Change and Conflict in the U.S. Army Chaplain Corps since 1945

Change and Conflict in the U.S. Army Chaplain Corps since 1945

Anne C. Loveland

Legacies of War / G. Kurt Piehler, Series Editor

THE UNIVERSITY OF TENNESSEE PRESS / KNOXVILLE

The Legacies of War series presents a variety of works—from scholarly monographs to memoirs—that examine the impact of war on society, both in the United States and globally. The wide scope of the series might include war's effects on civilian populations, its lingering consequences for veterans, and the role of individual nations and the international community in confronting genocide and other injustices born of war.

First Edition.

The paper in this book meets the requirements of American National Standards Institute / National Information Standards Organization specification Z39.48-1992 (Permanence of Paper). It contains 30 percent post-consumer waste and is certified by the Forest Stewardship Council.

Library of Congress Cataloging-in-Publication Data

Loveland, Anne C., 1938–
Change and conflict in the U.S. Army Chaplain Corps since 1945 / Anne C. Loveland. — First edition.
p cm. — (Legacies of war)
Includes bibliographical references and index.
ISBN 978-1-62190-012-2 (hardcover: alk. paper)
1. United States. Army. Chaplain Corps—History—20th century.
2. United States. Army. Chaplain Corps—History—21st century.
3. United States. Army—Chaplains—History.
4. United States. Army.—Religious life.
5. Soldiers—Religious life—United States.
I. Title.

UH23.L69 2013
355.3′470973—dc23
2013022460

To the memory of

my husband, Otis B. Wheeler,

and

my parents, Edith Ellen and John Wayne Loveland

Contents

ILLUSTRATIONS

FOREWORD

Historians writing accounts of British North America and the early years of the American republic see religion as a major political, cultural, and intellectual influence. There is a burgeoning interest in considering the religious life of the Civil War soldier. But historians who write about the twentieth century often ignore the role of religion in American life and imply modernization produced growing secularization. In *Change and Conflict in the U.S. Army Chaplains Corps since 1945,* Anne Loveland offers a compelling account of the major organizational, theological, and cultural shifts that have taken place in the army chaplaincy since the end of World War II. It is a work that has much to offer to military historians, religion scholars, and those seeking to understand the broader changes within American society since 1945.

Military chaplains hold a unique position within the U.S. Army in the twentieth and twenty-first centuries. They are commissioned officers, but by law and custom, do not carry weapons or lead soldiers into combat. Chaplains must gain the endorsement of their religious denominations before joining the chaplaincy and are expected to follow the tenets of their faith. At the same time, chaplains are also required, like all soldiers and officers, to take an oath at enlistment that requires them to protect and defend the U.S. Constitution, as well as to adhere to the lawful orders given to them by their civilian and military leaders. Army chaplains represent one of the clearest instances of direct state support of religion by the federal government. As army officers, chaplains receive their salary and benefits while in service, not from their religious denomination, but from the federal government.

The chaplaincy, as Loveland shows, changed significantly over the course of the post–World War II era. Like the broader American society, the army during the Cold War, embraced religion and support for the chaplaincy remained strong in the late 1940s and 1950s. Ideologically, national leaders saw that faith in God made America distinctive and separated us from Godless Communism. During the 1950s military leaders tasked chaplains with the explicit role of ideological warriors designed to indoctrinate soldiers through the Character Education programs.

During the Vietnam War, many chaplains responded to the call to help bolster morale among troops serving in Indochina. In the end, chaplains could do little to reverse the fraying of the army reflected in the growing problem of drug abuse. Concerns were raised in some quarters about the failure of chaplains to speak truth to power and their commitment to prophetic ministry. As Loveland's account shows, a chaplain when learning from a serviceman about the My Lai Massacre failed to ensure the chain of command made a proper investigation.

The composition of the chaplaincy changed significantly after the 1960s. Mirroring the rise of evangelicals in the wider society, evangelical ministers made up an increasing percentage of the chaplains with the U.S. Army. The evangelical embrace of the chaplaincy would be heightened by the decline in the number of mainline Protestant ministers and Reform rabbis who became chaplains that partly reflected disillusionment with the Vietnam War. Equally significant is the expansion of the chaplaincy to include chaplains outside of the Judeo-Christian tradition with the appointment of Muslims, Buddhists, and even Hindu chaplains.

Running throughout this study is a fundamental and almost irresolvable tension—reconciling the demands of the military norms with those of a chaplain's faith. This tension would lead evangelicals to object to efforts by the U.S. Army to promote a uniform Protestant worship service and Sunday school curriculum. More recently, it has prompted protests from some evangelicals that commanding officers and supervisory chaplains have ordered them not to "pray in Jesus' name" at command ceremonies that encompass a broad audience of soldiers of different faiths. As Loveland documents, the chaplaincy and the religious life of the American soldier became engulfed in the culture wars of the decade of the new century.

Anne Loveland's incisive work will encourage scholars to rethink the role of religion in one of America's most important and oldest institutions. This is a history of the army chaplaincy, and the Legacies of War series would be enriched by a comparative study of the U.S. Navy chaplaincy, as well as works that consider the role of chaplains in the armed forces of other countries. Equally vital is scholarship examining the religious life of the average soldier and how it has changed over time. We hope *Change and Conflict in the U.S. Army Chaplain Corps Since 1945* is the first of many works that examine the relationship between war and religion in the Legacies of War series.

G. Kurt Piehler
Florida State University

ACKNOWLEDGMENTS

Writing a book can be a lonely, even isolating, experience. I was fortunate to have a loving husband and dear friends who showed an interest in my work. Thank you, Otis, and Gresdna Doty and Jim Traynham, Katie P. Howard, Ann Sumner Holmes, Virginia Cronin, Rosi McGowan, and Cathy Steles.

I began working on the general topic of religion in the military in the 1980s. Along the way, I learned a great deal from members of the chaplain services. On a visit to the Army Chaplain School in the 1980s, I met with Army Chaplain Branch historian William J. Hourihan, who provided me with a useful introduction to the chaplains' work. During the past three decades I have gained information and insight from a number of army and navy chaplains regarding their varied ministry to the military: John W. Brinsfield Jr., Billy W. Libby, Galen H. Meyer, Arnold E. Resnicoff, Charlotte E. Hunter, Timothy J. Demy, and Gary R. Pollitt. Attorney Arthur A. Schulcz Sr. enhanced my understanding of the problem of religious pluralism in the Navy Chaplain Corps. When it came time to look for photographs, I was lucky to have the assistance of Marcia G. McManus, director-curator of the U.S. Army Chaplain Corps Museum, Fort Jackson, South Carolina, who did a wonderful job selecting a number of photos from the museum archives to illustrate diverse aspects of army chaplains' ministry.

Several individuals and organizations have encouraged my work on religion in the military by inviting me to contribute essays for publication or to present conference talks. In this regard, I wish to thank Lewis Perry and Karen Halttunen, Doris L. Bergen, Mark Silk, Kristen J. Leslie, and the Feinman Center for American Jewish History at Temple University.

Over the course of many years, my Louisiana State University History Department colleague Gaines M. Foster and I have read and commented on each other's manuscripts. Now serving as dean of the College of Humanities and Social Sciences, and despite a heavy workload, Gaines did a painstaking, thought-provoking review of my manuscript that prompted me to reconsider and clarify some of my thinking about the chaplains and thereby improve the book.

While revising the manuscript and seeking a publisher, I was fortunate to have the assistance of a number of other individuals. Clair Willcox, editor-in-chief of the University of Missouri Press, persuaded me to cut and condense an unwieldy manuscript of one-thousand-plus typewritten pages into a more streamlined five hundred pages. Then, when I sent a proposal to the director of the University of Tennessee Press, Scot Danforth, he immediately expressed interest in reviewing the manuscript for publication and accomplished that task in record time. His cheerful, encouraging e-mails and his sense of fairness made the review process much less of an ordeal than I had anticipated. Several readers of the manuscript made very helpful suggestions for improvement, which I have tried to honor. I am also indebted to UTP editorial assistant Thomas Wells and manuscript editor Gene Adair for being so prompt, cordial, and intelligent in advising me during the publication process, and to freelance copyeditor Karin Kaufman for her work on the manuscript.

Last but not least, I wish to thank the staff of the Interlibrary Borrowing Department of the LSU Middleton Library for their expert and gracious assistance in obtaining research material I needed to write this book, as well as earlier books. They have been an indispensable resource during my tenure at Louisiana State University.

Introduction

Army Chaplains in Cultural Transition

During the sixty-seven years following World War II, U.S. Army chaplains and the Army Chaplain Corps confronted many challenges and great change. Their "cultural transition" involved three intersecting cultures: the civilian religious community, the military, and the chaplaincy. Change in one culture produced change in the others; thus the chaplains' cultural transition was a two-way process. They effected change in the civilian and military cultures just as new developments in those sectors prompted chaplains to develop new ways of conducting their military ministry.

The influence the civilian religious culture exerted on the chaplains is apparent throughout the period. Chaplains were, of course, largely a product of that culture. They had been educated in civilian colleges, universities, and seminaries. They were ordained by its churches, denominations, and faith groups. They were required by the military to be "endorsed" by a civilian ecclesiastical agency, certifying that they were clergy and possessed all the necessary religious qualifications for serving as military chaplains. So it is not surprising that chaplains often reflected the thinking of the civilian culture. In the 1950s, when army chaplains developed a program called Character Guidance, they based its theological, moral, and civic rationale on the American religious consensus of that decade—what I call the "religion of democracy." That consensus broke down in the early 1960s; two new "cultural strategies" emerged, one conservative and sectarian, promoted mainly by conservative Christian evangelicals, the other liberal and pluralist, promoted mainly by mainline Protestant leaders, each group striving for cultural dominance. In the 1960s and subsequent decades, chaplains frequently adopted one or the other of those strategies in addressing problems they encountered within the U.S. Army and the Chaplain Corps. They also felt the impact of the two strategies when conservatives or liberals criticized chaplains as individual clergy or the programs they administered. Indeed, the criticism generated in the civilian religious community was a major factor in prompting chaplains to reevaluate and revise their concept and practice of military ministry in the latter part of the twentieth and the first decade of the twenty-first century.

The 1960s was a watershed decade for the Chaplain Corps and the U.S. Army, as well as the civilian society. On the one hand, it signaled increasing religious diversity, secularization, and sectarianism in the American religious culture, matters chaplains had to deal with in the military context. On the other hand, the decade marked the beginning of the Vietnam War, perhaps the single, most influential factor in army chaplains' cultural transition.

Chaplains' almost unanimous and outspoken support for the war made them a likely target of antimilitary, antiwar protest during the Vietnam era. The anti-chaplain critique formulated by antiwar critics in the civilian religious community had a profound effect on chaplains' thinking, not only during but also long after the war had ended. Chaplains' experience in the Vietnam combat environment also proved decisive. During the war they developed a new "ministry of presence" for soldiers.

In its aftermath they created a new "battle-focused" moral advisory role to commanders and pondered a "soldier-focused," less militaristic approach to morale building among the troops. Going beyond their traditional priestly and pastoral roles, they conceived and implemented a new "institutional," secularized ministry to the military organization. Reorienting their role as religious and moral educators, they moved from teaching mandatory Character Guidance to all army personnel to offering instruction in ethics, moral leadership, and world religions in the army service schools. Throughout the period, but especially beginning in the 1980s, they addressed the challenge of religious pluralism and "religious accommodation" within the army, revising old guidelines and formulating new ones to ensure the First Amendment right of free exercise for all army personnel. In the process, they traded an exclusivist religious orientation for an inclusivist, pluralistic vision, officially recognized in chaplain field manuals and army regulations.

Some of the changes chaplains made in their military ministry were in response to crises or challenges that arose during the period 1945–2012. In the late 1940s, army chaplains proposed what became Character Guidance as a remedy for the soaring rate of venereal disease among soldiers in occupied Europe and Japan. During the Vietnam era, the chaplains initiated or cooperated with army leaders in the development of programs to resolve an unprecedented, multifaceted crisis in the army caused by a decline of discipline and morale among enlisted personnel, incompetent leadership on the part of the officer corps, and the erosion of the professional military ethic. During the first decade of the twenty-first century, chaplains faced the challenge of deciding how they might best help the army to resolve yet another, and very different, kind of crisis. Chaplains themselves experienced two crisis situations: one the result of the Vietnam era antichaplain critique, the other a consequence of increasing religious pluralism, secularization, and sectarianism within the Chaplain Corps, as well as in the army and the civilian religious community.

In the course of addressing these crises and conducting their ministry to the military, the chaplains enhanced their professional status in the army. Indeed, this was part of their calculation. Throughout the sixty-seven-year period covered in this book, army chaplains sought two main objectives: ensuring the moral, religious, and spiritual welfare of all army personnel and becoming, as they put it, an "integral," well-respected component of the army organization and its mission. In realizing these goals, they discovered a means of exerting a profound, and in some cases unique influence on soldiers and commanders, the army as an institution, and the military community.

1

Teaching the Religion of Democracy

In 1946, the U.S. Army faced an unprecedented breakdown of discipline and morality among American soldiers in occupied Europe and Japan. The main problem was widespread sexual immorality and a "staggering" rate of venereal disease among the troops, which received considerable coverage in U.S. newspapers. The civilian community became alarmed, and the Veterans of Foreign Wars and a number of American religious leaders issued scathing commentaries on the crisis.[1] Army chaplains also voiced concern. Some wrote to their endorsing agents, and one chaplain presented his thinking in an interview with an investigator for the U.S. House of Representatives Military Affairs Committee. Other chaplains expressed their views within the military command structure. On January 11, 1946, members of the Tokyo-Yokohama Chapter of the Army and Navy Chaplains' Association sent a letter to Gen. Douglas MacArthur, Supreme Allied Powers Commander, in which they announced that they were "strongly of the conviction that prostitution as it relates to the U.S. Occupation Forces in Japan is producing moral degradation that is exceptionally widespread and unusually ruinous to the character of American troops."[2]

Like the members of the Army and Navy Chaplains' Association, many of the civilian commentators blamed the VD crisis on military policies: tolerating, even sanctioning, soldiers' access to houses of prostitution and encouraging sexual promiscuity by issuing contraceptives.[3] There was general agreement that the mandatory sex hygiene lectures the soldiers received and the prophylactic measures made available to them (condoms and chemical treatment) were inadequate in preventing sexual immorality and venereal disease. As one commentator wrote, "Effective moral and religious training" was "the only secure and permanent method."[4]

Army Chief of Chaplains Luther D. Miller not only appreciated the seriousness of the problem but was also aware of his fellow chaplains' dissatisfaction with the way it was being handled. In December 1946, he sent a memorandum to the director of personnel and administration of the War Department General Staff outlining a program to combat venereal disease in the army.[5] The War Department had recently established an experimental universal military training unit at Fort Knox, Kentucky, and the chaplains were giving the trainees a series of "Citizenship and Morality Talks." Miller suggested implementing the chaplains' lecture program throughout the army, and he included copies of the talks that had been prepared for the Fort Knox unit. "It is believed that these lectures, delivered once a week, will be of inestimable value and benefit to all Regular Army personnel and trainees in the development of a moral and spiritual background upon which to develop the deeper aspects of morale," he wrote.[6]

Secretary of War Robert Patterson responded to Miller's overture in late January 1947 in a letter titled "Discipline and Venereal Disease." Noting that the annual VD rates in the army were "higher than at any time in the past thirty years," Patterson authorized the new character education program Miller had recommended for the regular army. "The Corps of Chaplains bears a special responsibility for the moral and spiritual welfare of troops," Patterson declared. "To aid the chaplain in meeting this responsibility, commanding officers will allocate appropriate periods in the regular training schedule for instruction in citizenship and morality which all personnel will attend." Patterson charged the Office of the Chief of Chaplains (OCCH) with preparing the instruction.[7] The exchange between Miller and Patterson initiated the development of the army program that became known as Character Guidance (CG), officially recognized in Department of the Army Circular 231, issued in August 1948.[8]

The Rationale for Character Guidance

In his memorandum to Secretary Patterson, Miller proposed the chaplains' lectures as a solution to the army's VD problem. But he envisioned Character Guidance as much more than that—a comprehensive program of instruction in religion, morality, and good citizenship. In three articles published in chaplain and military journals, he presented his thoughts. He estimated that a million or more men would be coming into the army each year, and 50 percent of them would have no religious background. The other 50 percent would be nominal church members, but many of them would also benefit from having their faith strengthened and enlarged by the chaplains' lectures. Character Guidance would also improve military efficiency and enhance combat effectiveness. "Native bravery and intensive training cannot alone produce the best type of soldiers," Miller observed. For the army "to perform its mission with honor to the nation, it must have moral and spiritual nurture and inspiration of the highest type." And "character rooted in the knowledge and worship of God makes for fighting power and staying power." As for promoting good citizenship, he pointed out that "an adequate program of education for citizenship in a democracy must place the emphasis upon the character as well as intelligence; and in the realm of character, religion has always had the greatest influence."[9]

Army leaders agreed with Miller. They had developed a new perspective on military training that assumed an indispensable role for religious, moral, and citizenship education. In an address to First Army chaplains, Maj. Gen. John M. Devine, commanding general, Ninth Infantry Division, admitted that in the past, the army had operated on too narrow a concept of military training, thinking that it should include "only physical development and the teaching of strictly military subjects." At best, commanders had paid "lip service" to the idea that training in moral and spiritual values was desirable. Now, he asserted, army leaders recognized that "a true concept of training must be based on the nature of man and must recognize all his needs," including "moral, ethical, and spiritual guidance," in order to produce "a reliable, self-respecting, sincere, and loyal citizen-soldier."[10] Gen. Jacob L. Devers, chief of Army Field Forces, presented a similar view in the *Army Information Digest*. The army's training responsibility was "not limited to teaching military techniques," he declared. "To make the man a better fighter, we must make him a better citizen. We must help him relate the task at hand to . . . the greater democratic ideals of our Nation."[11]

President Harry S. Truman also advocated religious and moral training in the armed forces and appointed a number of civilian advisory committees to aid him in promoting it. The first of these, the so-called Compton Commission, was formed to advise him on universal military training, a program under consideration by Congress that Truman himself favored. After the passage of the Selective Service Act in June 1948, the commission turned its attention to the thousands of young men who would be drafted into the U.S. Army. Its successor, the Committee on Religion and Welfare in the Armed Forces, established in October 1948, also focused on that group.[12]

Both committees were primarily concerned with "the moral and spiritual well-being of American youth." In considering their needs, they emphasized their young age (well over 50 percent would be under twenty-one) and their moral and emotional vulnerability in the military environment.[13] The Compton Commission especially stressed the need for more chaplains (it suggested a ratio of one chaplain for every five hundred to six hundred trainees), who would not only conduct religious services but also spend one or more hours each week counseling the men on moral and ethical matters. They should also work closely with medical officers in conducting sex hygiene lectures for the trainees, the commission suggested. Continence rather than prophylaxis should be emphasized—chaplains and medical officers should "show the moral reasons for avoiding exposure to venereal diseases rather than the methods for evading the consequences therefrom."[14]

The Committee on Religion and Welfare promoted the work of chaplains as well as Character Guidance.[15] It emphasized the "special obligation" on the government, the military, and the civilian community to provide for young soldiers' spiritual and moral well-being. "They have left behind their daily contact with family, friends, and church groups," the committee explained. "Their attitudes, their character, their moral and spiritual fiber will be fashioned for an appreciable time under the influence of military discipline." The goal should be not only to offer a religious and moral environment that would sustain these young people during their period of service but also to ensure that they returned to civilian life as good as or better people than they were when they entered the military. By virtue of their large numbers, "these returning servicemen and women will have a profound effect upon our social fabric," the committee declared, "and, therefore, support for their religious and moral well-being is an inescapable obligation that faces the whole nation."[16]

President Truman's advisory committees spoke to, and to some extent for, individuals and groups in the civilian community who feared and denounced the demoralizing influence the military environment exerted on young soldiers. In the mid- to late 1940s, a small but vocal group of pacifists and liberal mainline Protestants were engaged in protesting the "militarization" of America; their targets included the proposals for universal military training and, later, the enactment of Selective Service. The religious and moral "hazards" of camp life, encountered by young people "at one of the most impressionable periods of their lives," were among the chief reasons these antimilitarists gave for criticizing military training. They pointed to bad company, profanity, drunkenness, gambling, and other temptations (sexual promiscuity and VD were not mentioned in so many words). They also contended that "mass regimentation" and indoctrination in "blind unquestioning obedience" would undermine initiative, self-reliance, and thoughtful judgment; they certainly would not contribute to "the building of moral fiber and the making of good citizens." A representative of the Federal Council of Churches declared that universal military training "would tend to destroy . . . all those freedom-giving and character-building institutions of the Christian tradition."[17] However misinformed or exaggerated such assertions may have seemed to military and government leaders, they posed a significant public relations problem for the army. If only for that reason, they needed to be addressed, and the Truman advisory committees were especially helpful in that regard.

"Faith of Our Fathers"

Although some people referred to CG as "the chaplains' program," it was, in fact, according to U.S. Army regulations, a command responsibility. Chaplains were to assist commanders in promoting "healthy moral, social, mental, and spiritual attitudes" in army personnel.[18] The Office of the Chief of Chaplains was assigned the task of developing the instructional materials for the program. Chaplain Martin H. Scharlemann, an instructor at the Army Chaplain School who had prepared the chaplains' talks for the Fort Knox unit, spent five years writing the lectures for the CG program. Sixty of them, each designed to last fifty minutes, were eventually published in six volumes under the general title *Duty-Honor-Country*.[19] The OCCH also provided chaplain instructors with elaborate lesson plans that explained how to present the lectures.

In the late 1940s and 1950s, attendance at the Character Guidance presentations was mandatory for all army personnel. During the eight weeks of basic training (BT), personnel received four hours of CG instruction, and during the eight weeks of advanced individual training (AIT), two hours. All other units and organizations received monthly Character Guidance instruction.[20] In the 1960s the requirement changed slightly. Basic training featured four hours of CG; following basic training, all personnel below grade E-6 (i.e., below staff sergeant) received one hour of such instruction monthly. Officers and enlisted personnel of grade E-6 and higher were to receive a "monthly briefing" on the content of the instruction being presented to lower ranking personnel.[21]

What did chaplains teach these soldiers? In 1962, a member of the OCCH staff offered a succinct description of the purpose and content of the CG lectures:

> The moral principles and values which [Character Guidance] seeks to promote and inculcate are not those of any particular religious

> denomination. Rather they are the principles and values of personal moral conduct which reflect the mature judgement of the vast majority of Americans and which are clearly delineated in the history, traditions, and culture of the United States. They are . . . clearly expressed in the words and writings of the Founding Fathers, in the historic documents of the American political system, and in the words and examples of those historical figures who played a leading role in the establishment and growth of the country. The program is theistically oriented in the sense that American traditions and institutions are founded upon a Judao-Christian [*sic*] concept of God and natural moral law.[22]

The chaplain's statement, intended as a description of Character Guidance, also evoked the religious and cultural consensus of the American people in the 1950s and early 1960s—what I call the "religion of democracy" or the "national faith."[23] Since chaplains drew heavily on that consensus in their CG lectures, it will be helpful to present a brief discussion of its content.

The religion of democracy was a "cultural strategy" consciously created by American clergy, educators, public moralists, and political leaders for various purposes: to inculcate certain religious, moral, and civic values and to sanctify the American way of life and enhance national cohesion and identity.[24] The strategy had two components. One was religious—not Protestantism, Catholicism, or Judaism, the three major faiths of America in the 1950s, but an amalgam of the three, which minimized the differences between them. The religious sociologist Will Herberg pointed out the existence of this component in *Protestant-Catholic-Jew,* published in 1955. He observed that most Americans in the 1950s viewed those three faiths "as three diverse, but equally legitimate, equally American, expressions of an overall American religion, standing for essentially the same 'moral ideals' and 'spiritual values.'"[25] It is important to keep this religious component in mind; it was what distinguished the religion of democracy from so-called civil religion.[26]

The other component of the national faith was a highly spiritualized conception of the origins and nature of American democracy, of "Americanism" and the so-called American way of life, and of the communist threat and the means of overcoming it. It emphasized the founding fathers' faith in God and the consequent "religious foundation" of American democracy. President Dwight D. Eisenhower, a principal spokesman of the national faith, declared that "the founding fathers wrote [their] religious faith into our founding documents . . . they put it squarely at the base of our institutions."[27] Other advocates of the religion of democracy spoke of America as a "covenant nation" and "a nation under God."[28] The Judeo-Christian tradition was said to be the main source of America's ideals and principles.[29] Just as faith in God was declared to be central to the eighteenth-century founding, it was also considered a basic element of true Americanism. "Without God there could be no American form of government, nor an American way of life," Eisenhower asserted. "Recognition of the Supreme Being is the first—the most basic—expression of Americanism."[30]

In addressing the communist threat, spokesmen for the religion of democracy portrayed the conflict between the United States and the Soviet Union as a struggle between "the God-fearing power of democracy and the God-hating power of communism."[31] President Truman, among others, emphasized the "moral and spiritual" threat posed by the communist ideology, which he said challenged "our belief in God," "our democratic form of government," "all the values of our society," and "our way of life."[32] Both President Eisenhower and the chairman of the Atomic Energy Commission, David E. Lilienthal, agreed with John Foster Dulles's repeated warnings against relying solely on "material things" such as military weaponry and industrial power in combating the communist threat. Americans must recognize the "priority of spiritual forces" and "the faith which in the past made our nation truly great," the secretary of state declared.[33] In 1953 Eisenhower asserted his "unshakable belief that is only through religion that we can lick this thing called communism."[34]

This emphasis on spiritual power notwithstanding, the religion of democracy supported military preparedness and the use of military might against America's enemies. It endorsed a spiritualized form of militarism. Since many Americans associated militarism with Japan and Nazi Germany, the advocates of the national faith made it more palatable by infusing it with traditional American values, ideals, and religious beliefs. Military preparedness was said to be "within the tradition of our founding fathers," a means of preserving "our heritage of freedom," and supported by "the teachings of Scripture."[35] In 1951, Secretary of Defense George C. Marshall defended Selective Service by quoting George Washington and declaring that the idea of the citizen's duty to participate in the common defense was "deeply ingrained in the American tradition."[36] As the Cold War intensified, government, religious, and military leaders increasingly relied on the language and doctrines of the national faith in justifying America's massive military system—as a means of defeating "militant international communism" as well as preserving "the ideals of individual liberty and spiritual values under God, . . . the genius and animating spirit of our nation."[37]

Teaching the National Faith

The imprint of the national faith on Character Guidance lesson plans and lectures is unmistakable. In their classes with soldiers, chaplains invoked the language and doctrines of the religion of democracy for a variety of purposes: to encourage faith in God and instill morality, patriotism, and good citizenship, as well as to inculcate the military virtues of duty, honor, and sacrifice.

"One Nation Under God," the first of four lectures delivered to soldiers in basic training in the 1950s and 1960s, clearly reveals the influence of the national faith. The lesson plan for the lecture listed its two objectives: "To help the individual to understand the effect that faith in a Supreme Being has had on the origin and development of our country" and "To lead the individual to a recognition of the importance of the spiritual element in his training." Most of the information the chaplain presented aimed at proving that "*we as a nation are* DEPENDENT *upon and* RESPONSIBLE *to Almighty God.*" According to the lesson plan, he was to point out, among other things, that "our Founding Fathers believed that the American form of government, and the American way of life were both impossible without God." In concluding the lecture, he was to "EMPHASIZE: That we must cultivate within ourselves the religious beliefs and attitudes that were a part of those who built our nation." The manual also suggested delivering a final exhortation, warning the soldiers to "be careful that you don't get so involved with rifles and grenades and maps and tactics, and all the other details of training, that you forget what we are as a nation, and what you are really to do and to be as soldiers." The chaplain was to tell the soldiers that they, more than other Americans, needed to understand that "we are 'A Nation Under God,'" because they were particularly called upon to "sacrifice" for their country.[38]

Other CG lectures also stressed the importance of religious belief. Chaplains taught soldiers that "the worship of God is a requirement of moral living,"[39] that "service to the nation is most effective only when religion becomes part of individual life,"[40] and that believing in and worshiping God was part of the American way of life. "Public institutions and official thinking reflect a faith in the existence and importance of divine providence," one lecture observed.[41]

Chaplains taught soldiers "basic morality."[42] They emphasized that moral principles came from God, "the final source of authority,"[43] and therefore were universal, unchanging, and eternal. These absolute moral principles were also said to be a component of the "traditional religion and morality which have always been basic to American life"[44] and woven into "the moral fabric of the American way of life."[45] They constituted "the undergirding principles of both our Government and our society."[46]

A stated objective of the CG program was "to develop the kind of soldier who has sufficient moral understanding and courage to do the right thing in whatever situation he may find himself." An article published in the December 1948 issue of the *Chaplain* noted that "formerly chaplains tried to persuade men not to fall into evil ways and explained the consequences of wrongdoing. Now they attempt to build the character of officers and enlisted men in such a way that they will not want to do wrong."[47] The new approach was evident in the way the chaplains addressed the problem of venereal disease. As John Willoughby has observed, the army command had treated VD as a medical problem to be solved by prophylactic measures—a "relatively amoral" way of controlling the epidemic.[48] By contrast, chaplains, in their CG lectures, stressed "the moral reasons for avoiding exposure to venereal disease rather than the methods for avoiding the consequences."[49] They also showed a forty-minute training film on venereal disease control, *Miracle of Living*, which emphasized the "moral approach."[50]

Besides "basic morality," chaplains taught soldiers the personal and civic virtues associated with good citizenship and Americanism. The personal virtues chaplains emphasized were integrity, self-reliance, courage, obedience, fair play, and persistence. The civic virtues included respect for authority, appreciation for the dignity of human beings, awareness of the individual's moral and social responsibility, and patriotism. According to a CG manual, they were "to give the American soldier a deeper knowledge and love of the country he serves, and a better understanding of why it is worthy of his service, and, if need be, of his very life."[51]

Inculcating patriotism involved teaching anticommunism, with an emphasis on the religious and ideological differences between democracy and communism. A lecture titled "Our Moral Defenses" described the Cold War as a contest of ideas "in which nothing less than the souls of men are at stake." It explained that Americans believed in the "freedom 'idea,'" which originated in "an acceptance of that Divine Providence in Whose name the signers of the Declaration of Independence pledged each other their lives, their fortunes, and their sacred honor." This conviction was said to be "basic to our whole theory of liberty." Karl Marx and his followers denied not only the existence of God but also another assumption of the "freedom 'idea,'" the existence and validity of certain moral principles established by God. Unlike the communist ideology, the "freedom 'idea'" viewed conflicts between human beings as a consequence of individuals' failure to obey God's laws rather than the result of different classes' relation to the means of production and distribution—and that "order and justice," not class struggle and revolution, constituted the divinely ordained objectives of society and history. "It ought to be clear from all this how vastly superior the freedom 'idea' is to the doctrine of Communism," the lecture concluded. "However, freedom succeeds only where its underlying moral principles are understood and appreciated. . . . There can be little doubt of the fact that the present ideological conflict is total war. . . . Its outcome will depend on what you and I believe and do with ourselves. Our defenses are moral. They consist of a belief in and the practice of what is right according to the Moral Law."[52]

Promoting Patriotism and Combat Effectiveness

Although the imprint of the religion of democracy remained strong throughout the history of Character Guidance, some of its content changed in response to issues arising from the Korean War and the intensification of the Cold War. In the aftermath of Korea, chaplains paid increasing attention to inculcating patriotism, largely in response to the public furor over the "brainwashing" and indoctrination of American prisoners of war (POWs). Twenty-one of the POWs had refused repatriation to the United States, and others were alleged to have collaborated with enemy propagandists or committed serious offenses against their fellow prisoners. Many American citizens, government officials, and military leaders concluded that the POW's had succumbed because they were not sufficiently patriotic or committed to American ideals, and that the armed services and the public schools shared the blame for their lamentable behavior.[53] That was the view presented in the mass media by popular writers such as Eugene Kinkead, author of the bestselling *In Every War but One,* and Virginia Pasley, who wrote *21 Stayed: The Story of the American GIs Who Chose Communist China—Who They Were and Why They Stayed.*[54]

Perhaps unwittingly, President Eisenhower helped reinforce the notion that ignorance of American ideals explained the POWs' behavior. At a news conference in October 1953, when a reporter raised the issue, Eisenhower said, "When you take the meager education that we give to our people, sometimes as to what their obligation is to a free form of government, what it means to support it, what it means to keep it and pass it on, you sometimes wonder that there weren't more of our people that succumbed, at least temporarily."[55] The following year, in an address to the Military Chaplains Association, Eisenhower emphasized the role chaplains should play in fortifying soldiers against enemy indoctrination. They must "make sure every individual knows what his country stands for, and therefore what is the basic cause for which he fights." He exhorted the chaplains to continue "your work among our armed services to help raise and keep up to the highest possible pitch the morale and the spiritual strength that we so badly need, as we defend freedom against totalitarianism in this world."[56]

In 1955, the Defense Advisory Committee on Prisoners of War issued its report. Although it minimized the extent of POW misconduct and lamented "the adverse publicity" and "misconceptions" surrounding the issue, it portrayed the American prisoners as seriously handicapped "when plunged into a Communist indoctrination mill." Many of them "had heard of Communism only as a name" and knew too little about the ideals and traditions of the United States. That was why they "couldn't answer arguments in favor of Communism with arguments in favor of Americanism."[57]

To remedy this problem, the Defense Advisory Committee recommended the formulation of a code of conduct for American servicemen. Promulgated by President Eisenhower in an executive order on August 17, 1955, the Code of Conduct for Members of the Armed Forces of the United States consisted of six articles. It began with the statement "I am an American fighting man. I serve in the forces which guard my country and our way of life. I am prepared to give my life in their defense." It ended with "I will never forget that I am an American fighting man, responsible for my actions, and dedicated to the principles which made my country free. I will trust in my God and in the United States of America." In articles two through five the American serviceman pledged never to surrender of his own free will; to endeavor to escape if captured by the enemy and not to betray his fellow prisoners; to refuse to give any information during interrogation other than name, rank, service number, and date of birth; and to make no statements disloyal or harmful to his country and its allies.[58] It was a prescription for behavior in combat and in the event of capture, but its language and format suggests it was also intended to function as a pledge of loyalty, even, to use Adm. Arthur W. Radford's term, a "creed."[59]

The Defense Advisory Committee also recommended enhancing moral and citizenship training, in the civilian community as well as in the armed forces.

"Pride in a country and respect for its principles—a sense of honor—a sense of responsibility—such basics should be established long before 'basic training,' and further developed after he enters the Armed Forces," the committee declared. Both the military and civilian communities shared the responsibility for building "spiritual and educational bulwarks against enemy political indoctrination."[60]

Chaplains incorporated the lessons drawn from the POW scandal into the Character Guidance program. A 1961 CG manual cited research that revealed that the POWs who capitulated to enemy indoctrination did so because of a lack of faith. Not only were they devoid of confidence in themselves, their country, and the American Way of Life, but they also lacked confidence in "divine providence" and "personal faith in God."[61] The manual pointed to Character Guidance as a remedy. It was "designed to give the American soldier a deeper knowledge and love of the country he serves, and a better understanding of why it is worthy of his service, and, if need be, of his very life." In addition, chaplains worked "continually to deepen and strengthen the knowledge of that traditional religion and morality which have always been basic to American life."[62]

Chaplains began to teach the code of conduct in their monthly classes. A 1957 army regulation on code of conduct training listed eleven objectives. Most of them involved strengthening the individual soldier's "survival ability," developing his understanding of U.S. military law and the Geneva Conventions as they pertained to prisoners of war, and providing information about the physical and mental challenges he would face as a POW and how to cope with them. However, three of the training objectives called for political, moral, and religious indoctrination: "develop resistance to enemy political and economic indoctrination through education in the basic truths and advantages of our democratic institutions as opposed to the fallacies of communism"; "motivate the individual toward our national objectives"; and "develop the moral fiber and religious motivation of the American soldier to fortify him with the weapons of faith and courage."[63] Here was an opportunity for chaplains to show how CG could help army leaders achieve those objectives. The Office of Chief of Chaplains issued new instructional materials to assist chaplains in discussing the code.[64] They were to point out to soldiers "the part played by the mind and spirit in our national security," the importance of keeping alive "the mental and moral 'will to resist,'" and a POW's obligation "to remain faithful to the United States and to the principles for which it stands." Chaplains should also urge each soldier, "throughout his captivity," to "look to his God for strength to endure whatever may be-fall [him]."[65]

By the early 1960s, the Character Guidance program had undergone a significant evolution. In the late 1940s and 1950s, chaplains sought to build soldiers' character and citizenship fitness and to improve their moral behavior. Then, in the mid- to late 1950s, they began to emphasize their role in producing good soldiers. The 1961 CG manual described the program as "a moral management tool for the Army commander" that provided "a direct route to morale, efficiency, and discipline in the command."[66] It contended that CG enhanced combat efficiency: "Resolute character and strong moral conviction" counted as much as military skills and weaponry in ensuring good performance on the battlefield.[67] Chaplains also pointed out their special ability to use religious and spiritual values as "motivational factors" in inspiring soldiers to fight "under any circumstances and in the face of any enemy."[68]

Chaplains Playing the "Backfield"

In a 1947 article in the *Army and Navy Journal*, published eight months after the implementation of Character Guidance, Chief of Chaplains Luther Miller pointed out the significance of that action: "The Army chaplain is no longer playing guard; he is in the backfield. Commanders more and more are making up plays with the chaplain carrying the ball." Since CG was a formal, compulsory component of the military training schedule, the chaplain was directly involved in military training in a way that he had not been before. Miller predicted that in the future, chaplains would be increasingly "called upon as a specialist in citizenship and morale as well as an authority in religion."[69] He and many other army chaplains were convinced Character Guidance would enhance their status and influence in the army. They were right. The new program placed chaplains in a far better position than that of previous decades.

In the late nineteenth century, according to the military historian Edward M. Coffman, "many if not most officers were contemptuous or suspicious" of chaplains.[70] Conversely, chaplains thought the army had little interest in promoting religion and even less in supporting the chaplaincy.[71] Their role was not well defined, and commanders often assigned them "collateral duties" that were irrelevant to their religious and moral ministry. During World War I, President Woodrow Wilson and his Secretary of War relied less on chaplains and more on civilian groups in the Commission on Training Camp Activities (CTCA) to provide for the moral and religious welfare of soldiers.[72] Many commanders continued to treat the chaplains as "handymen," assigning them various collateral duties, such as "unit historian, librarian, post exchange officer, bond sales officer, band director, athletic officer, morale officer, . . . graves registration officer, [and] education officer."[73]

The chaplains' status improved during World War II, mainly because Army Chief of Staff Gen. George C. Marshall supported their work.[74] He was also persuaded by the chief of chaplains to relieve chaplains of the irksome collateral duties.[75] Nevertheless, some World War II chaplains complained about commanders' lack of respect. Chaplain James L. McBride observed that many commanding officers were not "aware of the importance of religion in motivation to action." They did not believe, as he and other chaplains did, that "religion can and does make men's souls strong for battle." And since most of the officers were not active church members and showed little interest in formal religion, they either "overlooked" the chaplain entirely or failed to provide the facilities and other support he requested. Other chaplains complained that their commanders did not understand the chaplain's ministry or "failed to accept or carry out their responsibility for the moral and spiritual welfare of the men under their command."[76]

Chaplains' problems with commanding officers continued in the aftermath of the war. In 1947, two civilian clergymen who visited some four hundred chaplains serving American occupation forces in Europe and the Pacific reported that many officers gave "only token recognition to the chaplain but no personal allegiance to honor what he stands for" and that few attended chapel services or sought to "strengthen or assist personally in the chaplains' work."[77]

Implementation of Character Guidance greatly improved chaplains' standing with commanders. Suddenly, or so it seemed, they became quite vocal and active in support of the army's religious and moral programs. When Secretary of the Army Frank Pace Jr. sent the commanding generals of the army copies of Defense Secretary Marshall's 1951 memorandum "Protection of Moral Standards," he added his own enthusiastic endorsement of character education. He not only applauded the fact that the army had "pledged itself to a consistent and concerted effort to develop sound character in its soldiers" but also urged "all officers and noncommissioned officers [to] familiarize themselves with the Army Character Guidance Program." He called for "renewed effort and emphasis at every level to instill these spiritual and moral qualities in all officers and enlisted personnel so as to make them more effective members of the Army and more substantial citizens."[78] During the late 1940s, and into the 1950s and early

1960s, other well-known and respected military leaders spoke and wrote in praise of the chaplains and Character Guidance for inculcating spiritual values, moral principles, and good citizenship.[79] In the late 1940s and especially in the 1950s, commanding officers, and the officer corps in general, not only supported CG but also cooperated with chaplains in developing other religious and moral programs for soldiers.

Winning Civilian Support

One of the reasons military leaders praised and cooperated with the chaplains' ministry, especially Character Guidance, was because it garnered support for the army in the civilian religious community—support it badly needed during the VD crisis and the debates over universal military training and Selective Service. One of the purposes of the Fort Knox experiment had been to counteract civilian concern about soldiers' immoral behavior and the negative effects of military training. The War Department had mounted a massive public relations campaign in behalf of the pilot program, even engaging chaplains in the publicity effort.[80] Character Guidance advanced that campaign, and on a much larger scale, over a longer period of time, and attended by much less controversy.

During and after the Korean War prominent civilian religious leaders joined the chorus of military men praising the religious vitality of overseas troops, the high level of worship attendance at military installations, and the great improvement in the moral behavior of soldiers.[81] In 1956, in the annual report of the Lutheran Service Commission, Secretary Carl F. Yaeger declared that although U.S. newspapers still featured disturbing accounts of soldiers' off-base activities, there were "encouraging signs of a lifting of the level of the conduct of American military personnel." In particular, he noted that "officers and men of the U.S. Armed Forces in increasing numbers are daily exemplifying the best in Christian living. Many are vigorously resisting the temptation to make of their service life a period of spiritual inactivity or a long 'lost weekend.'"[82]

The most remarkable change of attitude in the civilian religious community was a more positive view of military service. Although some religious groups and individuals continued to oppose the peacetime draft during the 1950s,[83] others came to accept it as a necessary component of national security. Some went even further, presenting an idealized view of military service that made it not just a civic responsibility but also a sacred Christian duty. Thus William V. Kennedy, writing in the *Catholic World,* declared that the idea that every able-bodied man was responsible for the defense of home, community, and nation was "as fundamental to the American way of life as it is integral to the Christian moral code."[84] Even some liberal, mainline Protestants, who had been active in the movements against universal military training and Selective Service, conceded the possibility of viewing military service positively, "in Christian terms," even as an element of the Christian vocation.[85]

Mission Accomplished

Character Guidance also benefited chaplains. One of their longstanding goals was to improve their professional standing in the army by making their ministry an integral part of the military institution. Character Guidance was the first of several programs developed to achieve that objective.[86] During the 1950s, they added new emphases more directly related to the military mission, such as teaching the code of conduct and bolstering soldier morale—instilling the attitudes and capabilities necessary to combat effectiveness.

In the aftermath of World War II, army chaplains' situation had looked quite unpromising. Much changed in the decade and a half that followed, largely as a result of the development of Character Guidance. Army Chief of Chaplains Gerhardt W. Hyatt (1971–75) recalled that in 1952, when he began working in the Office of the Chief of Chaplains, "it appeared to me that [the Chaplain Corps] was not taken seriously as an agency of the Army," that "we were tolerated because title X required the Army to have chaplains," but "we were not really integrated into the Army as a professional branch." Army Chief of Chaplains Patrick J. Ryan (1954–58) pointed to a significant change as early as 1956, when he declared that the army's religious program was "in a stronger position than at any time I can remember in my twenty-eight years of service." And in 1961, Navy Chief of Chaplains George A. Rosso observed that all of the chaplaincies had progressed from being "tolerated" to being "accepted" and, finally, to being "respected."[87] Given their achievements of the 1950s, army chaplains had every reason to contemplate the beginning of a new decade with great confidence. Instead, as we shall see in the next few chapters, the 1960s proved to be a difficult time for chaplains.

2

THE SIXTIES WATERSHED

The decade of the 1960s was a watershed in the history of the Army Chaplain Corps, just as it was for American society and culture.[1] The American people became more religiously diversified, and their government, society, and culture became more secularized. The consensus forged by the religion of democracy broke up, and two new, competing strategies claimed the culture-shaping role it had performed in the 1950s. The Vietnam War provoked intense debate within a civilian community that had, for the most part, supported military preparedness and the Korean conflict during the previous decade. All of these developments registered a profound impact on U.S. Army chaplains and their ministry, beginning in the 1960s and continuing into later decades.

The "Seismic Shift"

"Seismic shift" was the term theologian-historian Martin E. Marty used to describe the reconfiguration of American religion in the 1960s.[2] The main shock came from the dramatic decline in the membership of the mainline Protestant denominations. In 1966, their total membership reached its peak as a percentage of the population, representing 56.2 percent. But then, during the period 1965 to 1975, decline set in. For example, during that ten-year interval, the United Presbyterian Church in the USA experienced a membership decline of 12.4 percent; the United Methodist Church, 10.1 percent; the United Church of Christ, 12.2 percent; the Presbyterian Church in the United States, 7.6 percent; the Lutheran Church in America, 5.0 percent; and the Episcopal Church, 16.7 percent. By 1974, Protestants as a group represented only 50.4 percent of the total U.S. population.[3]

Another notable effect of the seismic shift was the increase in the number of Americans who appeared in the "Other religions" or "None" categories of religious preference surveys. "Other" meant other than Protestantism, Roman Catholicism, and Judaism. This category included Buddhism, Islam, and Hinduism, various "Eastern religions," and cults and sects such as Zen Buddhism,

Baha'i, Transcendental Meditation, Yoga, Hare Krishna, and the Nation of Islam (Black Muslims). During the late 1950s through the mid-1970s, the percentage of "Others" in the total population increased steadily: to 1 percent in 1957, 3 percent in 1967, and 4 percent in 1972 and 1976. These percentages translated into significant numbers. By the mid-1970s, various polls and surveys revealed that "a projected six million persons were involved in Transcendental Meditation, five million in yoga, . . . and 2 million in Eastern religions." The "None" category was applied to individuals who said they had no religion or who identified themselves as atheists or agnostics. In 1952, the "Nones" represented just 2 percent of the total U.S. population. In 1957, the percentage rose to 3, then fell back to 2 percent in 1962 and 1967. In 1972, however, it jumped to 5 percent, and in 1975 it rose to 6 percent.[4]

Discovering Secularization

In the 1950s, religion suffused American culture. Political campaigns, discussions of U.S. foreign and military policy, the Pledge of Allegiance, even U.S. coinage and postage stamps bore the imprint of the religion of democracy.[5] In the 1960s, Americans "discovered" secularization. Sociologist Peter Berger noted its reciprocal relationship with religious pluralism. "In modern society," he observed, "secularity and pluralism are mutually reinforcing phenomena. Secularization fosters the civic arrangements under which pluralism thrives, while the plurality of world views undermines the plausibility of each one and thus contributes to the secularizing tendency."[6]

In the 1960s, few doubted that America was becoming a more secular society. So many facts and findings seemed to confirm that view. In addition to the increase in the number of Americans who identified themselves as "Nones" and "Non-affiliated," a 1968 Gallup survey revealed that church attendance by adults had declined to 43 percent of the population, the lowest point recorded to date.[7] Not surprisingly, public opinion polls showed that many Americans perceived a significant erosion of religious faith, whether in their own lives or in society as a whole.[8] In 1968 religious sociologists Rodney Stark and Charles Y. Glock announced the "dawn of a post-Christian era." While many Americans were "still firmly committed to the traditional supernatural conceptions of a personal God, a Divine Saviour, and the promise of eternal life," the trend, they declared, was "away from these convictions." What they called "a demythologized modernism" was "overwhelming the traditional, Christ-centered, mystical faith."[9] A decade or so later, however, religious scholars recognized that secularization had not made as much headway in America as some had predicted. Its main influence was in other areas of American life—in higher education, for example, and in the government. The religious culture and American society as a whole had not been greatly transformed.[10]

Nevertheless, secularization and the reconfiguration of American religion shaped a religious culture very different from that of the 1950s. The significant number of Americans who identified themselves as "Other religions," "None," or "Non-affiliated" belied that decade's Protestant-Catholic-Jew paradigm. Religion no longer exercised so pervasive an influence. And the perceived consensus among and within the three major faith groups disappeared, as each of them began to play a more distinctive culture-shaping role.

During the 1960s, as sociologists Wade Clark Roof and William McKinney observed, the Roman Catholic Church "moved into the mainstream, socially, culturally, and religiously." Its membership increased steadily between the 1950s and 1980s, and in the aftermath of Vatican II, the church became "much more pluralistic, more voluntary, more American." During the 1960s, Roman Catholics were especially visible in the civil rights movement and the protest against the Vietnam War. By the 1980s, the National Conference of Catholic Bishops (NCCB) had "emerged as a kind of collective conscience for the nation," treating the "great issues facing the nation and the world" and "addressing themselves not just to Catholics but to all Americans."[11]

Judaism (Reform, Conservative, and Orthodox) had many fewer members than Catholicism, but during the 1960s and later decades, Jews wielded a significant influence on American culture, generally in favor of liberal causes.[12]

The culture-shaping influence of Protestantism declined, partly because of its membership losses but mainly as a result of its division into three different groups: liberal mainline Protestants, conservative mainline Protestants, and evangelicals. The liberals and the conservatives disagreed on matters of theology and worship within the mainline denominations. The evangelical Protestants mounted an aggressive, sectarian movement outside those denominations. At a time when mainline Protestant membership was declining, the evangelicals experienced impressive growth. During the period 1965–75, for example, membership in the Southern Baptist Convention rose by 18.2 percent, in the Seventh-day Adventist Church by 36 percent, and in the Assemblies of God by 43 percent.[13]

Competing Cultural Strategies

The division of Protestantism contributed to the emergence of two new, competing cultural strategies, each of which presented a distinctive "moral vision" for the American people and the nation.[14] One, promoted by the evangelicals, was conservative and sectarian; the other, advanced by liberal mainline Protestants, was liberal and pluralist. Depending on the issues involved, the conservative strategists found allies among conservative mainline Protestants and conservative Catholics and Jews; the liberal strategists received support from liberal Catholics and Jews, as well as the so-called secular constituency.[15]

The Conservative, Sectarian Strategy

Conservative evangelicals, including pentecostals and fundamentalists, were the principal formulators of the conservative, sectarian cultural strategy. For many years, the National Association of Evangelicals (NAE), founded in 1942, was their chief representative and consensus builder, "the voice of evangelical Christianity speaking for morality and righteousness in matters of public concern."[16] The adjective "sectarian" denotes the strategy's exclusivist slant.[17] "Conservative" suggests, among other things, the "restorationist" motif so prevalent in evangelical pronouncements on American religion, culture, and society.

Evangelicals regarded the Bible as the inerrant, inspired Word of God, and the infallible, all-sufficient authority in faith and life. They believed that the moral principles presented in the Bible were absolute. They insisted that salvation came only to those who had been "born again," having professed personal faith in Jesus Christ as their savior from sin. They contended, moreover, that their particular faith was the only true faith, the only "true Christianity."

Evangelicals were devoted to spreading the one, true faith—in their phrasing, "evangelizing the world for Christ."[18] They justified their compulsion by citing the Great Commission handed down by Jesus in Matthew 28:19: to "go . . . and teach all nations, baptizing them in the name of the Father, and of the Son, and of the Holy Ghost." The Lausanne Covenant, formulated in 1974 by the International Congress on World Evangelization, provided a more modern definition of evangelism, widely accepted by evangelicals. "To evangelize is to spread the good news," it declared. Evangelizing involved "the proclamation of the historical, biblical Christ as Saviour and Lord, with a view to persuading people to come to him personally and so to be reconciled to God."[19] Evangelicals adamantly defended their right to evangelize, on the grounds that it was authorized by both God and the First Amendment to the U.S. Constitution.

The conservative cultural strategy was, first and foremost, an effort to address what evangelicals and many of their allies saw as a fateful moral crisis in America. They dated it from the 1960s, when, they said, Americans "discarded the Bible" and a rising tide of immorality, secularism, and godlessness engulfed the nation.[20] The Supreme Court school prayer and Bible-reading cases, decided early in the decade, marked the beginning of what evangelicals and other religious conservatives saw as a concerted effort to "marginalize" religion in America.[21] They were convinced that the "forces of secularism" were taking over, that "a new religion of secularism was evolving" as Americans "drifted away" from "Judeo-Christian values."[22] Their awareness of increasing religious diversity, especially the increase in non-Christians, also shaped this perception of America as a secular, "godless" nation.

The remedy for America's rejection of God and godly principles was obvious. "We desperately need a genuine revival of spiritual righteousness in our land," Jerry Falwell declared.[23] Of course "genuine" meant a revival of the one, true faith, conservative evangelical Christianity. Promoting such a revival became one of the main objectives of the conservative, sectarian cultural strategy, to be accomplished largely by means of individual and mass evangelism—one-on-one "sharing of the faith," the preaching of televangelists like Falwell, and the huge, citywide "crusades" conducted by Billy Graham. To reverse the moral and spiritual decline of America, evangelicals also called for the restoration of the ideals and principles of the founding fathers. "Either we will return to the moral integrity and original dreams of the founders of this nation," Pat Robertson warned, "or we will give ourselves over more and more to hedonism, to all forms of destructive antisocial behavior, to political apathy, and ultimately to the forces of anarchy and disintegration that have throughout history gripped great empires and nations in their tragic and declining years."[24]

The restorationist theme derived from evangelicals' particular, sectarian vision of the American past, which incorporated elements of the religion of democracy. They contended that America had been founded as a Christian nation in covenant with God, that the founding fathers were all religious men who venerated the Judeo-Christian tradition, and that the Declaration of Independence and the Constitution recognized the existence of a "Supreme Being" and were based on biblical principles.[25] God blessed America as long as it honored him and his "inerrant Word," but then, according to the evangelical narrative, in the 1960s Americans turned against God, violated his laws, and sank into immorality and secularism. Unless they returned to God and the Bible, and restored America to its original, Christian character, his judgment would surely be upon them and the nation would perish.

The Liberal, Pluralist Strategy

The chief formulators of the liberal, pluralist cultural strategy were the clergy, and especially the leadership, of the mainline, liberal Protestant denominations. Official pronouncements of their positions and objectives came from the various denominational bodies, as well as from the National Council of Churches (NCC) and the World Council of Churches. As mentioned earlier, Jews and liberal Catholics supported and helped promote the liberal strategy, as did many of the "Nones" and "Non-affiliated."

In analyzing the liberal theological orientation of late-twentieth-century mainline Protestantism, commentators stress its openmindedness.[26] Among the characteristics Robert Wuthnow cites are "greater room for interpretation of doctrinal creeds, self-conscious syncretism of symbolism from several of the world's religions, a more privatized form of religious expression, [and] mixtures of social scientific and theological reasoning."[27] Richard Hutcheson, writing in the late 1970s, contended that "openness, tolerance, inclusiveness, [and] acceptance of diversity" had been "characteristic of the mainline churches ever since the twenties," and he noted that although their doctrinal statements established "general parameters," they were "seldom applied in a way excluding anyone." Like Wuthnow, he also called attention to their incorporation of a "secular-scientific world view."[28]

While the liberal strategy reflected many of the traditional emphases of American Protestantism, it was very much a product of, and a response to, the 1960s religious reconfiguration. Although the significance of the conservative evangelical movement did not register immediately, it soon became a major concern. So did the mainline denominations' membership decline, which continued during subsequent decades. The increasing religious diversification of the American people also forced liberal Protestants to face the challenge of American religious pluralism. In January 1961, in a *Christian Century* article titled "Protestantism Enters Its Third Phase," Martin E. Marty pointed to the inauguration of "the first non-'Protestant' president of the United States" as a symbol of "the end of Protestantism as a national religion and its advent as the distinctive faith of a creative minority." For more than a century it had exerted considerable numerical, moral, religious, and political influence in America—commentators spoke of "the Protestant hegemony" and "the Protestant establishment." Now, Marty contended, Protestants must confront, and accept, the fact of religious pluralism in America.[29]

Thus the "pluralist" component of the liberal strategy was a recent addition. Mainline Protestants had long accepted doctrinal diversity *within* the Protestant fold, but they had not, as a rule, extended such openmindedness to Catholics, Jews and other faith groups—or to nontheists—before the 1960s. "Conscious theological pluralism" was "something relatively new to Protestantism," Hutcheson recalled in the late 1970s. But, he noted approvingly, the mainline churches were now "pluralistic by design," welcoming members who differed in "faith, style and opinion" and cultivating "ethnic, cultural, and economic diversity," which, he admitted, "most mainline churches have in short supply."[30]

During the religious revival of the 1950s, liberal mainline Protestants had engaged in evangelistic activity. But according to historian Thomas Berg, their "new embrace of religious pluralism" in the 1960s prompted them to reconsider its propriety and to reject it as a component of their cultural strategy.[31] Hutcheson also regarded the liberals' "discomfort" with evangelism as a consequence of their commitment to religious pluralism. "To offer Jesus to others as a savior from sin is to suggest that the objects of evangelism are sinners," he explained. Accepting religious pluralism meant being "willing to be inclusive rather than exclusive," an orientation "soundly rooted in the biblical image of the one body with a diversity of members and gifts." Liberals' acceptance of "the relativism of the modern scientific world view" and its distrust of "absolutes" was also a factor. Hutcheson described evangelism as "an activity of absolutists who are convinced that 'there is salvation in no one else, for there is no other name under heaven given among men by which we must be saved' (Acts 4:12)." Thus "to evangelize . . . is to imply that the views of others who have not accepted or do not agree with our own beliefs about Jesus Christ are not valid—or at least less valid than our own." Liberals were unwilling to take such a position, "even by implication."[32]

Like the evangelicals, liberals grounded their strategy in a particular view of American history. But their "usable past" differed dramatically from that of the evangelicals. Rather than viewing America as a Christian or Protestant nation, liberals emphasized the cultural and religious diversity that had long existed in

America, beginning with the first settlers who came "in search of religious freedom." Historians writing in the 1960s emphasized the multiplicity of religious traditions that flourished during the colonial and revolutionary eras. They called attention to religious dissent, disestablishment, and voluntaryism, and to the men and women who sought to preserve and expand religious freedom and diversity.[33] In 1966, in an article in the *Journal of Church and State,* Franklin H. Littell disparaged the idea of America as a "Christian nation" as a "myth" or "ideological construct" presented "in defiance of the historical record." Littell was a prominent, well respected religious historian, and this was a scholarly article, but he spoke as a liberal strategist when he described the perpetrators of the myth as "the champions of 'Christian America,' represented most vocally these days by the Radical Right of our spiritual underworld."[34]

Like the evangelicals, liberals made the framing of the Constitution the centerpiece of their "usable past." They, too, revered the founding fathers. But they valued them for creating a secular, as opposed to a "Christian," government, an accomplishment Littell described as America's "most important single contribution . . . to human liberty and law."[35] Whereas the goal of the conservative strategy was to restore that Christian orientation, the liberal objective was to conserve and enlarge the founders' legacy, adapting it to the changing political, cultural, and religious context in America.

Unlike conservative evangelicals, who regarded "the forces of secularism" as a key factor in America's moral crisis, liberals viewed "the secular" in a much more positive light. The inspiration came from the "new," "secular," or "radical" theology formulated by a number of liberal theologians in the 1960s. These "new theologians" were much influenced by two German Protestant theologians, Dietrich Bonhoeffer and Rudolf Bultmann. Echoing Bonhoeffer, they spoke of a "world come of age," in which human beings recognized that "we must live in the world as if there were no God." Heeding Bultmann's call for the "demythologizing" of Christianity, they advocated a "secular" interpretation of the gospel, stripped of otherworldliness and the preoccupation with personal salvation, focused instead on the problems of this world. This "rediscovery of the secular" concentrated liberal Protestants' attention on the economic, social, political and religious "structures" that, in the words of a 1965 NCC statement, "help determine men's lives."[36] The new theologians were very critical of what they called the "culture religion" of America, which sanctified the political, social, and economic status quo.[37] They urged the creation of a "servant church" that would focus its attention on current domestic and international problems, such as racism, poverty, war, the arms race, and environmental pollution. They exhorted the clergy, especially, to engage in "prophetic" ministry, challenging the myths, ideologies, and institutions of society in the light of biblical faith.[38]

Chaplains and the Competing Strategies

Although U.S. military chaplains conducted their ministry in the armed forces, they were neither ignorant of, nor unaffected by, the religious culture of the civilian community. They were not only aware of what was going on in it, they also "belonged" to it. They were members of its various churches, denominations, and faith groups, had been educated in its theological seminaries, and were certified by civilian endorsing agencies. Many chaplains held and publicly expressed views similar to those promoted by the cultural strategies, and the civilian strategists occasionally took a keen interest in the chaplaincies, their programs, and individual chaplains.

In the 1960s, strategists and chaplains on both sides of the religious divide became involved in controversies over three army chaplaincy programs: the General Protestant Service, the Unified Protestant Sunday School Curriculum, and Character Guidance. First Amendment, church-state issues were at the heart of all of the controversies, a reflection of the intense debate going on in the civilian religious community regarding the school prayer and Bible reading decisions the U.S. Supreme Court had rendered in the early 1960s.[39]

The Sectarian Challenge in the Army Chaplain Corps

The sectarian thrust that powered the 1960s conservative cultural strategy was already a conspicuous element of evangelical Christianity in the 1940s, when the National Association of Evangelicals began endorsing evangelical clergy for the military chaplaincies. The NAE had three main objectives. One was to offset the large number of Catholic chaplains in the armed forces. Another was to annul the liberal, mainline Protestant orientation of various chaplaincy programs. Third, and most important, the NAE regarded the armed forces as a great mission field and chaplains as "missionaries to the military." Evangelical chaplains were committed to converting as many military personnel as possible to evangelical Christianity, in accord with the Great Commission.[40]

Like civilian evangelicals, evangelical chaplains believed that their faith was the only "true Christianity." Neither Roman Catholicism nor mainline Protestantism qualified as such, because, among other things, they did not recognize the "new birth" as "God's method of salvation."[41] Another manifestation of evangelical sectarianism was intense dislike of "ecumenism," a doctrine and practice evangelicals associated with the mainline Protestant denominations, the National Council of Churches, and, especially, the World Council of Churches. Some evangelicals included the Orthodox churches and the Roman Catholic Church in the ecumenical movement. They opposed ecumenism partly because it rejected their conviction that theirs was the "true faith" but also because they associated it with "doctrinal deviations" such as relativism, universalism, and humanism.[42]

In the 1950s and 1960s, these sectarian beliefs prompted evangelical chaplains and their civilian allies to criticize two Army religious programs: the General Protestant Service (GPS) and the Unified Protestant Sunday School Curriculum. The GPS was a regular Sunday worship service conducted by a Protestant chaplain and adapted to the needs of personnel who came from a variety of denominations and churches. According to the 1958 and 1964 chaplain field manuals, Protestant chaplains were required to offer a "general service of worship which is acceptable and meaningful to the maximum number of Protestant personnel in the command." However, "no chaplain is required to conduct or participate in any service, rite, or sacrament contrary to the requirements of his denomination."[43]

Writing in the *Military Chaplain* in 1958, Chaplain Roy J. Honeywell emphasized the chaplain's obligation to the men attending the GPS, who represented a broad spectrum of Protestant belief, from liturgical to evangelical and from liberal to conservative. To offer them a sectarian service would be inappropriate. Although Honeywell recognized the chaplain's legitimate concern for members of his own denomination, he insisted that the chaplain's "debt to the few must not outweigh his obligation to the many," which required him to present "a living message which relates helpfully to the problems of great numbers of men."[44]

Most evangelicals regarded the GPS as more or less ecumenical and opposed it on that ground. They could have declined to participate, as the chaplain manuals stated. They did not challenge it directly. Instead, some, perhaps many, discovered or created opportunities for incorporating their sectarian beliefs and practices into the worship service—for example, preaching the doctrine of the "new birth" or introducing the altar call. They were encouraged in such stratagems by articles published in the *Chaplain* and various denominational magazines in which their fellow chaplains testified to their own experience preaching "doctrinal sermons" during the GPS or ending it with an altar call or invitation.[45]

The army implemented the Unified Protestant Sunday School Curriculum in 1954 to ensure continuity of religious instruction for the children of military personnel who attended chaplain-organized Protestant Sunday schools. Because the curriculum was intended to represent a cross section of the Protestant denominations, chaplains were required to use interdenominational or nondenominational rather than sectarian or denominational instructional materials.[46] Just as some evangelical chaplains found ways of incorporating their sectarian beliefs and worship practices into the GPS, others dodged the requirements of the curriculum by using instructional materials of their own choosing or presenting the prescribed ones in accord with their own theological stance. With the support of civilian evangelicals, they also protested against the curriculum.

In attacking it, the evangelicals employed two arguments. In the one, they contended that the materials were theologically and doctrinally inconsistent with the Bible and infused with liberalism and ecumenism. The NAE went so far as to charge that chaplains were being required to use Sunday school materials containing "heresy." The other argument was based on the First Amendment to the Constitution. The evangelicals asserted that in requiring evangelical chaplains to participate in the Unified Protestant Sunday School Curriculum program and use only the prescribed instructional materials, the army and the chief of chaplains were "forcing their particular brand of theological beliefs and church polity on others of different beliefs," thereby violating chaplains' right to the free exercise of religion. Appealing to the establishment clause, some evangelicals cited the Supreme Court's school prayer and Bible-reading decisions, despite the fact that they were generally regarded as anathema among evangelicals.[47]

The Office of the Chief of Chaplains defended the unified curriculum against evangelical protests. It stated that although it encouraged its usage, no chaplain was compelled to use it or need fear punishment for refusing to do so. The OCCH also noted that nothing prohibited the establishment of denominational Sunday schools where the need existed.[48] In 1964, the chiefs of the three chaplaincies asked the Defense Department's judge advocate general (JAG) to respond to assertions by the NAE and a U.S. senator that it was "illegal for the Armed Forces to select the religious literature to be used in the Armed Forces Sunday Schools." In the opinion it sent to the Armed Forces Chaplain Board, JAG argued that the content of the curriculum was "not propounded by the Government" but was selected by "ordained ministers of the church acting solely in their capacity as such, unhampered by any Government restriction." Seeing no violation of the

First Amendment establishment clause, JAG advised chaplaincy and military authorities to view the dispute over the curriculum as "a religious matter into which, instead of compliance, it would actually be a violation of the Constitution to interfere."[49]

In the 1970s, evangelical chaplains continued to criticize and evade the requirements of the GPS and the Unified Protestant Sunday School Curriculum. The OCCH continued to uphold the requirements. The stalemate ended in the 1980s, when the Army Chaplain Corps began formulating new policies aimed at supporting free exercise of religion for all military personnel, including chaplains. Even so, evangelical sectarianism did not cease. In the 1990s and first decade of the twenty-first century, as we shall see in later chapters, it focused on other, even more difficult issues.

Secularizing Character Guidance

The Character Guidance controversy lasted almost ten years, from 1960 to April 1969. The military people involved in it included two army chiefs of chaplains, Charles E. Brown Jr. (1962–67) and Francis L. Sampson (1967–71), the secretary of the army, the army chief of staff and the deputy chief of staff for personnel (DCSPER), the Office of the Judge Advocate General and the general counsel of the army, and the secretary of defense. At first the controversy was an "in-house" matter. Then individuals and groups in the civilian religious community, divided between the two cultural strategies, began to participate. Taking the side of the liberal strategy were the American Civil Liberties Union (ACLU), several liberal Protestant and interfaith organizations, three chaplain-endorsing bodies, and editorial writers of the *Nation,* the *New York Times,* the *Washington Post,* and *Commonweal.* Spokesmen for the conservative strategy included an Indiana congressman, the head of the National Association of Evangelicals, the editor of *Christianity Today,* the commander in chief of the Veterans of Foreign Wars, and the executive director of the Military Chaplains Association.

The controversy started when OCCH began receiving negative commentary regarding the CG program, from army personnel as well as civilians. Some commanders expressed disapproval of mandatory training in what they called "civil religion." A number of enlisted men complained to the secretary of defense and members of Congress about being "*forced* to go to a class in religion" and being told by a chaplain "that in order to believe in democracy you must believe in God."

Several liberal Protestant denominations, speaking individually or as members of a National Council of Churches Study Conference on Church and State, questioned various aspects of the CG program, including "denominational instruction in [religious] doctrine and ethics" and mandatory attendance. In December 1962, Lawrence Speiser, director of the Washington office of the ACLU, wrote a letter to Secretary of the Army Cyrus R. Vance, charging that trainees at Fort Devens, Massachusetts, were being subjected to "religious indoctrination" during Character Guidance lectures.[50]

Hoping to forestall further complaints, in March 1963 Vance informed army commanders that CG sessions were to be used exclusively for discussion of assigned topics and that such instruction should not take place in chapels or chapel facilities except in cases of military necessity. Chief of Chaplains Brown decided in June to issue a similar warning to his chaplains. In his monthly newsletter, after citing Vance's directive, he told the chaplains that to ensure the "non-religious nature" of Character Guidance training and to prevent its being confused with religious instruction, they were specifically prohibited from using CG lectures "to deliver a sermon, to announce religious services, to upbraid troops for nonparticipation in chapel programs, to show religious films or to expound their own theological views." Only the scheduled topic was to be discussed, and only approved Department of the Army training materials were to be utilized.[51]

In referring to the "non-religious nature" of Character Guidance, Brown expressed the position he and his successor, Francis Sampson, maintained throughout the controversy. Brown conceded that the program was "theistically oriented," but he maintained that it was "not a religious program" since it had not been devised to support any religious doctrine or institution and since it offered instruction not in "religious principles" but in "ethical, moral and psychological principles" underlying "traditional American concepts of personal integrity and responsible social conduct." He insisted that the army was "both free and obligated to uphold the basic moral and spiritual principles on which this Nation is founded," one of which was "a belief in God." In 1966, he did withdraw the "One Nation Under God" lecture from the program, citing two concerns: it might be construed as "trespassing on the sphere of religion" and the topic violated the First Amendment. (The second concern was apparently inspired by a "legal opinion" presented by the Office of the Army Chief of Staff.) In June 1968, the army issued a revised *Character Guidance Manual,* which stated explicitly that

CG "does not teach religion"—on the same page where it declared that "religion of a soldier's personal choice is recognized as a basis for the strongest moral motivation either in peace of war."[52]

On April 15, 1968, ACLU director Speiser wrote a letter to the under secretary of the army stating the organization's view that the Character Guidance "program as currently conceived and conducted raises problems under the First Amendment of the Constitution."[53] Five days later, having received a memorandum from the general counsel of the army regarding Speiser's letter, Army Chief of Staff Harold K. Johnson assigned the matter to DCSPER, with JAG and the chief of military personnel as the principal action officers. Then followed eleven months of deliberation by the military leaders and the Office of the Chief of Chaplains. Francis L. Sampson had become chief of chaplains in August 1967, and he provided a detailed description of the exchange of views in the annual *Historical Reviews* for 1967–68 and 1968–69.

During the deliberation, Sampson viewed Army Chief of Staff Harold K. Johnson and Vice Chief of Staff Ralph E. Haines Jr. as allies. He had good reason to do so. Besides the fact that they were men of strong religious convictions and active in evangelical endeavors, they expressed views sympathetic to Sampson's position during the controversy. The judge advocate general and the general counsel were another matter. In May 1968, the acting JAG sent his recommendation to DCSPER. Having analyzed the legal issues the ACLU had raised, he concluded that the CG program did pose "some troublesome constitutional issues." However, he thought the program could be "successfully defended if all religious passages are deleted from the training materials, reliance on religious texts and illustrations is discontinued, and additional guidance and adequate supervision are furnished to chaplains in the conduct of the program."[54]

Chief of Chaplains Sampson set forth his position in several memoranda to DCSPER. The OCCH was not opposed to policies prohibiting chaplains from inculcating religion in CG classes, and it believed that the existing advisories against "preaching" in such classes were appropriate and sufficient. However, it adamantly objected to the proposal to discontinue "reliance on religious texts and illustrations" in CG materials and lectures. The office specifically protested any prohibition of "the use of references to and illustrations from any religious literature as long as such citations are not used as authoritatively propounding a truth or doctrine."[55] Thus the inclusion and use of "religious references" in CG lectures became the main issue in the controversy.

In June 1968, Sampson received what he believed to be the "final word" on the controversy, a staff study prepared by DCSPER and approved by Chief of Staff Johnson. According to Sampson, it allowed chaplains to "incorporate references to religious materials as a supplement to published lesson plans," but warned them that they "must avoid 'preaching' or using the religious material 'in any manner which implies exclusive authority, priority, or validity for that particular source.'" Sampson wrote in the 1968–69 *Historical Review* that chaplains "could live with" those instructions "without compromising their professional interest in the Character Guidance Program."[56]

In early December 1968 came unsettling news. Sampson learned of a December 6 letter the general counsel of the army had written to the ACLU, stating a position "quite different" from that of the DCSPER staff study. The general counsel had declared that the religious references in CG materials were sufficiently numerous to indicate that "inadequate attention" had been paid to ensuring that the CG program was "wholly secular in its approach to training our personnel in matters of duty, honor, and patriotism." He said that the army was now eliminating "all passages with religious connotations," and he assured the ACLU that the problem would not recur. He added that "the Army has emphasized that its 'chaplains, in conducting this program, are performing a military function on behalf of the command and are not to use the program in any way as a religious training program.'" On December 11, the secretary of the army disapproved the DCSPER staff study recommendation allowing CG instructors to incorporate "religious references" into their lectures. He, too, emphasized the importance of the chaplain's acting "strictly in his role as a staff officer performing a military function for the command." Including religious references in his lectures would not only call attention to his religious role, but could also "lead to the charge that the Character Guidance program is a compulsory religious training program rather than a proper secular program related to training Army personnel on matters of duty, honor, and patriotism."[57]

Sampson responded by sending DCSPER a "Chief of Chaplains Position Paper on Character Guidance Instruction." He declared OCCH's strong opposition "to any inclusion of religion or religious dogma" in the CG classes, since that would violate the rights of the soldiers who were required to attend them. But OCCH was equally opposed to "any attempt totally to prohibit the use of religious references, illustrations, or materials," so long as they were not used to imply "exclusive authority, priority, or validity." Such a prohibition would, among other things, deny the chaplain instructor "recourse to the historical-religious or cultural-religious foundations of civilization as . . . reflected in the great literature of the ages" and would even preclude the use of many relevant quotations and illustrations from "the founding documents of the United States."[58]

A new round of deliberations ensued and continued into March 1969. Sampson issued a revised position paper in which he elaborated on his concern that chaplains would not be able to use references from "historical documents of the United States" such as the code of conduct and the Declaration of Independence. He declared that to eliminate such references because the chaplain was the instructor or because the quotations might be "constitutionally suspect" would be "carrying the striving for 'secularism' in Character Guidance instruction to the point of absurdity." In effect, Sampson was defending the teaching of the religion of democracy. "Intentionally avoiding any reference to the historical-religious or cultural-religious foundations which shaped our national heritage is not neutrality, but premeditated sterility, as well as the propagation of historical inaccuracies," he stated. Sampson also drew up a statement of "implementing guidance." JAG scrutinized it and Sampson had to accept some revisions in its wording. On the use of what Sampson called "historical-religious and cultural-religious references," for example, JAG required the following statement: "Historical and cultural references which have incidental religious significance will be used in a strictly secular sense and only where necessary for an understanding of the subject matter of a particular Character Guidance lesson plan."[59]

When DCSPER accepted Sampson's revised position paper, along with the "implementing guidance" as reformulated by JAG, Sampson could claim a partial victory in his fight against the secularization of Character Guidance. After all, JAG and the general counsel had initially demanded the elimination of all religious references, connotations, passages, texts, and illustrations.[60]

But the controversy did not end. It entered a new phase in mid-March, when the *Nation* ran an editorial thanking the ACLU for "persuading the Pentagon" to eliminate religious indoctrination that violated the Constitution; and Congressman William G. Bray (R-Ind.) complained in the House of Representatives that the chief of chaplains was "knuckling under" to "unjustified meddling on the part of the ACLU." A few days after Bray's harangue, on March 28 and 29, the newspapers broke the story of the army's directive. They reported that the army had ordered chaplains to eliminate all references to God and religious philosophy in the Character Guidance lectures in response to an ACLU complaint that the mandatory lectures were being used for religious indoctrination.[61]

The cultural strategists pounced upon the issue. Liberals praised the army's new policy in support of "the American tradition of separation of church and state" and "the democratic principle of religious freedom." Conservatives censured it as "only one of many attempts to remove God from our national life." Calling for a congressional investigation, the head of the National Association of Evangelicals asserted that "for the government to tell a chaplain, who is an ordained clergyman, what he may or may not say with reference to God in any lecture is to make a farce of the First Amendment."[62] On March 29, Secretary of Defense Melvin R. Laird intervened in the controversy, ordering a Defense Department review of the army directive. In a news release explaining his action, he declared that, as Secretary of Defense, he would "insure [*sic*] that this department abides by the law at all times." He also emphasized the "special obligation" on "our commanders" to "present an inspiring program of character guidance to members of the armed forces, particularly to the thousands of young men and women who enter the service each year." On April 2, he held a meeting in the office of the assistant secretary of defense, Roger Kelley. Representatives of the three chiefs of chaplains attended, along with various U.S. Army and Defense Department officials. They decided that the Armed Forces Chaplains Board should study the CG program and present its recommendations by April 11. At the conclusion of the meeting, Kelley ordered the preparation of a news release for the defense secretary to give to the media the following day.

In the April 3 press release, Laird made his position on the Character Guidance issue perfectly clear: "I want to state that there will be no prohibition against the use of 'God,' 'Supreme Being,' 'Creator,' 'faith,' 'spiritual values,' or similar words. Reference to these terms are [*sic*] appropriate for inclusion in the Character Guidance Program." Having overruled the army directive, he went on to insist that "all three military departments consistently have adhered to the position that the espousal of religious dogmas or particular sectarian beliefs is not the purpose [of] and has no place in the Character Guidance Program."[63]

Laird's pious grandstanding gratified the conservatives. He was celebrated as the man who had "restored God" to Character Guidance.[64] But his pronouncement did not reinstate CG's original, emphatically religious orientation. The program had already been somewhat secularized, as a result of Sampson's begrudging concessions. And the final recommendations presented in the Pentagon and Armed Forces Chaplains Board reviews allowed chaplains to use "religious references" in accord with the restrictions formulated by the judge advocate general.[65] The gradual, one might even say inexorable, secularizing process continued. Our Moral Heritage (OMH), the program that replaced CG in 1970, exhibited its predecessor's "somewhat secularized" orientation. Human Self-Development (HSD), instituted in December 1971, was a "wholly secular" program—and almost as controversial as Character Guidance.[66]

The controversies involving the General Protestant Service, the Unified Protestant Sunday School Curriculum, and Character Guidance all reflected two of the major concerns of the 1960s: the implications of the establishment and free exercise clauses of the First Amendment and the impact of secularization. The military and civilian communities continued to debate those issues during the latter part of the twentieth and the first decade of the twenty-first century.

The next chapter considers another, quite different controversy of the 1960s, a reaction to the other major shock of that decade—the Vietnam War. It pitted a segment of the civilian religious community, including advocates of the liberal cultural strategy, against the chaplains. The main issue was their support for and participation in the war. Of all the controversies of the 1960s, it had the most profound and lasting effect on chaplains and their ministry.

3

CHAPLAINS UNDER FIRE

Writing in the *Christian Century* in 1973, Navy Chaplain Richard G. Hutcheson wondered whether the military chaplaincy would turn out to be "one of the casualties of the Vietnam war."[1] Indeed it was. It did not receive a mortal blow, but during that conflict chaplains suffered criticism more serious than any they had ever encountered. Individuals and groups in the civilian religious community censured them for supporting and participating in the war, as well as for engaging in a military ministry. Opposition to the war was the main impetus behind this critique, but it was also informed by antimilitarist thinking and the secular theology.

Civilian and Chaplain Support for the War

As is well known, the civilian religious community was divided on the war, as were the denominations, churches, and faith groups that constituted it.[2] The prowar camp presented a number of arguments defending the war. The most common was the one put forward by President Lyndon B. Johnson and his administration—Americans were fighting in Vietnam to help the South Vietnamese resist communist enslavement and to sustain "the principle for which our ancestors fought in the valleys of Pennsylvania," that every people had the right to "shape its own destiny."[3] Conservative religious leaders portrayed the war as a struggle against communism, invoking the "domino theory" and the doctrine of containment, as well as the Declaration of Independence. Fundamentalists cited the Bible in preaching that Vietnam was a "holy war." Without going that far, the more moderate evangelicals appealed to Mark 12:17 and Romans 13:1–7. A very few religious leaders and ethicists, including the National Conference of Catholic Bishops, grounded their defense in the just war tradition.[4]

The great majority of army chaplains believed in the war.[5] In 1971, *New York Times* reporter Ralph Blumenthal interviewed some fifty chaplains during a four-week tour of military bases in the United States, and all of them expressed support. Indeed, Army Chief of Chaplains Francis L. Sampson told Blumenthal that fewer than a dozen chaplains had left the military because they opposed the war.[6] Unfortunately, sources from the Vietnam era do not provide much information on the reasons chaplains gave for supporting the war.[7] Some were more inclined to castigate the civilian antiwar movement than to explain why they believed the intervention and the war were justified, whether from a strategic or moral point of view.[8] The views chaplains did express were similar to those of prowar civilians. Many, if not most, agreed with or adopted the view expressed by their chief of chaplains. In the *Historical Review* for 1969–70, Sampson asserted that the United States had no choice but to honor the commitment it had made to maintain the independence of South Vietnam, and he praised President Johnson for refusing to "sell our national soul at the price of political expediency."[9] Later, interviewed by Blumenthal, he said that violence was an evil, "but the man who doesn't resist the violence becomes culpable, it seems to me. In the case of nations, innocent friends who are being attacked by aggressive nations, it seems to me if we made a commitment of friendship to a country, we owe it our support." Blumenthal also asked Deputy Chief of Chaplains Gerhardt W. Hyatt for his view of the war. He replied, "A man of discernment has to give his government the benefit of the doubt." Hyatt was a member of the Lutheran Church–Missouri Synod, and he seemed to be paraphrasing Martin Luther, who had argued that a Christian should not serve in an unjust war, but if he could not decide whether a cause was just, or if he lacked information, should support his government's position.[10]

Some army chaplains supported the war as part of the anticommunist crusade, insisting that the United States had a moral obligation to help South Vietnam resist communist "enslavement."[11] The head chaplain of the Fourth Infantry Division at Fort Carson told Blumenthal, "I believe there's an unjust aggressor keeping [the South Vietnamese] from their God-given freedom."[12] A Lutheran periodical quoted Chaplain LaVern W. Gardai justifying the war "on the grounds that it's a fight against Communism which is 'ungodly,' and a fight for freedom, which is 'connected with all the positive aspects of religion.'" In the *Army Digest* another chaplain invoked the religion of democracy: The desire to assist South Vietnam in obtaining the religious and political "blessings" Americans enjoyed was "implicit in the American way of life" and an outgrowth of America's "magnificent heritage of freedom."[13]

Chaplain Dissenters

A small minority of army chaplains publicly opposed the war, although how many others were critical of the war but remained silent, for whatever reasons, will never be known.[14] Chief of Chaplains Sampson did say, in 1971, that fewer than a dozen chaplains had resigned from the military because they opposed the war.

Various sources offer glimpses of chaplain dissent. The Office of the Chief of Chaplains included a brief note in the 1969–70 *Historical Review* about a chaplain who expressed disagreement with U.S. government policy and army operations in Vietnam and sought release from active duty. The OCCH had "recommended approval as an exception to policy, on the ground that he could not function effectively as a chaplain."[15] In *A Rumor of War,* Philip Caputo remembered a chaplain who addressed anguished questions to him in the officers' mess about whether "these boys are dying for a good reason" or "getting killed for nothing."[16] In 1989, in an article published in *Stars and Stripes,* Chaplain Gerald Mangham said that during the war, apparently while serving in Europe, he delivered sermons in which he criticized the U.S. intervention in Southeast Asia, although he said he avoided the political arena for fear of jeopardizing his ministry.[17]

There is more information on a few chaplains who questioned the war. In 1971, Chaplain William J. Hughes wrote in the Methodist *Christian Advocate* about the dilemma he had faced in trying to serve as a man of God in a war in which "the moral issues are ambiguous at best." He noted that many of his fellow chaplains viewed the war as a just cause in which the United States was assisting a weak nation to resist conquest and enslavement. For them, the issue and their duty were "clear-cut." But he knew of other chaplains who "wrestle daily with their consciences and pastoral responsibilities." When he went to Vietnam, he was opposed to U.S. involvement, and his opposition intensified during his tour of duty. Nevertheless, he had deliberately refrained from expressing his views in his preaching because, he said, "I genuinely felt that I and my men should not, indeed could not, oppose this war from within the area of conflict." At the same time, he had done what he could "to bring our involvement here to a quicker end." He had prayed for "our leaders" and the negotiators in Paris. He had "written and spoken to both prominent and ordinary citizens with the hope that I might encourage them to continue their opposition." He apparently discussed his views and actions with other chaplains, including whether he should stay in the army as a chaplain. "Some feel there is no room for me within the military," he wrote. "They feel in all honesty that I am hypocritical and false." Some had encouraged him to resign.[18]

Chaplain Billy W. Libby told of the lengthy period of ethical deliberation that preceded his public declaration of opposition to the war. It began in the early 1960s, when he was stationed in Europe. From that vantage point, he watched the U.S. involvement in Southeast Asia expand, all the while reading voraciously on the subject and "praying for God's will." In November 1967, he accompanied the 101st Airborne Division to Vietnam. "My year in the War was simultaneously the most rewarding and the most shattering of my time in the Chaplaincy," he remembered. "I came away convinced that the death and destruction, given the ambiguous and often contradictory objectives of the conflict, raised profound theological issues about the Creator and the Creation."

During the two years following his tour in Vietnam, Libby continued to pray about the war, still deliberating. He took a special interest in soldiers who were applying for conscientious objector status and insisted that they be treated fairly and in accord with military regulations. He apparently refrained from expressing his own doubts about the war. Then, in February 1972, having served eighteen months in the same troop unit and congregation, Libby announced in a sermon that "for me, the war is no longer morally defensible." He quoted similar statements by the United Methodist Council of Bishops, the general conference, and his own annual conference, as well as those of other Protestant denominational bodies. The reaction to his sermon surprised him. The members of his congregation included his unit commander and family, three battalion commanders, and many soldiers. "We discussed the sermon as part of my ministry in that particular congregation and in the light of my self-understanding as a Methodist minister," he wrote. The members of the congregation expressed a variety of opinions, but no one left the congregation. The response from his fellow chaplains, however, was "heavily negative." Only three or four openly supported him. One, a Lutheran, insisted on the right of conscience, even though he disagreed with Libby's position. Another, a member of the Disciples of Christ, was in the process of resigning from the chaplaincy because he, too, opposed the war. "The general response from Chaplains was to distance themselves from me," Libby wrote. "I had rocked the boat too much, and they did not want to identify with me." In June 1972, he was passed over for promotion to lieutenant colonel. The commanding general had written a letter, which he attached to Libby's officer efficiency report, in which he quoted from Libby's sermon, declared his views to be out of place in the army, and recommended that he be assigned to nonpolicy jobs that would not involve contact with soldiers.[19]

During the Vietnam era, few Americans knew about Hughes, Libby, or other chaplains who publicly declared their opposition to the war. Libby's views did not receive attention outside the military, and his *Military Chaplains' Review* article was published after the war ended. Hughes's statement appeared in 1971 in a limited-circulation denominational journal. He did, however, receive mention in a 1972 *New York Times* article as "one of the rare chaplains who have publicly questioned the morality of the war in Vietnam."[20] But even if chaplains' opposition to the war had been known—even if a greater number had spoken out against it—chaplains probably would not have escaped the criticism heaped upon them by antiwar people in the civilian religious community. Their dual role as clergy and military officers rendered them particularly vulnerable, and they were singled out for a special kind of censure.

The Origins of the Antichaplain Critique

Most of the criticism directed at the chaplains came from individuals who were associated with an organization called Clergy and Laymen Concerned About Vietnam (CALCAV), formed in January 1966. Its religious and ideological orientation was emblematic of the various strains of thought that contributed to the emergence of the antiwar movement of the 1960s. A brief review of the history of the movement will help in understanding why CALCAV, in opposing the Vietnam War, focused so much of its attention on military chaplains.

The earliest protests against U.S. intervention in Southeast Asia came from pacifists and liberals (both religious and secular) who had been involved, in the 1950s and early 1960s, in the campaigns against militarization, conscription, and the arms race and for nuclear disarmament. In 1963–64, the National Committee for a Sane Nuclear Policy, the Fellowship of Reconciliation, the American Friends Service Committee, and the Women's International League for Peace and Freedom held antidraft rallies, issued statements questioning U.S. policy in Vietnam, and offered assistance to conscientious objectors.[21]

Religious turmoil in South Vietnam, the assassination of President Ngo Dinh Diem (1963), the bombing of North Vietnam (1964), and the landing of American combat troops (March 1965) prompted other segments of the civilian religious community to criticize the intervention. The Protestant journals *Christian Century* and *Christianity and Crisis* and the Roman Catholic *Commonweal* published articles critical of the war.[22] Two ad hoc ecumenical groups protested the war in 1963 and 1965. Also in 1965, two liberal Protestant denominations, the United Church of Christ (UCC) and the American Baptists, expressed "doubts about military escalation" and offered "cautious alternatives without criticizing the government."[23]

According to Mitchell K. Hall, the individuals who formed CALCAV were frustrated by the noncommittal stance of most other denominations and faith groups and were morally offended by government officials' attempts to discredit the growing antiwar movement. In October 1965, some one hundred Protestant, Catholic, and Jewish clergy in New York City held a meeting to discuss U.S. policy in Asia. The event led to the founding of CALCAV, which soon became the leading independent religious antiwar group in the debate over Vietnam.[24]

Most of the members of CALCAV were theological and social liberals—mainline Protestants, Roman Catholics, and Jews. The most prominent of these were Harvey G. Cox Jr., John C. Bennett, John B. Sheerin, Abraham J. Heschel, Richard John Neuhaus, William Sloane Coffin, Richard Fernandez, and Robert McAfee Brown. Two pacifists, William Robert Miller and Gordon C. Zahn, assisted in disseminating the CALCAV critique.[25] Besides CALCAV, a number of religious bodies expressed views critical of the chaplains.

For CALCAV, the Vietnam War was preeminently a religious and moral issue. In *Vietnam: Crisis of Conscience,* published a year after its founding, the organization declared, "Our nation is embroiled in a conflict in Vietnam which we find it impossible to justify, in the light of either the message of the prophets or the gospel of Jesus of Nazareth."[26] CALCAV's primary focus was on the conduct of the war—what it termed "the immorality of the warfare in Vietnam," as evidenced by the high number of civilian casualties, the commonplace use of napalm and white phosphorous, the forced evacuation of towns and villages, the defoliation of crops, and the torture of prisoners to secure information.[27]

Members of CALCAV spoke often of the "moral outrage" and "horror" of the war, and of the feelings of shame, pain, guilt, and sorrow they experienced in contemplating it. They believed the war had provoked "a crisis of conscience" in America. The nation as a whole, its churches and synagogues, every American citizen—all needed to confront "the terrible realities" of the war. The religious community, in particular, had shown itself unwilling to provide guidance on "this primary and inescapable moral problem of our time." The professed mission of CALCAV was to "goad" the religious community "into responsibility." In full-page newspaper ads, books and articles, as well as interfaith conferences and demonstrations, it urged individual clergy and their religious bodies to speak and act against the war in such a way as to arouse a public outcry strong and broad enough to "make itself felt in Washington." Nothing less would satisfy. At a time when many churches and synagogues refrained from taking a position on the war,

CALCAV contended that silence and inaction were "immoral," even constituted "betrayal." It is important to note that in charging the religious community with particular delinquency in regard to Vietnam, CALCAV members, as clergy and laymen, admitted their own culpability. As Robert McAfee Brown pointed out in one of the essays in *Vietnam: Crisis of Conscience,* "We of the churches and synagogues, who should for years have been sensitizing the conscience of the nation, bear a particularly heavy burden of guilt for what has happened so far, and therefore bear a particularly heavy responsibility for righting that wrong and initiating a new direction."[28]

In 1971, CALCAV published *Military Chaplains,* a collection of essays on various aspects of the chaplaincy, some by civilian clergy and laymen, others by chaplains, all of them quite critical. Richard Fernandez and Harvey Cox, codirectors of CALCAV, explained why the organization had decided to publish the book. Fernandez wrote that it was an effort to explore issues "laid bare and exposed" by the Vietnam War: how "one 'does religion' within the military structure," "how one can be within a particular structure of ministry while at the same time remain faithful to the biblical tradition," and the "areas of conflict which plague the chaplaincy."[29] Cox, the editor of the book, pointed to the organization's desire to understand "why the whole American religious community failed so utterly to perform its religious and moral task" regarding Vietnam. The authors of the essays regarded the chaplaincy "primarily as a crucial case of how churches and synagogues relate themselves to the state and to its military arm." However, Cox insisted that CALCAV and the contributors were not "singling out the chaplain as dramatically different from the rest of us" and that "the question of how one speaks truth to power is not a question that chaplains alone must grapple with." Indeed, he added, "in many ways *all* professional ministers share in some way the chaplain's dilemma." Chaplains simply represented "a more severe case of what in some way infects us all."[30]

The Role-Conflict Thesis

The antichaplain critique advanced by CALCAV and other religious organizations was based on two premises: first, that chaplains experienced a conflict between their roles as military officers and clergy, and when confronted with a choice between loyalty to military values or their religious commitment, they invariably resolved it, consciously or unconsciously, in favor of the military values; and second, that all clergy, whether civilian or military, had a "prophetic responsibility."

The so-called role-conflict thesis was presented by Waldo W. Burchard in his 1953 Ph.D. dissertation and was summarized the following year in "Role Conflicts of Military Chaplains," an article published in the *American Sociological Review* (*ASR*)."[31] (Most people who knew of or cited Burchard's work had read this article, or a reprint of it, rather than the dissertation.) Burchard designed his research and interpreted its results in such a way as to prove what he already believed—"that the New Testament offers no ideological support for military activity" and that "it is impossible for the Christian in military service to put [its doctrines] into practice." He contended that the answers the chaplains gave to his questionnaire showed conclusively that they deferred to "military values" whenever they were confronted with a conflict arising from what he called "the Christian philosophy."[32]

The Burchard thesis seemed especially appealing to people who were antimilitary, antiwar, antichaplain, or all three. In the late 1950s and early 1960s, it found favor among prominent sociologists of religion.[33] In the mid-1960s, Pierre Berton summarized it in *The Comfortable Pew,* a book that was popular among the "new theologians." According to Berton, "Burchard discovered that the views of the chaplains he surveyed differed in no way from those of other officers on the deep question of the morality of modern warfare." After describing some of the views Burchard reported, Berton observed that "this outlook is very similar to the one that formed the core of Adolph Eichmann's defence during his trial in Israel."[34]

Like Berton, the antiwar critics assumed the validity of the role-conflict thesis. They found it particularly useful in explaining and censuring chaplains' involvement in basic training, counseling conscientious objectors (COs), and moral and religious instruction.[35]

Thus, in 1970, in the *Christian Century,* Robert E. Klitgaard charged chaplains with sanctioning, even "sanctifying," basic training, which he described as a "dehumanizing" ordeal that involved verbal and physical abuse, demeaning harassment, and moral indoctrination. New recruits, Klitgaard explained, often experienced a "moral crisis" during basic training. They confronted, in many cases for the first time, "the whole question of war and organized violence." But the chaplain failed to provide the ethical and spiritual guidance they desperately needed. "This is the man who is there to help you with your problems," Klitgaard wrote. Yet in carrying out his official functions in basic training, "he bestows the aura of religious respectability, however unwittingly, on dehumanizing practices.

In effect he sanctifies warmaking and professional criminality without really coming to grips with the trainee's moral dilemma. And he 'legitimizes' Vietnam by not discussing it, by almost necessarily avoiding the issues of conscientious objection and participation in an unjust war (could there be such a thing?)." No wonder that, in the minds of trainees, the chaplain came to be "seen as a sort of moral underpinning of the military system."[36]

In criticizing chaplains for avoiding the issue of conscientious objection, Klitgaard pointed to one of the chief concerns of the antiwar critics—the chaplain's role in CO proceedings. During most of the 1960s, the military defined conscientious objection according to fairly restrictive guidelines, none of which, it should be noted, recognized the validity of objection to a particular war, that is, selective objection or, more loosely, war resistance. The Selective Service Act authorized exemption from combat status only for a person who, "by reason of religious training and belief," conscientiously opposed participation in war in any form.[37] Beginning in the mid-1960s, chaplains were instructed by their chiefs of chaplains, the judge advocate general, and other authorities to follow guidelines set forth in the 1965 Supreme Court decision *United States v. Seeger,* in which the Court argued that "religious training and belief" included "all sincere religious beliefs which are based upon a power or being, or upon a faith, to which all else is subordinate or upon which all else is ultimately dependent."[38] In 1970, chaplains received further instructions to broaden their interpretation of the phrase "religious training and belief" in accord with a new Supreme Court decision, *Welsh v. United States,* which stated that "religious belief" included strongly held beliefs "which are purely ethical or moral in source and content" and not based on expediency.[39]

To obtain CO status, a serviceman had to go through a lengthy procedure that included notifying his commander of his intention to seek such status, preparing and submitting a written application, and being interviewed by a military psychiatrist, a middle-ranking officer knowledgeable about CO policies and procedures, and a chaplain. The decision on the application, which could not be appealed within the military, was rendered by a review board made up of representatives from the offices of the chiefs of chaplains, the judge advocate general, the surgeon general, and the chief of personnel.[40] According to the military, the chaplain's task in the CO procedure was not to act as a counselor to the applicant, shepherding him through the process of obtaining CO status. Nor was he to recommend approval or disapproval of the application. He was to assist the review board in making its decision by interviewing the applicant and then sending a report to his commanding officer explaining the source of the applicant's belief and presenting a judgment as to its "sincerity."[41]

In an article published in *Commonweal* in 1969, Gordon Zahn declared that the chaplain was "so closely identified with the military establishment in rank, in uniform, and in point of view" that he was "the last person to whom one might turn for an answer to a crisis of conscience."[42] Three years later, in *Military Chaplains,* he wrote that most chaplains probably would not go out of their way to prevent an applicant from obtaining CO status, "though, of course, some do." The more likely problem was the chaplain's "unconscious bias." Zahn contended that the chaplain would have difficulty understanding the applicant's objections to war, and that he would be inclined, as an "expert" on the matter, to try to convince him "that he has misinterpreted or misapplied the moral and theological principles which have brought him to those 'incorrect' conclusions."[43]

Two mainline Protestant churches expressed concern about the chaplain's role in the CO process. The United Presbyterian Church in the USA and a task force of the United Church of Christ recommended that the Department of Defense make the chaplain interview voluntary rather than mandatory, leaving it to the applicant to decide whether to have it and whether the chaplain should submit a report.[44] Among the reasons the UCC task force presented was concern about chaplains' "subtle bias" against CO applicants. "A conscientious objector . . . confronts the chaplain at the core of his ministry with a statement challenging its legitimacy," it explained in its report. Given the chaplain's "close identification with the military and its mission," he was unlikely to render an impartial judgment as to the applicant's "religious sincerity and moral integrity."[45]

Antiwar critics also saw evidence of chaplains' deference to military values in the way they conducted various army programs aimed at enhancing soldiers' religious faith, morality, and morale. In *Military Chaplains,* sociologists Peter Berger and Daniel Pinard were quite critical of the way chaplains provided religious instruction. Unlike Character Guidance lectures, which were mandatory, religious education classes were voluntary, and the chaplain was supposed to offer instruction in the religious tradition to which he subscribed. Instead, Berger and Pinard asserted, chaplains taught a "military religion" that portrayed military service as a religious as well as a civic duty; described authority, in all its forms, as "good and God-given"; defined conflict between the United States and communist countries as a "holy war"; and advocated values that enhanced soldiers' military efficiency.[46]

The UCC 1973 task force report focused much of its attention on what it called the chaplain's "morale maintenance" role. Although it granted "the importance and need for morale in the military enterprise, and even the covert role religion may play in achieving it," it declared that "turning the chaplain into a morale officer is an abuse of his or her clerical calling." It regarded morale building as "a military rather than a religious function" and expressed concern about the "danger of using religion as a rationale for the righteousness of our cause, sacrificing faith to the fighting spirit."

Character Guidance and code of conduct training posed a similar problem. The UCC task force objected to the chaplain's being required to teach a compulsory character education program "run by the commander . . . and intended to increase military efficiency, discipline, and well-being." It was even more critical of the military's expectations regarding code of conduct training. After quoting a chaplain field manual on the objectives of such training, the task force declared, "Such statements of military theology demonstrate quite clearly that it is the duty of the chaplain to underpin patriotism with religion, to infuse the killing of the enemy with the nuances of a crusade, and to sacralize dying with the assurance that one's life is given for God as well as country." Assigning chaplains such a duty constituted "an abuse of the chaplain and his ministry, exploiting his primary mission for military ends and compromising his liberty under the gospel." Without "that freedom of the pulpit," the chaplain became "a mere house priest."[47]

Urging Chaplains' Prophetic Responsibility

The UCC task force's reference to the chaplain's "liberty under the gospel" points to the other premise of the antichaplain critique, that chaplains had a "prophetic responsibility." The task force explained the phrase, saying, "We reserve his right to speak for God against the nation and its policies under the gospel."[48]

The idea of the clergy's prophetic responsibility was to a great extent a legacy of the "new" or "secular theology" of the 1960s. In *Military Chaplains,* Harvey Cox and other contributors pointed to the "exciting rediscovery" in recent decades, within Christianity and Judaism, of "the critical-prophetic dimensions of biblical faith."[49] One of the questions they sought to answer in the book was "How does a chaplain proclaim a prophetic gospel when he is wearing the uniform of the military, is paid by the state, and furthermore is dependent on his superior officers for advancement?"[50]

Judging from what the antiwar critics wrote and said on this question, the answer was, basically, "He doesn't" or "He can't." In their view, chaplains abdicated that responsibility—during soldiers' basic training, in counseling CO applicants, in providing moral and religious instruction, in building troop morale, and especially in supporting the war in Vietnam.

On this score, Gordon Zahn was the chaplains' harshest critic. In 1969, he published two articles, one in *Judaism* titled "The Scandal of the Military Chaplaincy" and the other in *Commonweal* titled "What Did You Do During the War, Father?" He explained that he used the term "scandal" in "its theological sense of *skandalon*: leading others into immorality or, at least, blinding them to the immorality of acts in which they are, or may become, engaged." His main charge was that chaplains failed to provide soldiers with "adequate moral guidance . . . on issues relating to the nature of a given war, its objectives, and its means." This, he insisted, was one of their major responsibilities—"awakening and instructing the consciences of the men so that they could not be trapped into performing patently immoral acts of war." Indeed, at one point in the *Commonweal* article, he suggested that such guidance might even include counseling "disobedience." In Zahn's view, by abdicating this responsibility, chaplains not only deprived soldiers of the moral guidance they desperately needed but also lent support to an immoral war. Even the chaplain who did not play the role of "morale-booster," encouraging men to fight and kill, was guilty of "scandal," according to Zahn. "His presence on the scene is certain to be interpreted as a token of spiritual compatibility, if not of outright endorsement," he declared.[51]

While Zahn criticized chaplains for refusing to provide soldiers with moral guidance on the war, the United Church of Christ task force censured them for shirking that responsibility in carrying out their role as advisers to commanders and their staff in Vietnam. Its report included a lengthy statement defining chaplains' prophetic responsibility. Among other things, they were "to be a Christian sign of contradiction within the military and the State, fulfilling the prophetic calling of discerning God's judgment on the policies and practices of the nation." Specifically, they were "to speak for the church by serving as counselor and critic to the commander and his staff on all matters of religion, morality, and ethics in the life and work of the military."

In the opinion of the task force, chaplains had failed on both counts. On the one hand, it censured chaplains for "remaining silent long after the truth about the war was known and their own churches had condemned it as immoral and unjust." On the other, it faulted them for failing to exercise a "restraining role" on commanders in Vietnam. The task force noted that chaplains and their defenders justified officer rank for chaplains on the ground that it gave them access to and influence over the commander. But the conduct of the war suggested otherwise. "We have witnessed such a profound lack of restraint in the military's policy, practice, and weaponry that credulity fails in the face of overwhelmingly contrary evidence," the task force declared. "We challenge the chaplaincy, from the Chiefs on down, to demonstrate in the case of that war—with saturation bombing, free fire zones, napalm and guava bombs, the torture of prisoners, political assassination squads, and My Lai—just how military chaplains have influenced commanders for moral ends in order to humanize the war."[52]

"Where Were the Chaplains?"

Of all the antiwar criticism directed at the chaplains, the most severe focused on their involvement in atrocities committed by American troops in Vietnam. This was partly owing to the antiwar critics' preoccupation with the conduct of the war. In January 1968, almost two years before the media broke the news of the My Lai massacre and its cover-up, CALCAV had published its own revelation of war crimes in Vietnam. *In the Name of America* was a compilation of news reports and magazine articles, many based on eyewitness accounts, that documented what CALCAV called the "consistent violation" of international agreements relating to the rules of war and the breakdown of "moral constraint" by U.S. and allied military personnel in Vietnam.[53] Thus the critics already knew or suspected that war crimes were endemic in Vietnam.

Nevertheless, they were shocked by My Lai, and their articles on the subject resound with anguish—and anger.[54] "Where were the chaplains?" Robert McAfee Brown asked in his essay in *Military Chaplains,* referring not just to My Lai but also to the "hundreds" of other such incidents reported in the CALCAV book. "On an issue of the greatest moral sensitivity, on which the representatives of the churches could have been expected to speak forthrightly and prophetically, why were the chaplains so silent?" In Brown's opinion, their silence was especially serious because "in this instance a special prophetic responsibility rested upon the military chaplaincy" and chaplains failed to exercise it. We would "expect a military chaplain to be leading the outcry when murder is committed in the name of freedom," he declared.[55]

Gordon Zahn also censured the chaplains for their "crashing silence."[56] In *Military Chaplains* he observed that only when journalists questioned them directly did chaplains say anything about My Lai. "Without such outside initiatives one suspects there would have been no public expression of moral indignation on the part of our uniformed clergy," he wrote. What they did say, "once the well-kept secret was exposed," he found unacceptable. He quoted two chaplains' statements published in the *Boston Pilot* in late December 1969. One came from the U.S. Army Republic of Vietnam (USARV) deputy command chaplain, who declared, "We do not debate the morality of the war in general or the morality of any particular war. Our job is to look after the spiritual welfare of the men." In the other, the command chaplain of the Military Assistance Command Vietnam (MACV) described the massacre as "a breakdown, a moral collapse," but cited, apparently in exoneration of the chaplains, the Character Guidance lectures in which they "set ethical standards" and the sermons in which they preached "Christ crucified." The deputy command chaplain's statement confirmed Zahn's contention, in "The Scandal of the Military Chaplaincy," that chaplains deemed "questions bearing on the morality of the war" to be "outside of their area of responsibility." And like other antiwar critics who regarded Character Guidance as a program for military indoctrination, Zahn doubted its effectiveness in preventing past or even "future atrocities."[57]

Nor did Zahn believe the chaplains who claimed that they were unaware of the massacre. Did they not hear "the widespread rumors that made 'Pinkville' a matter of common knowledge?" The answer, in Zahn's opinion, was obvious. "They must have known or they must have heard." And if they did protest, "assuming any did protest," it was "registered 'through channels' with care being taken that it did not become a matter of public record."[58]

The Indictment of the Chaplains

The phrasing antiwar critics used in censuring chaplains for their involvement in My Lai and other war crimes helps us to understand the thinking behind their general indictment of the chaplains. Recall that Robert McAfee Brown asserted that in the case of the My Lai massacre "a *special* prophetic responsibility was not exercised."[59] Zahn, discussing an incident in Vietnam that involved a violation of international law regarding prisoners of war, cited the "*direct*" and "*immediate* obligation" of a chaplain to speak out against such a practice, the more so "because the chaplain has the direct responsibility for the spiritual welfare and moral guidance" of the troops.[60] He pointed out the chaplains' particular responsibility to awaken and instruct the consciences of their men to prevent their being "trapped into performing patently immoral acts of war."[61] Harvey Cox described chaplains as "those most consistently responsible for interpreting the biblical faith to the military."[62]

Why did the critics hold chaplains to a "special prophetic responsibility"? The critics agreed that the churches and synagogues of America bore a "particularly heavy burden of guilt" and "a particularly heavy responsibility" regarding the war.[63] Chaplains were members of and endorsed by those same churches and synagogues. Presumably, they were equally culpable with the civilian members and clergy. But in *Military Chaplains,* Cox described them as "a more severe case" of the failure of the "whole American religious community . . . to perform its religious and moral task" regarding Vietnam.[64]

What, then, made chaplains "a more severe case"? Cox pointed out two components of the religious community's failure: "First, it failed to warn us that there were no compelling moral reasons to *get into* the war (in fact the moral arguments clearly indicated the opposite). And, then it failed to restrain the means utilized once the decision to wage war was made."[65] Basically, the religious community's failure was its silence on the intervention and the conduct of the war. Chaplains, by contrast, were not silent on the war. They publicly supported it. Apparently, in the mind of the critics, that constituted a worse failure of conscience and responsibility than "mere" silence.

Another reason the critics viewed chaplains as "a more severe case" was because they performed their ministry in and to the military. As we have seen, many of the critics regarded the chaplains' presence in the military as a symbolic endorsement not just of the war but also of the military as an institution. At the same time, however, the critics regarded that presence as an opportunity, unavailable to the civilian clergy, to exercise the prophetic responsibility more immediately and more directly within the military. Not only did chaplains have access to the soldiers—in the barracks, at the post, even in the combat environment—but as members of the commander's staff, they also were in a position to offer advice, and criticism, regarding command decisions and policies. Given such access, their failure to restrain troops and commanders from engaging in or ordering immoral acts of war seemed especially reprehensible. Several statements by antiwar critics reveal the rationale behind this kind of thinking. In a "Commentary by Religious Leaders on the Erosion of Moral Constraint in Vietnam," the introduction to *In the Name of America,* the twenty-nine signers insisted that all Americans were in some sense responsible for the Vietnam War and the crimes Americans committed in it. Just how responsible depended on what the individual had done or not done. One of the criteria by which the religious leaders judged culpability would seem especially applicable to the chaplains: "The citizen who knows of wrong committed in the name of his country, and remains silent, is thereby implicated in the perpetuation of those wrongs."[66] In an article in the *Nation,* Richard Falk, professor of international law at Princeton University, invoked "Nuremberg principles" in arguing that "the circle of responsibility is drawn around all who have or should have knowledge of the illegal and immoral character of the war" and that "anyone who believes or has reason to believe that a war is being waged in violation of minimal canons of law and morality has an obligation of conscience to resist participation in and support of that war effort by every means at his disposal." Falk also argued that while Vietnam raised "the issue of personal conscience . . . for everyone in the United States," it was "raised more directly" for those serving in the armed forces.[67]

Chaplains were not only *in* the military, they were *of* the military. Most of the critics of the chaplains, especially those who invoked the role-conflict thesis, were strongly antimilitarist. "There is nothing so vile as the arrogance of the military mind," Rabbi Abraham Heschel wrote in *Vietnam: Crisis of Conscience.*[68] In similar, highly charged language, Cox revealed his own antimilitary bias in discussing the reasons why CALCAV had decided to address the subject of chaplains in *Military Chaplains.* He declared that the time had come for a "thorough rethinking of the place of the military . . . in American life." Over the last thirty-five years it had grown to "grotesque proportions," commandeering "far more than a mere lion's share of our national goods," reaching into nearly every household through the draft, spreading "its metallic claws around the globe to hundreds of bases and bivouacs." Its "gargantuan scale and vast complexity" defied control. Any discussion of the military chaplaincy would have to consider the "crucial difference between, on the one hand, serving in an army we perceive principally as a means of promoting self-defense and securing domestic tranquility; and on the other hand, serving in an army that holds distant peoples (and discontented groups at home) in the mailed fist of imperial domination." The central question to be answered was, "Are there any armies in which ordained priests, ministers, and rabbis should *not* serve? Or are even the 'worst' armies (in the sense of the purposes for which they are used) equally in need of spiritual ministrations?"[69]

The invective antiwar critics heaped on chaplains reflected the hostility they felt toward "the military machine" and the militarization of America—hostility rendered more intense by virtue of America's involvement in an immoral, unjust war. "Warmonger" and "a greased cog in a machine for killing" were mild in comparison with the other epithets the critics used in disparaging the chaplains, including "religious eunuch," "military lackey," "house priest," "kept" clergy, and "mistress of the military." Zahn accused chaplains of "scandal," and *Time* magazine quoted Lutheran pastor Richard John Neuhaus saying that clergy who volunteered for the chaplaincy "expose themselves to 'spiritual prostitution.'"[70]

Despite Cox's disclaimer in his introduction to *Military Chaplains,* the critics did single out chaplains for especially harsh criticism. They declared that all Americans were "responsible," to a greater or lesser degree, for the war being fought in Vietnam. But when they assessed the chaplains' culpability, they used stronger language. Chaplains, they said, "legitimized" the war—in effect, were complicit in it. Their indictment included numerous charges: publicly supporting the conflict, endorsing it by their "presence" in the armed forces and in the Vietnam combat environment, oiling the military machine by sanctioning a dehumanizing program of basic training, discouraging conscientious objectors, preaching and teaching a "military religion," and, most grievously, failing to prevent or speak out against atrocities and war crimes committed by U.S. troops. All of these charges impugned chaplains' moral and religious integrity.

The Chaplains' Rejoinder

The antiwar critics' indictment was unprecedented—more publicized, more harsh, and more serious than any criticism chaplains had ever received from the civilian religious community. Individual chaplains and their civilian supporters felt compelled to respond publicly. They addressed both components of the antichaplain critique, the role-conflict thesis and the concept of prophetic responsibility.

Chaplains and their supporters conceded that some chaplains did experience role conflict as a result of their dual status as clergy and military officers. But they insisted that the "tensions" or "temptations" chaplains experienced were not significantly different from those encountered by civilian clergy in a parish ministry or in the administration of a denomination or faith group. They especially disputed the antiwar critics' assumption that *all* chaplains were subject to role conflict and that they *invariably* gave priority to military values. Chaplains and their supporters had more confidence than the antiwar critics in chaplains' ability to overcome such conflict and to remain faithful to their religious commitment.[71]

Chaplains responded in various ways to the charge that they were abdicating their prophetic responsibility. Some simply repudiated it, based on their own experience. In an interview published in the *Lutheran Witness,* Deputy Chief of Chaplains Gerhardt W. Hyatt declared that "the myth that the commanders impose undue influence on the content of a chaplain's ministry has no foundation." He observed that "problems" occasionally arose but were "resolved by the statutory and regulatory protection which is weighted heavily in favor of the chaplain."[72] Another Lutheran chaplain, John Himes, wrote that the "prophetic voice" was heard "surprisingly frequently"—and "where it is most effective, within the command structure where the decisions are made rather than on the streets."[73]

Two army chaplains took issue with Robert Klitgaard's assertion in his *Christian Century* article that chaplains legitimized the Vietnam War by not discussing it and by avoiding soldiers' questions about conscientious objection and the morality of war. William E. Austill declared it an "outrageous statement." He contended that "every chaplain is available for discussion of the issues perplexing the soldier." William L. Wells pointed out that he himself had participated in "a number of discussions on these subjects" in chapel-sponsored groups, which were attended on a voluntary basis.[74]

The chaplains' rejoinder included more than expressions based on personal experience. It focused on the fundamental issue in the controversy over prophetic responsibility. The antiwar critics insisted that the right and duty to exercise that responsibility was absolute—for all clergy, whether civilian or military, on any topic, and in any context. By contrast, chaplains and their supporters not only recognized but also approved and accepted limitations on that right and duty.

In 1970, the General Commission on Chaplains and Armed Forces Personnel, one of the largest of the chaplain-endorsing agencies, issued a document titled "Chaplains' Guidelines for Free and Responsible Expression of Conscience in the Military."[75] The document observed that chaplains were "commissioned under an agreement between the church and government" and that both institutions expected the chaplain "to conduct his ministry in accord with the dictates of his conscience and the requirements of his denomination, as he deals creatively with his obligations and opportunities in the military community." The commission did not minimize the difficulty of meeting that expectation. "These dual responsibilities to church and state must be held in a balance which is precarious at best," it declared. "This delicate tension must always be recognized."

The commission suggested that, almost inevitably, a chaplain would, at some time or another, be subject to a "conflict of conscience." Section IV of its document read as follows: "If in the course of his military ministry this inherent tension should become conflict, the chaplain must use every resource of both church and state to seek reconciliation." But there must be no compromise of "religious liberty." If the chaplain found that reconciliation was not possible, then "the dictates of his faith and denomination determine his course of action." The commission's emphasis on maintaining "religious liberty" and adhering to "the dictates of faith and denomination" suggests that resignation was not inconceivable. It recognized that in some cases, a chaplain would be unable to effect a reconciliation between the demands of conscience and denomination on the one hand, and the requirements of the military institution on the other. He would have to resign his military ministry.[76]

Recognition of the limitations the military institution imposed upon a prophetic stance was more explicit in an article published in a Yale Divinity School (YDS) journal in 1970, written by Sam Lamback, Dan Garrett, and John Brinsfield. Brinsfield was a YDS graduate; Garrett and Lamback were enrolled there when the article was published. The three men had recently served in Vietnam as "line officers at the company level." One of the subjects they discussed in the article was what they called the "intrinsic limitations defined by the nature of the ministerial setting" in which chaplains worked. Although they contended that chaplains could raise "moral questions" about "the day-to-day life in the military setting," and even about the conduct of the war in Vietnam, they observed that a chaplain who forcefully questioned "the 'raison d'être of the military per se" would "undoubtedly . . . be standing on 'thin ice.'" In view of the "intrinsic" limitations, they concluded that the chaplain "must be willing to accept a limitation on his prophetic role in order to minister in and to the military."[77]

Most chaplains seem to have accepted the trade-off. As mentioned earlier in this chapter, very few chaplains (including those in the other chaplaincies as well as the army's) vigorously opposed the war within the military or resigned their ministry in protest. Their acceptance of limitations on the prophetic role is also evident in the self-imposed restrictions some of them adhered to when speaking about the morality of the war.

Chaplain William Hughes, one of the few chaplains who publicly dissented against the war, deliberately refrained from expressing his antiwar opinions while ministering to combat troops in Vietnam. Chaplain William Wells, describing chaplains' discussions of the war with soldiers, noted that they "generally avoid the whys and wherefores of Vietnam in classes in which attendance is mandatory." He himself believed that approach was "both proper and constitutional." In his opinion, it was the duty of the president and Congress to determine policy in Vietnam, explain it to the American people, and appeal for their support. He thought it would be "basically wrong" for the military, including chaplains, to speak publicly for or against "any political, social, economic or foreign policy of our government." Chaplain Clarence Reaser took a similar position in two articles published in 1969 and 1972. In the earlier one, he wrote that if a chaplain thought the war was wrong, "he should speak on the subject to civilian communities, to legislators—to people, in short, who have the means to take the action he recommends." However, he questioned the judgment of a chaplain who expressed his views "on the evils of the Vietnam war" to military personnel, especially those serving in Vietnam, since they did not make policy or exert sufficient influence to change it. In the later article, he declared that the role of chaplains was "neither to attempt to alter administration policy nor to defend it."[78]

There are at least two reasons why chaplains accepted limitations on the prophetic role. One was that although the antiwar critics regarded the prophetic role as the sine qua non of a godly ministry, chaplains traditionally preferred the pastoral role and continued to do so in the Vietnam era.[79] Indeed, Chaplain Reaser flatly declared that the chaplain's main assignment was to speak to "the fears and problems and frustrations of men who are facing the basic issues of life and death," addressing "their deepest needs which go far beyond the question of whether or not the United States is being moral in its involvement in Vietnam."[80]

Another reason chaplains accepted the "intrinsic" limitations on the prophetic role was because they believed that in spite of those restrictions, they enjoyed what the General Commission on Chaplains and Armed Forces Personnel called "a measure of freedom" in offering moral guidance and even criticism regarding military policies. For example, the commission observed that the chaplain, in his pastoral ministry to soldiers in the combat environment, "can effectively and concretely emphasize the moral limitations upon the use of force" asserted in official military doctrine. In a more general way, he could be "a demonstrably effective counterforce to the thrust toward brutalization, cynicism, and contempt for human life that combat tends to entail."[81] Recall that Lamback, Garrett, and Brinsfield pointed out that chaplains could raise "moral questions" about "the day-to-day life in the military setting" and even about the conduct of the war in Vietnam. They offered the example of a chaplain assigned to an infantry battalion who was in a position, as a member of the commander's staff, to "follow closely" the unit's tactical planning. His "first-hand knowledge" of the means employed in waging war, combined with his ready access to the commander, provided him with "an unusual opportunity" for offering moral guidance on such questions as the use of "indirect fire capability"—and "with the potential for immediate correction."[82]

Some chaplains appealed to or used the language of the secular theology in justifying their accommodation to limitations on the prophetic role. (This despite the fact that the antiwar critics invoked the same theology in support of their absolutist view of the role.) As we saw in the previous chapter, the new theology taught that the task of the church, its ministers, and believers lay within the secular world and in the midst of its problems. Some chaplains were quite receptive to this "turn to the secular" and its implications for their military ministry. Indeed, they saw and described their ministry as a "secular form of service" or a "secular ministry." Thus Chaplain Albert Ledebuhr wrote that "the church is not an isolated island divorced from the realities of daily life" and that "the minister must serve the people where they are." Indeed, "the ministry of the church is at its best where the clergyman, armed with the gospel, comes face-to-face with people where they work, sleep, eat and play." In Ledebuhr's view, this was exactly how the chaplain conducted his ministry to the military.[83]

In 1973, in a *Christian Century* article, Chaplain Richard Hutcheson invoked the idea of chaplains' military ministry as a secular form of service in such a way as to defend chaplains' accommodation to the limitations of the military institution and at the same time disparage the antiwar critics and their antichaplain critique. "Is there a theology of involvement in the world?" he asked.

> Is there a place in the church's mission for Christian professionals within the secular structures, sharing the lives of those served, participating in the ambiguities and contradictions of their existence, seeking to minister to and even sometimes to modify the structures of which they are a part? Or does the mission of the church require its professionals to sit above the social and institutional structures of the world as "come-outers," unsullied by participation, free to urge disobedience, passing judgment as outsiders?

In discussing the issue of prophetic responsibility, Hutcheson admitted that a chaplain did not have complete freedom—to "pass judgments on the military command structure and its mission," for example, or to "make common cause with dissidents" or urge disobedience of orders. But, as other chaplains pointed out, he did have that "measure of freedom"—to "proclaim an authentic religious message within the military," "to engage with the power structure to affect the quality of life in it," and even to seek to "modify" the military structure.[84]

The antichaplain critique was devastating, but it had a tonic effect. It forced chaplains to reevaluate the military ministry in the light of the antiwar critics' criticism, the new secular theology, and the Vietnam War. Many of the changes chaplains made in their ministry in later decades were rooted in that reevaluation process—the new emphasis on "institutional ministry"; new doctrine that assigned chaplains a role in providing moral guidance *and criticism* of command policies, in garrison as well as the combat environment; and the development of a new role for chaplains as instructors in ethics (including the ethics of warfare), moral leadership, and world religions. In the next chapter, which recounts chaplains' experience in the Vietnam combat environment, we shall see how it, too, raised important questions about the purpose and conduct of military ministry, with important ramifications in later decades.

4

Navigating the Quagmire

While chaplains in the United States were embroiled in religious and ideological conflict with the antiwar critics, combat chaplains in Vietnam faced a very different challenge—ministering to soldiers involved in counterinsurgency warfare.[1] It was similar to the challenge Chaplain Richard Hutcheson described in speaking of the "theology of involvement." Ministering in a secular environment, "sharing the lives of those served, participating in the ambiguities and contradictions of their existence," chaplains "in country" did not enjoy the moral certainty of either prowar chaplains or their antiwar critics in the States. They not only came to recognize the ambiguities and contradictions of the war in Southeast Asia but also became aware of ambiguities and contradictions in the military ministry. Thus many combat chaplains were led to examine—or reexamine—their understanding of war in general and the Vietnam War in particular, of the nature and purpose of military ministry, and of their own moral accountability.

A Ministry of Presence

The 1964 and 1967 U.S. Army chaplain field manuals devoted a special chapter to "The Chaplain in the Combat Zone." Although chaplains should not and were not required to bear arms, they were expected to "accompany their troops into combat." In describing chaplains' responsibilities, the manuals emphasized their "personal presence." They were to provide "frequent religious services" in "the field" as well as engage in religious education. Pastoral care was especially important. The manuals advised chaplains that "continuous, person-to-person visits, spiritual care of the wounded, counseling, sharing of privation, and spiritual support in the hour of need make the chaplain a symbol of the concern of both God and the nation for the soldier under stress of combat." Such pastoral care was the essence of what came to be known as the "ministry of presence."[2]

For combat chaplains, the ministry of presence meant coming under fire with the troops, risking land mines and booby traps, seeing their men's torn flesh, severed limbs, and shattered bodies, comforting the wounded and dying on the battlefield, and helping to carry them to the medevac helicopter. Chaplain Jerry Autry recalled that "few days passed when soldiers in my unit didn't take fire or become involved in some kind of enemy activity."[3]

Chaplains did not have a great deal of control over the type of assignment they received in Vietnam, so some became combat chaplains as the result of arbitrary assignment. Others, by their own testimony, sought out a combat ministry. Some wanted to serve the troops in Vietnam that "had it hardest and hurt the most." Chaplain Claude Newby wrote that that was where he "belonged"—with "the grunts." Chaplain James D. Johnson explained that "acting and interacting" with soldiers in the field, "wherever they are, whatever they are doing," was also a way of demonstrating that "I care enough to be with them."[4] Other chaplains confessed to ulterior motives for seeking a combat ministry. Some wanted to be "where the action was" or to be "tested" by war. Others were escaping a humdrum civilian pastorate, or they calculated that assignment to an infantry battalion would advance their career in the army. Some assumed that "spreading the Gospel" would be easier in the combat zone, where troops would be most vulnerable and in need of the "spiritual comfort" that chaplains could provide.[5]

The chaplain field manuals were somewhat vague on the matter of where, exactly, the chaplain should locate himself in the combat zone. "Generally," they declared, "in combat and combat support battalions, the chaplain is located in the vicinity of the battalion aid station, so that he can rapidly respond to the need for ministering to the wounded and dying." But, they added, "he must continue to serve the needs of the men who are carrying the battle who are not casualties." The manuals concluded with an open-ended recommendation: "The chaplain should, with the concurrence of his commander, locate himself where he can best accomplish his mission. This may or may not be at the battalion headquarters."[6]

Some army chaplains used the term "circuit riding" in describing how they functioned in the combat zone.[7] Chaplain J. Robert Falabella, assigned to three battalions of the Twenty-fifth Infantry Division, wrote that he went "out to the field" for five days or so each week. Returning to the division base on Saturday, he would check on the wounded in the evacuation hospital and then, on Sunday morning, conduct three religious services at the base. That afternoon he would find transportation to begin another week in the field. Chaplain Jim Young, with the 101st Airborne Division, wrote that his troops were spread out among eleven fire bases, and visiting them involved a round trip of some three hundred miles.

Like most other chaplains, he sometimes traveled by helicopter, other times by jeep.[8]

Besides securing transportation and juggling base and field responsibilities, chaplains had to consider the needs of the men in the combat zone. Chaplain Newby, assigned to Headquarters, First Brigade, for duty with the Second Battalion, Eighth Cavalry, wrote that he spent little time in the battalion rear area: "My goal was to spend two or three nights in 10 with each infantry company." Which company depended on the battalion mission and current operations. He usually went to the company "most overdue for a visit," but sometimes he focused his attention "where the hurting was greatest, in consequence of recent events, current action, or because imminent action had high casualty-producing potential." He estimated that companies rarely waited longer than two weeks for a visit. Newby said that he "regularly . . . went along on platoon and company-size operations, which I considered maneuvers, not patrols, where the chaplain belonged." However, observing "unwritten, unofficial chaplaincy 'doctrine,'" he generally did not intentionally accompany squads.[9]

When Chaplain Joseph P. Dulany was with the First Cavalry Division (Airmobile), he generally ministered to company-sized units, and only occasionally to platoons. While visiting a company, he participated in whatever activity engaged it. At the end of the mission, or during breaks, he made himself available for counseling and talking with the men and also conducting religious services. He generally did not visit a company when it was in contact with the enemy, but if a unit came under fire while he was there, he "was present to comfort the wounded and pray for the dead." In most cases, however, he said he "flew in after a contact to counsel the men and conduct services."[10]

Other chaplains did go on combat operations with their men, and they reported that the troops often responded with surprise and appreciation, perhaps because they did not expect it.[11] Some chaplains who wanted to go into the field encountered resistance from senior or supervisory chaplains and commanders. Johnson believed that they based their opposition on the model presented in previous wars. Combat in the two world wars and Korea "proved that the chaplain's best location was at the battalion aid station," usually "positioned a mile or two to the rear of the front lines," he explained. They were also concerned that accompanying the troops in combat would expose the chaplain to "daily dangers such as snipers, booby traps, and firefights." His supervisory chaplains asked him, what if he were "wounded or killed, how might that impact the troops?"[12]

Autry's commander voiced other objections. "To be honest, I never understand the chaplain's place in combat," he told Autry. He said Autry would be "better off" back at the landing zone, "ready to see our soldiers at the aid station." And, he declared, "I don't want you out here because we don't have time to look after you. The troops are too busy to be concerned with anything but their jobs." Autry wrote that when he heard the commander's statement, he felt "marginalized." He observed that chaplains heard such comments "all the time," which he attributed to the fact that "those in command don't see what we do as essential or important." While some chaplains conformed to their senior chaplains' or commanders' restrictions, he himself was determined to be with the troops. "I was a minister, my life was dedicated to service, and I wanted to be where I could do the most good," he declared.[13]

Johnson reacted in a similar way. Both his division and brigade chaplains directed him to remain on the Mobile Riverine Force aid boat during combat operations. "But, they're not on the ground where I am," he observed. He decided that being wounded or killed during combat was "a risk I . . . must take if I am to do the ministry God has called me to do." He *would* go on combat operations with the troops, whenever possible. He *would* "be where they are, regardless." His supervisory chaplains were "not on the ground where I am." Johnson felt vindicated when he saw how his unit responded to his decision. The first time he went out on an operation, a dozen or more officers and soldiers commented that they could not believe he was doing so, and several asked him why. "This is obviously new to them, and they all seemed impressed," he wrote. "This makes me feel good."[14]

According to historian Henry F. Ackerman, chaplains regarded the ministry of presence as the most important ministry they performed in Vietnam.[15] One of the reasons they gave was the feeling of emotional and spiritual "connectedness" they had with the soldiers they served. It came partly from sharing the conditions in which the troops lived and worked, partly from receiving their approval and appreciation. "There was never a time before or since [Vietnam] that I was so integrated into a group of men," declared Chaplain Robert L. Campbell. "There was a sense in which I was one with them as a 'suffering servant.'" Jerry Autry, remembering "the hard-to-believe acceptance of the troops, their utter willingness to share themselves, their tremendous desire to have me with them," wrote that he "suddenly . . . felt useful, less marginalized."[16]

Chaplains and Troop Morale

According to both the 1964 and 1967 chaplain field manuals, the chaplain, as a member of the commander's special staff, was supposed to provide him and his staff with "advice, information, and plans on matters pertaining to religion, morals, and morale as affected by religion." In a section on the chaplain's "mission" in the army, the two manuals declared that he "has a leading role in the deliberate and systematic cultivation of moral and spiritual forces in the Army." This included "stimulating and guiding the growth of the soldier's spiritual and moral sense of obligation," which would, among other things, enable him to be "a devoted defender of the nation."

In the United States, chaplains fulfilled their morale responsibility in the lectures they presented in the Character Guidance program and in training soldiers in the code of conduct. As we saw in chapter 1, the 1961 CG manual declared that one of the purposes of the program was "to give the American soldier a deeper knowledge and love of the country he serves, and a better understanding of why it is worthy of his service, and, if need be, of his very life." Similarly, one of the stated objectives of code of conduct training was to "develop the moral fiber and religious motivation of the American soldier to fortify him with the weapons of faith and courage."

The 1964 and 1967 chaplain field manuals noted that "combat conditions generally preclude a formal program of character guidance instruction" but insisted that the chaplain "can advise his commander and assist the staff by planning and leading group discussions in the positive aspects of moral courage, the spirit of sacrifice, [the] sense of duty and integrity." The chaplain should also exemplify "the noble qualities of character which success in combat demands" during his "daily contacts" with the men and his visits to small groups.[17]

In an address to the 1968 convention of the Military Chaplains Association, Chief of Chaplains Francis L. Sampson listed the qualities he believed essential to chaplains' "effective ministry," several of which had a bearing on the chaplain's morale role. One was "an intense loyalty and love for his country." Another was "a great respect for the military" and "what it stands for." Chaplains should know and convey to the troops that the military "is not an aggressive force, seeking enemies to destroy," but is dedicated to preserving peace and national security and implementing "national policy as it pertains to treaties with friendly nations who, of themselves, are not strong enough to repel aggressive powers from without." There was "nothing un-Christian in these missions," Sampson insisted.[18]

Whether because of their own inclinations, official guidance in the manuals, or guidance from the chief of chaplains, some chaplains serving in Vietnam vigorously promoted support for the war among the troops. Some soldiers who had seen combat in Vietnam expressed very negative views of these chaplains. In an investigation sponsored by Vietnam Veterans Against the War, a soldier who had served with the 101st Airborne Division recalled his battalion commander assuring his men that "there'll be enough VC [Vietcong] to go around," and the chaplain chiming in with, "It's better to give than to receive and do unto others before they do unto you."[19] Other sources provide additional glimpses of zealous morale-boosting chaplains in Vietnam. Michael Herr, in *Dispatches,* wrote of a chaplain who began one of the many sermonettes aired over Armed Forces Radio by praying, "Oh, Gawd, help us to learn to live with Thee in the struggle against Thine enemies." The *Washington Post* published a news story in 1970 about a chaplain who, when asked by Col. George S. Patton III to pray for a large body count, offered the following petition: "Oh, Lord, give us the wisdom to find the bastards and the strength to pile on."[20]

The memoirs combat chaplains published during and after the war offer a different perspective on chaplains' morale-building efforts in Vietnam. Being with the troops in the combat zone, chaplains were able to judge their morale by their talk and demeanor—and their "weary and hollow eyes." Even in the early years of the war, chaplains detected a general decline in morale. The men arrived in country believing in the American intervention, Chaplain Earl C. Kettler wrote, but after being there for several months, they tended to "do a lot of staring into space." Over time, as Chaplain Autry observed, the troops "lost faith in their cause." He and other chaplains attributed the demoralization partly to civilian protest against the war, but mainly to the merciless grind of combat, especially "the continual loss of men through death and serious injury."[21]

Chaplains came to realize that for most of the men, surviving the war and going home to the United States was the main goal and that soldiers fought not so much for country or for God as for each other.[22] Chaplains also understood, probably better than other officers, the nihilism that gripped so many soldiers, exemplified in Ron Kovic's often quoted declaration, "There is no God for me after Vietnam" and in the proverbial "It don't mean nuthin.'"[23] Chaplain Larry Haworth wrote that although he himself "knew, felt, and observed God's presence in trying and difficult situations," he well understood why so many soldiers "honestly ask the question, 'Where was God?' and they honestly say that they did not feel him there." It was, Haworth pointed out, because so many of their experiences were "so horrible, traumatic, and unspeakable that many men and women honestly believe[d] that God had abandoned them."[24]

Chaplains may have been able to recognize, understand, and empathize with the troops' low morale because they themselves became disillusioned with the war. "Any confidence I had in the Vietnamese effort has completely faded," Chaplain Kettler wrote in a letter to his wife during his 1964–65 tour of duty. "The war here isn't worth one American life," he declared in another. He confessed to having become a "cynic about Vietnam." The weakness of the South Vietnamese army was one reason; he also cited the failings of American political and military leaders. In one letter he criticized a speech of President Johnson for its lack of "conviction" and failure to explain why the United States was in Vietnam. In several others he deprecated administration and military leaders' optimistic pronouncements about the progress of the war and the high morale of the troops.[25]

Chaplain Falabella, whose tour of duty included the 1968 Tet Offensive, wrote of becoming "more and more disillusioned with the competency of the American military and political leadership." Chaplains Dulany and James Johnson also criticized what they regarded as mismanagement of the war, putting most of the blame on "the politicians who keep this war going." They were "out of touch with the impact of the real war," Johnson declared. Dulany faulted military leaders for not having the courage to tell the politicians "to leave the tactical management of the military situation to the commanders on the ground." Chaplain Newby decried the influence of "civilian leaders," especially their "limited war objectives and the strategy of gradual escalation," as well as the growing number of "media attacks on the Army, the fighting soldier and the chaplains who served him."[26]

Given the state of their own morale and their assessment of troop morale, it is not surprising that some combat chaplains eschewed bellicose morale boosting.[27] Chaplain Falabella described two occasions when he sought to bolster morale. The first was an evening mass held after he and his men had spent a long, exhausting day in the field. They had already endured one downpour. Now, as he started the service for the "weary, soaked GI's," it began to rain again.

> In exasperation rather than with any irreverence, I prayed, "Father, can't you see how tired these boys are? How wet? They couldn't even get a

> hot meal tonight because of the rain. Now they are here at mass to pray to you and it's starting to rain again! What kind of a Father are you? Are you going to let it rain at mass after all they have been through today?"

Falabella sought to raise the men's spirits partly by presenting himself as their advocate, taking their side even to the point of irreverence, but mainly by calling upon God to recognize and alleviate their distress. The prayer was an appeal to God, but it also showed how much Falabella sympathized with and cared for his men. The fact that, as Falabella noted, "the rain did stop for the remainder of the service," surely heightened the men's perception of a caring God and chaplain.

On the other occasion, Falabella did seek to strengthen the men's combat motivation. They had been "going at it for some time," he remembered. "They were overworked and near exhaustion." Their faces looked "tired and haggard," and they were clearly under "tremendous strain." Asked by the commanding officer to "go around and speak with the troops," Falabella walked the entire perimeter, talking with groups of men:

> As I went around I recall commenting to the men on the tough going everyone was having recently. When they would mention our losses and seem overly despondent, I would jump on them about the fact that it was about time the Mech started being its old self; that they had been pushed around long enough. It was time to start getting mad. . . . "They've been sniping at you, placing booby traps and mines against you; mortaring you. Now, the enemy is getting brazen enough to come right up to you. It's about time we started to do something besides feeling sorry for ourselves."

Falabella explained in his memoir that the enemy had the advantage that night. The battalion was "in a critical posture," not having had time to prepare for a fight. "If the enemy did attack, however, I felt convinced our men would fight like their old selves, despite their weariness and set-backs, but it would be costly." He felt sorry for having spoken harshly to them: "There was nothing more I wanted to do emotionally than openly sympathize with them in their complaints over their bad luck. But I could not enjoy the luxury of those emotions when their lives were in critical danger." The harsh words were necessary. Some targeted the enemy, but it was the ones he directed against his men that counted, carefully chosen for the purpose of strengthening unit cohesion—by reminding them of their "old self"—so that they would be able to fight effectively and survive.[28]

The same concern for and identification with the soldiers' psychological and emotional state may be seen in two other chaplains' descriptions of their morale role. In these cases, the chaplains dealt with soldiers who had refused to go into combat.

At his commander's request, Chaplain Johnson interviewed three men who were in confinement for having refused combat. For thirty minutes he listened to them explain why they had done so. In his memoir, he wrote, "What can I do, encourage them to change their minds so they can get shot or killed? I tell them they'll be court-martialed if they continue to refuse to go to the field. We have a matter-of-fact conversation about their options. Their rationale is that going to LBJ [Long Binh Jail] is better than getting killed." Following the interview, Johnson returned to his ship and held a memorial service for a soldier who had been killed in combat. Afterward, he wrote that he could not help thinking that "if this memorialized soldier had refused to go to the field, he'd probably still be alive."[29]

Chaplain Dulany told in his memoir of being called to the battalion aid station to see about some men who had withdrawn from a combat situation and disobeyed their commander's order to go back into battle. "The commander seemed unable to understand why these men refused to obey his order and instead 'fell out,'" despite the fact that they "exhibited classic symptoms of battle fatigue," Dulany wrote. He and a Captain Cave attended to their needs. Dulany counseled them and gave them an opportunity to "tell their stories." Cave administered a sedative and sent them to a tent for a night's rest, after which they returned to duty.[30]

Chaplains Bearing Arms

Just as their identification with the soldiers influenced chaplains' approach to morale building, it also shaped their thinking about the bearing of arms. In a chapter titled "The Chaplain in the Combat Zone," the 1964 and 1967 chaplain field manuals declared that the "*chaplain is a noncombatant*. He should not bear arms; he will not be required to bear arms. He is protected by the provisions of the Geneva Convention in this role."[31]

According to Chaplain Autry, most chaplains in Vietnam did not carry weapons. Some cited army regulations and the Geneva Convention as a reason. Some contended that carrying a weapon would violate the tenets of their faith or calling.[32] However, other chaplains did carry a weapon of some kind, either concealed on their person or stowed in their jeep. In fact, some of these chaplains were "assigned" or urged to obtain a weapon and, if necessary, given "unofficial" training in using it. The .45-caliber pistol seems to have been the principal side-arm. Chaplain Dulany kept an M-79 Grenade launcher with flechette rounds between the front seats of his jeep.[33]

Some chaplains arrived in Vietnam in agreement with the regulation against bearing arms and then came to oppose it. Their reports of the debates they carried on in their minds and with other chaplains regarding the regulation give an idea of the various factors that ultimately influenced them to make the change. Some of them seized upon the phrase "will not be required to bear arms" while ignoring the one that said "he should not bear arms." More commonly, chaplains contended that in Vietnam, the noncombatant protection afforded by the Geneva Convention was irrelevant, since the Vietcong did not observe the convention or regard chaplains as noncombatants.[34] Other chaplains explained their decision to carry a weapon by saying that they wanted to avoid being a "liability" or a "drag" on troops involved in combat.[35] When Chaplain Autry accompanied troops on combat operations, he noticed that they seemed "uncomfortable" when he was in the field with them. He guessed they felt he had to be protected. Autry decided that the soldiers should not be worrying about him; they needed to focus on their own safety. "When I began carrying a weapon, they were less concerned because I was not someone they had to look after," he wrote.[36]

Chaplain Falabella presented a detailed explanation of how his thinking about bearing arms changed during his tour in Vietnam. He initially believed carrying a weapon was "unfitting a man 'of the cloth.'" Upon arrival in Vietnam, and while awaiting assignment, he was briefed by the USARV chaplain, who told him, among other things, that "I should not forget that I was a chaplain and so conduct myself." Falabella asked what he meant by that and was told that a chaplain should not carry any weapons. He wrote that even then, the senior chaplain's advice conflicted with what he, Falabella, "understood to be the Army regulation[,] which simply stated that a chaplain would not be required to carry a weapon." He considered it "a fair and reasonable statement, leaving it up to the conscience of the chaplain to make the final decision." However, at that time he had no thought of opposing the regulation; the senior chaplain's advice seemed "realistic and sagacious."

For a while, Falabella was comfortable with not carrying a weapon. Then, prompted by conversations he had with soldiers, he began pondering the matter. The soldiers would ask about his views regarding the morality of killing and of the war. He would give them, and "the GI's would look at me and say, 'What you have said is interesting, Chaplain, but I do not see you carrying a gun.'" The more he thought about that, "the more it did seem to be a contradiction." He asked himself,

> Is there something morally wrong in carrying a weapon? Is it morally wrong for me to do what they are doing? . . . I am a man, a human being, just as they are. That I am a priest or preacher makes me by that title alone no more of a man of God than any other man or woman on earth, for we are all people of God.

Falabella concluded that since he did not believe it was wrong for the men to carry weapons and use them in Vietnam, there was no reason why it should be "wrong, morally or otherwise," for him to carry a weapon. "The environment I am in is no different from theirs."

Falabella believed that whether to carry a weapon was a decision to be made on moral grounds, a matter of conscience. He reached that decision during Tet, based on his "actual experience" in the combat environment. Having changed his thinking, he not only disparaged his former view as "naïve" but also felt misled by the USARV chaplain. Here Falabella expressed the sense of superior knowledge and realism that other combat chaplains sometimes exhibited, which they claimed was the result of being "on the ground," "with the men," in the combat zone. Initially, he had thought the senior chaplain "realistic and sagacious," but now he decided that "he no more knew what the real situation was 'out there' in the hedgerows, rice paddies, plantation fields and woods than most other high-ranking officers with their immaculately clean uniforms, shiny boots and air conditioned offices." The senior chaplain and the high-ranking officers, as well as the "visiting VIPs from Congress," never "lived in the field with the troops who were doing the actual fighting," he declared. "They no more knew the real situation than did those pseudo-intellectuals in the States."[37]

Combat Chaplains and the Morality of War

Chaplains in Vietnam had considerable difficulty responding to the questions their soldiers posed about the morality of killing and the war. Their own low morale and their disillusionment with the way U.S. political and military leaders justified and conducted the war were partly responsible. However, the main reason was that they were ill prepared, theologically and ideologically, to provide well-reasoned answers.

Because many of the denominations and faith groups that made up the civilian religious community were divided on the war, chaplains could not rely on them for guidance on the moral questions it raised. Older chaplains, who had attended seminary or engaged in a civilian ministry in the 1950s and early 1960s, remembered a time when, for the most part, the civilian religious community supported military conscription, the nation's wars, and its foreign policy. Now, in the late 1960s, that consensus had broken down and chaplains had to decide on their own what position to take on the intervention in Southeast Asia, the conduct of the war, antiwar protest, and conscientious objection.

In addition, most chaplains had received no formal instruction regarding the morality of war and related subjects. In the 1950s and 1960s, no theological seminaries offered the military chaplaincy as a major area of study, and only a few offered courses designed to prepare students for a military ministry—and those courses generally did not consider such subjects as the morality of war and the rights of conscience.[38] The three chaplain schools did not provide theological or ethical instruction in the morality of war. They assumed that chaplains had already received sufficient training in theology, ethics, and counseling and therefore focused on providing them with "military and professional orientation."[39] Thus the commandant of the U.S. Army Chaplain School pointed out that its main objective was to familiarize chaplains with the "framework" in which they would operate by helping them acquire "a reasonable amount of military information and background," a "thorough knowledge" of their "responsibilities, duties, and loyalties" as a staff officer, and "a reasonable understanding of how the combat arms fight" and of "the varied concepts of warfare in the modern age."[40]

By their own testimony, some chaplains had not thought very much, or very deeply—and others had not thought at all—about the morality of war in general or the morality of the Vietnam War in particular. Chaplain David E. Knight, who was thirty-seven years old when he went to Vietnam, remembered arriving with "lofty notions of the justness of our cause," seeing himself as "a soldier pitted against the godless forces of Communism." Chaplain Dulany recalled a shipboard discussion regarding the reasons for the American involvement in Vietnam in which he apparently did not participate but only listened. "At that time," he wrote, "I was unable to verbalize a coherent justification for our forces being in Vietnam and more directly for my participation in the exercise. I suppose that if pushed far enough I would have said that I was 'along for the ride.'" Chaplain Autry confessed that before going to Vietnam, he had "never seriously thought about whether the war was right or wrong. . . . It simply had to do with duty, although to be honest, I probably hadn't thought that through either."[41]

Chaplain Curt Bowers presented a detailed description of the problems he encountered during the twenty-one days he and his soldiers spent aboard a ship bound for Vietnam. The question the soldiers most often asked was "Can I be a Christian and still kill?" In trying to answer that question, Bowers discovered that there were "very few simple answers" to it. Searching the Bible for references to peace and war, to loving one's enemies, and to killing, "it seemed that the more we looked at the Scripture, the less simple the answers became with regard to the place of the soldier in the Christian community." Bowers reported the answers he gave to some of the soldiers' questions. "What about turning the other cheek, Chaplain?" Bowers wrote that "the only answer I could give them was that entering combat is a kind of turning the other cheek. You're letting the enemy smite your cheek rather than those you love." He quoted Romans 12:18: "If possible, so far as it depends on you, be at peace with all men." However, he added, "History shows that if we appease an aggressor, . . . we only encourage more aggression." Commenting on the "scary responsibility" of answering the soldiers' questions, Bowers noted that "we may not have arrived at any great or grand conclusions, but we did come face-to-face with some realities that life had never asked us to consider before this time."[42]

Once chaplains found themselves in the combat environment, questions about the morality of war could not be ignored. But Vietnam was not the best place for meditating on them. Chaplain James Johnson observed that in a threat-filled environment, it was hard to keep one's mind on "ethics" and not focus on "survival," and Chaplain Bowers pointed out that "there was not a great deal of time to think about the reason we were there. We were busy trying to find and encounter the enemy and gain the upper hand."[43] In their memoirs, some chaplains described their feelings of inadequacy.

Chaplain Joseph Dulany described how he tried to counsel a soldier who had mistakenly killed one of his fellow soldiers "and was having a very difficult time accepting what he had done." Dulany remembered that the soldier "wanted literally to crawl into a hole and pull the ground over himself." He wrote that "there was little that I could do other than to listen, pray and allow him to ventilate."[44]

Discussing his experience counseling soldiers in Vietnam, Chaplain William P. Mahedy expressed dissatisfaction with the training he had received in such matters. "Group dynamics, training as facilitators, sensitivity and awareness seminars, all helped us to be more caring and relevant but of course evaded real moral and religious issues," he wrote in his memoir of the war. He tried to be more than a "facilitator" but still felt somewhat inadequate. He cited, as an example, the feelings he had after counseling a young soldier who had approached him "somewhat sheepishly" with a question:

> "Hey, Chaplain," he asked quietly, "how come it's a sin to hop into bed with a *mama-san* but it's okay to blow away gooks out in the bush?" The unsatisfactory answer that I gave him that day was the only one that makes any moral or religious sense. It amounts to this: "You may kill others only because they are trying to kill you, for this is an act of simple, elemental self-defense." The answer is legitimate on one level but not on others. It is perhaps inescapable, given a combat situation, but it leaves much unresolved.[45]

Chaplain Falabella seemed more comfortable answering his soldiers' questions: "Will God forgive me for killing the people we are fighting? And what about the commandment, 'Thou shalt not kill'?" He based his comments on the difference between "killing" and "murder," and, like Mahedy, invoked the right of self-defense. "If someone attacked you[,] you had a right to defend yourself to the degree necessary." If, in defending yourself, "death occurred, the act causing it would not be murder, which is the unjust taking of another's life," but killing. Murder he defined as "clearly a moral evil as well as a physical evil," but killing, in such a situation, "while not a moral evil, is clearly a physical evil and naturally abhorrent to human feeling and sensitivities." He cautioned against confusing "the abhorrence of 'killing' . . . with the question of its morality[,] which deals with reason and not primarily feelings." Regarding the biblical commandment against killing, Falabella contended that "it seemed reasonable to consider that the concept of 'kill' . . . had to do with 'unjust' killing since not all incidents of killing in the Bible are condemned." Then, pointing to the exigencies of combat, he told his soldiers,

> You are clearly defending yourself since wearing a military uniform and bearing a weapon make you liable to enemy fire. Whatever else you may think of war theoretically, the fact is you are here on the field of battle and not safely seated in a classroom discussing subtleties about taking

> another's life. Unless you defend yourself you will not only jeopardize your own life but also the lives of your comrades, since fire power can make the difference in who will live and who will die. You say you do not want to kill. Who "wants" to kill and is truly a sane man? But if you do not fire your weapon with the intention of being effective, you could be a factor in your buddies' getting killed. In effect you would be killing your own people.[46]

Although the combat chaplains varied in the way they approached soldiers' questions about the morality of killing and war, they generally agreed that criticizing the U.S. intervention or the conduct of the war was inappropriate in a combat environment where soldiers were fighting and dying. This was the position approved and enunciated by the General Commission on Chaplains and Armed Forces Personnel. In "Dissent Over Vietnam," an editorial in the May–June 1968 issue of the *Chaplain,* Executive Director A. Ray Appelquist observed that as civilian criticism of the Vietnam War increased, the commission was increasingly being asked to comment on "the wisdom and morality of this involvement." In response to such requests, he wrote that he had "found it useful and necessary to cite the basic purpose of the churches in supplying chaplains to the armed forces." He contended that "while chaplains and chaplaincy executives are free to hold personal views, they are not officially charged with analyzing and commenting on national goals and policies." Appelquist also insisted that the position chaplains were adopting in Vietnam was consistent with, even required by, their "pastoral" responsibility. Serving in a pastoral role "does not constitute or imply an endorsement of war in general or of any war in particular," he wrote. "It reflects a Christian concern for all human beings, whether nobly or mistakenly involved in the immediate risks and burdens of war." In fulfilling the pastoral role, "chaplains choose to identify with men in uniform, and if necessary to suffer with them, in order to minister to them."[47]

Taking such a position rendered combat chaplains vulnerable to the civilian antiwar critics' charge that they were abdicating their prophetic responsibility. Some chaplains recognized an additional problem. Although the position seemed ethically appropriate to the ministry of presence, it may have undermined the chaplain's moral credibility with his men. In the United States, chaplains presented themselves, and were officially regarded, as authorities on religion, morality, and ethics. In Vietnam, when they evaded the moral and religious questions raised by the war, they appeared to be immune to such concerns, or unable or unwilling to address them. Chaplain Mahedy lamented the fact that soldiers looking for moral and spiritual guidance regarding the war did not find it, and he speculated that that was one reason for the resentment many veterans expressed toward the chaplains they had encountered in Vietnam.[48]

Chaplains and War Crimes

In his 1968 *Chaplain* editorial supporting combat chaplains' abstention from criticizing the U.S. intervention or the conduct of the war, A. Ray Appelquist added the following caveat:

> At the same time, as morally responsible men chaplains are not permitted to ignore the larger context in which they serve. It is to be expected that chaplains will be alert and sensitive to conditions of needless inhumanity and unlawful acts of war which might compromise their nation or undermine their own integrity and witness as ministers of the Gospel.[49]

Just as some chaplains were theologically and ideologically unprepared to address questions about the morality of the Vietnam War, war in general, and killing, so were some chaplains morally and psychologically ill equipped to deal with instances of brutality, atrocities, and war crimes committed by American troops. In a few cases the problem was ignorance of or failure to follow proper reporting procedures. In other cases, it was the chaplain's inadequate moral assessment of the incident. But the balance sheet also shows instances in which chaplains acted as Appelquist expected them to.

The most infamous of the war crimes that involved chaplains was, of course, the one known as My Lai. On March 16, 1968, in My Lai (4), a sub-hamlet of the village of Son My, some five hundred Vietnamese noncombatant men, women, and children were murdered by soldiers of Charlie Company, Task Force Barker, Eleventh Brigade, Americal Division. Commanders and their staffs within the division concealed the incident; not until November 1969 were the first accounts of the massacre published in the *New York Times* and other newspapers.[50]

On November 29, 1969, Army Chief of Staff Gen. William C. Westmoreland appointed Lt. Gen. William R. Peers to head an inquiry into the massacre and its cover-up. The Peers Commission drew up a lengthy, detailed report, dated March 14, 1970. Among other things, it revealed that two chaplains were involved in the initial response to the massacre. The division artillery chaplain, Carl E. Creswell, was told about the massacre by Warrant Officer Hugh C. Thompson Jr., a helicopter pilot who had witnessed the incident. Creswell advised him to lodge an official protest in command channels (which Thompson did) and said he would do the same through "chaplain channels." Creswell then informed his superior in the chain of command, Chaplain Francis R. Lewis, Americal Division, about the allegation. Lewis said he would forward it through the operations officer for the division. At the Peers inquiry, Lewis claimed he talked with three division staff officers about the massacre, but all of them denied any recollection of his doing so. The Peers Commission Report concluded that Lewis's discussions with the division officers were so informal and vague that they failed to register. Lewis also claimed to have spoken with the Eleventh Brigade chaplain, Maj. Raymond P. Hoffman, but at the inquiry Hoffman denied it. Hoffman did say that Chaplain Creswell told him he had reports that "our people had fired into women and children" and that Creswell continually "ragged" him and "pulled his leg" but never offered any specific information or made specific allegations and so he did not take him seriously.

Although the army did not file formal charges against Creswell and Lewis, the commission's report did criticize them, saying,

> It is clear from the actions—and the acts of omission—of Chaplains Lewis and Creswell, that while both were aware of the serious nature of the charges alleged by WO1 Thompson, neither took adequate or timely steps to bring these charges to the attention of his commander. It should have been evident to both of these chaplains that the idea of conducting an investigation of a war crime through chaplain channels was preposterous.

General Peers specifically criticized Lewis for conducting his own personal inquiry and then, when it proved futile, failing to ensure that an adequate investigation was conducted; and for limiting his action at division headquarters to "informal discussions with various staff officers." Peers faulted Creswell for failing to report the matter to his commanding officer or to the staff judge advocate or the inspector general and for reporting only to Chaplain Lewis and then, upon receiving no satisfactory response, failing to take effective action to ensure that an investigation would be conducted. Creswell himself rendered perhaps the most acute judgment on his conduct. He was quoted as saying, "In hindsight, I feel I should have done more. . . . God forbid that in a similar situation, any Chaplain should ever be content with the actions I took."[51]

When the fact of the massacre was revealed, in November 1969, the army, the Defense Department, and the Nixon administration declared My Lai an aberration.[52] Senior chaplains in Vietnam responded with a similar, albeit more nuanced, view. In late December 1969, the *Boston Pilot* published a news story with a Saigon dateline on the reaction of army chaplains. The author, Father Patrick Burke, S.S.C., wrote that chaplains condemned all atrocities "unequivocally." He quoted the USARV deputy command chaplain as saying, "We do not measure the gravity of an atrocity by its size. They are all morally wrong." In further discussion of the chaplains' views, Burke wrote that there was "general disbelief that the number allegedly killed is as big as reported" and that chaplains knew, "from personal experience," that "war is a dirty business and that civilians get caught in the middle." However, he insisted, "none of them doubt that the alleged incident at My Lai would be an exception, an isolated incident, but even so, not one to be condoned." The *Boston Pilot* story also included a quote from the MACV command chaplain, Father William R. Fitzgerald, O.M.I., the senior chaplain in Vietnam: "If the reports are correct, it means there was a breakdown, a moral collapse. In our character guidance lectures we set ethical standards and in church we preach Christ crucified."[53]

In the early 1970s, when the Office of the Chief of Chaplains received inquiries about atrocities committed in Vietnam, the stock response was to point out all that was being done to strengthen the moral fiber of the troops. The wording varied, but the message was always reassuring. Referring to soldiers' training in the laws of war, Chief of Chaplains Francis L. Sampson would point out that the army had "clear-cut rules on the conduct of warfare," which were "constantly reemphasized at every level of command both in training and in the combat zone." He would call attention to the army's "extensive" effort to provide a religious and moral ministry to American soldiers. He would declare that command support for that effort was "outstanding," that chaplains were "intensively" involved, and that the programs used in support of that ministry were "constantly being updated and restructured with the help of outstanding civilian and military experts." In sum, "every opportunity is utilized to reinforce the soldier's personal religious faith and convictions and to encourage the development and maintenance of the highest moral standards on the part of all personnel."[54]

Confidence in the military training and character education the troops received was one reason the chief of chaplains and other chaplains believed My Lai was an aberration. Another reason was their great faith in the basic goodness of the American soldiers fighting in Vietnam. In response to a letter writer who voiced concern about the moral corruption of young soldiers, Sampson urged him to trust in "the moral integrity of each American citizen to conduct himself with decorum and pride according to the dictates of his conscience."[55] Chaplains also had great faith in the salutary influence *they* exerted on the troops in the combat environment. The antiwar critics argued that their presence in Vietnam legitimized an immoral, unjust war, but some chaplains believed (or at least hoped) that it provided a countervailing influence against the dehumanizing effect of combat.[56]

One of the problems with emphasizing the example and presence chaplains provided for soldiers in the combat zone is that although chaplains tried hard to reach even the most remote outposts, the wide dispersal of troops in Vietnam made it physically impossible for them to be with very many men at any given time and place. Journalist Martin Gershen suggested that in the case of My Lai, the absence of a chaplain and his ministrations, at a crucial time in the life of Charlie Company, played a role in its commission of the atrocity. He points out the great emotional stress the members of Charlie Company had undergone during the month prior to the incident. They had lost forty-two men, representing "at least 40 percent casualties." The day before the massacre, a chaplain had visited the company and held a memorial service for the dead. But according to Gershen, "It was really too late." By that time, "Charlie Company was totally demoralized and psychologically destroyed. . . . They had become a platoon crazed with fury and frustration and fear; a platoon of a company that could no longer take it—and could no longer be trusted or responsible for its actions." Describing the thinking of the men "as they mourned the dead and feared for the living," Gershen wrote, "Something would have to be done . . . . The men had to destroy [the] enemy or die. It was kill or be killed. That's what they taught in the infantry."[57]

The problem with emphasizing the moral, religious, and legal instruction given the troops is that such instruction actually received minimal attention in basic training. Soldiers attended four hour-long Character Guidance classes and heard a one-hour presentation on the Geneva Conventions and war crimes regulations, but they received hundreds of hours of instruction on "how to obey and how to kill."[58] A soldier who testified before a hearing on war crimes remembered that the classes on decent treatment of civilians were "overshadowed by all of the other classes where the instructors constantly, you know, taught us 'blow them away, blow them away.'" The noncommissioned officers, most of whom had served in Vietnam, "would tell us stories of blowing away civilians, of what to expect in Vietnam, and they would always refer to the Vietnamese as gooks, and slant eyes and dinks." He said he and "most of the guys got the impression, you know, you cannot trust anybody, and as long as nobody is watching to be hard, tough, and have no feeling, and blow them away."[59]

Once in Vietnam, the experience of counterinsurgency warfare seemed to render irrelevant the distinction the laws of war made between noncombatants and the enemy. In a war where the enemy was "visibly invisible," it seemed the only way to survive was to make everyone the enemy.[60] As William Broyles remembered, in Vietnam "the lesson was driven home again and again: trust nothing, for the smallest child and the oldest woman can kill you. . . . We were trained to be soldiers, but we never learned who our enemy was. If anyone could be our enemy, then why not everyone? It was a small—if false—step, logically. But morally it led to the abyss—it led to My Lai."[61]

In the decades following My Lai, more information on the nature and extent of war crimes and atrocities committed in Vietnam has become available. There is still considerable debate on the subject, but a number of historians contend that My Lai was not unique, save perhaps in magnitude. Documents stored in military archives and evidence presented at public hearings and inquiries support Richard Hammer's contention that war crimes and atrocities were "endemic to the American military commitment in Vietnam."[62] Various sources, including chaplains' memoirs, give some idea of how chaplains dealt with such problems.

In *Casualties of War*, published in 1969, Daniel Lang told the story of the so-called Incident on Hill 192, the kidnapping, rape, and murder of a Vietnamese girl in 1966 by four U.S. infantrymen in the central highlands. All of the names in the book were pseudonyms. In fact, it was the story of Chaplain Claude Newby, who listened for two hours to a description of the incident from one of the soldiers who had witnessed it, believed him, and reported it to the Criminal Investigation Division office. Ultimately, the four soldiers were court-martialed.[63]

In a 1977 study of military chaplains, Clarence L. Abercrombie mentioned two instances in which chaplains criticized the way their commanders were conducting military operations. One chaplain reminded the commander that "our primary objective was to facilitate pacification, not kill 'Gooks.'" The other chaplain objected to "the burning and relocation of a [Vietnamese] village." Although none of the villagers was hurt and they were helped with the relocation, the chaplain explained that he "still felt it was cruel and unnecessary." Abercrombie also discussed four chaplains who said they had been instrumental in preventing the killing of enemy prisoners and another who had brought formal charges against the perpetrators of an incident similar to, but on a much smaller scale than, My Lai.[64]

In his 1989 history of army chaplains in Vietnam, Henry Ackerman described an effort by chaplains of the 198th Light Infantry Brigade, Twenty-third Infantry (Americal) Division to help their troops deal with the aftermath of the My Lai massacre. As one of them explained, the brigade had not been directly involved in the massacre, but the chaplains became concerned about the resentment, bravado, and anti-Vietnamese feelings the troops were displaying. At first they held group counseling sessions. Then, the chaplain wrote, "we realized we needed an experience rather than words." They organized a picnic for the soldiers and a group of young Vietnamese children from an orphanage, which proved to be a great success. As the chaplain pointed out, "It was crucial to us to show the troops that the Vietnamese were individual human beings who had also suffered in the war." After the picnic he said that he and his fellow chaplains began to see "a difference in how the troops acted in the field and how they saw themselves." They saw the Vietnamese "as individuals with human faces and personalities as only kids can personify."

Ackerman also recounted the experience of a division chaplain of the First Cavalry, in 1967, when one of his chaplains informed him of an incident in which soldiers had cut off the ears of North Vietnamese army corpses for "souvenirs." The division chaplain told his subordinate to "let it alone," that he would handle the matter. Then, he wrote, "I went immediately to the Chief of Staff and related the events. He was horrified. He told me, 'I'll handle this, chaplain, and many thanks.'" Two days later, the division commander held an "officers' call." According to the division chaplain,

> He directed that *every* officer be present, down to platoon leaders, at one of several sessions. There were *no exceptions*. He stated, I remember well, "What constitutes a crime in the United States of America, constitutes a crime in the 1st Cavalry. And I will prosecute anyone violating proper conduct to enemy personnel, living or dead. It's a short step from mutilating a corpse to mutilating a person."

While praising the commander for "heading off problems" and setting a high standard of conduct for the soldiers, the division chaplain concluded his account by pointing out the equally important "work of chaplains behind the scenes."[65]

While Lang, Abercrombie, and Ackerman document cases of chaplains reacting to potential or actual instances of brutality on the part of U.S. soldiers, the Vietnam memoirs written by combat chaplains provide a different kind of information, often revealed with remarkable candor. Shock, detachment, hesitation, indecision, ambivalence, empathy, self-interest, anger, and feelings of inadequacy—all of these emotions influenced their thoughts and actions in dealing with such instances.

Chaplain James Johnson described his reaction upon seeing his troops set fire to a hooch where they had discovered a homemade zip gun. He wrote that his conscience bothered him to see "someone's home . . . being burned." He remembered, as a child, a neighboring family huddled together, crying, after their house had burned to the ground. He tried to dismiss the matter by deciding that the hooch "probably belongs to a VC." Still, he pictured the VC family reacting to their loss the same way the neighboring family had. He wondered whether he should speak to the battalion commander about "hooch burning." If he did, "would the command see me as a rabble rouser?" That question led to others: "Is this just part of war? Do I just minister to the soldiers, or should I have a say in events such as this?" These, he admitted in his journal, were "not easy questions," and he had no "easy answers." While pondering them, walking around in the combat area, he suddenly realized that he made a perfect target for an enemy sniper. Awareness of that threat quickly relieved his mind of ethical questions and focused it on survival. "In this war," he observed, "morality has a way of becoming blurred in our quest to survive."[66]

Chaplain Dulany expressed similar sympathy for the families of dead North Vietnamese army soldiers when he saw a bulldozer pushing their bodies into a common grave. He "wondered how the Vietnamese families were alerted to the fact that a son, father, or husband had been killed in battle." In his Vietnam memoir he reflected on his response to incidents he had witnessed, or knowledge he had had of actions and behaviors "that could be considered as morally questionable," but which he did not think constituted "atrocities." The basic question was, "How could I have responded and maintained my stature and effectiveness as a unit chaplain?" In one instance, during an army search of a Vietnamese village for VC, weapons, and rice, the occupants were gathered into a group. South

Vietnamese National Police forcibly dragged some young women away from it and out of sight. "I thought they were taken away for interrogation," Dulany wrote. "A short time later piercing cries and screams came from the area where the men had taken the women. I was left to imagine what was taking place. I turned away without saying a word to anyone. Should I have investigated or at a minimum raised the issue with the commander of the unit responsible for the mission? Probably. But I remained silent."

Dulany cited two other instances when he remained silent. In one, he saw a two-and-a-half-ton U.S. military truck "turn abruptly from its normal lane of traffic and apparently intentionally run over an old woman squatting by the side of the road." He wondered, "Should I have chased this truck down and confronted the driver or at a minimum noted his tag and unit designation and reported the incident?" His answer: "Probably. But I continued on my way." In the other instance, in a tent at headquarters, he saw a Vietnamese soldier interrogating a suspected VC break several of the man's fingers and then his arm. "I turned away and acted as though I had missed the entire episode," Dulany wrote. Again, he posed the question, "Should I have spoken out? Probably. But I walked away without saying a word."

On another occasion, however, Dulany did act. Accompanying a maneuver company, he observed one of the soldiers wearing a "strange looking necklace" that turned out to be "a human ear, sun dried, . . . hanging from a string" around his neck. Dulany wrote,

> I was appalled and addressed the issue immediately. In my service that day and for many days afterwards, I talked about the value of life and questioned how a person could mutilate another human being's body and feel good about it. . . . I suggested further that actions such as these made one less than human. . . . I maintained that mutilation of a body was sacrilege.

Even so, looking back on his response, Dulany seems to have felt it was inadequate. He thought his censure of the mutilation of enemy bodies probably did some good, but he heard that such incidents continued. "Could I have done more? Probably, but I used the 'bully pulpit' available to me to address the issue. I used my position to speak of the dignity of human life."[67] Dulany did not say what the "more" might have been. He might have underscored the gravity of the offense by reminding his men that the maltreatment of dead bodies was defined as a war crime in *The Law of Land Warfare*, a U.S. Army field manual.[68]

Several combat chaplains recalled incidents involving VC suspects or prisoners. Johnson witnessed an interrogation of a suspect by a Vietnamese Kit Carson scout. He was talking rapidly to the detainee, slapped him suddenly, and then put a knife to his throat. Here is how Johnson described his reaction in his journal:

> My heart is pounding. I don't want to be witness to murder. Our Vietnamese interpreter quickly assures us that the scout won't cut his throat; he just wants the VC to think he will. . . . Nevertheless, I'm not going to allow neither [*sic*] this man nor anyone else to be executed if I can stop it. The morality of war is no different in this situation than anywhere else.

Then the scout blindfolded the VC. Johnson concluded his description of the incident, saying, "Apparently he's told that his head is going to be cut off if he doesn't talk. He begins chattering like an auctioneer." Johnson knew that killing the prisoner would have constituted a war crime under the Geneva Convention. What he did not seem to realize was that threatening the prisoner, first with the knife to the throat, then by telling him that his head would be cut off, constituted torture, which the convention also prohibited.[69]

Chaplain Autry described how he responded to another incident involving South Vietnamese soldiers and suspected VCs. An ARVN (Army of the Republic of Vietnam) platoon had captured "a couple of young boys who were definitely VC. . . . [They] had these young kids stripped down naked and stretched out and were beating the soles of their feet. They had already drawn blood." Autry wrote,

> I stopped it. The lieutenant didn't look pleased. I pulled him aside and reamed him out a little. I told him that it was not right, not good for our soldiers to see, that and if these VC knew something, which I doubted, we could get it another way. Americans, I said, didn't do things this way. He grinned like I was his dad scolding him.[70]

The descriptions presented above, of chaplains witnessing or becoming involved in "morally questionable" incidents or actual war crimes, reveal the many factors that shaped their response. Some seemed ignorant of proper military reporting procedures and/or the laws of war. One feared being typed a "rabble rouser" by the commander. Another could not decide whether to take any action. Some seemed to have a sense of inadequacy or futility, no matter how they responded.

Two other factors, possibly the most important of all, were the chaplains' firsthand experience of combat and their feeling of "connectedness" with the troops. Chaplains were only human, which is to say not immune to the morally corrosive effects of counterinsurgency warfare. Johnson's combat experience produced conflicting emotions about killing the enemy. In one journal entry he expressed disapproval of the emphasis "higher command" placed on body counts and of his own commander's being proud of his unit's record. "Proud of killing?" he wrote in his journal. "This isn't rabbit hunting. It seems almost as trivial as that. . . . I don't feel good that some human beings were killed today before my eyes, even though they would've surely killed us without blinking." But then, as an afterthought, he wrote, "I guess it's okay emotionally not to see the VC as people because to do so might drive us crazy. In later years, it does drive some crazy. It's easier to see this as a game with points scored, not people killed." Although, as we saw earlier, Dulany was disturbed when he saw a bulldozer pushing the bodies of dead North Vietnamese soldiers into a common grave, he had no problem with the body count. He saw it as "the accepted means for determining the success of a unit's performance in combat," and he noted, in a similarly dispassionate tone of voice, that "it was rumored that men from some units, other than ours, routinely shot Vietnamese farmers wearing the common black pajama garb and threw web gear on their bodies so they could be counted as VCs."[71]

In the combat environment, some chaplains encountered moments of self-knowledge, during which they experienced feelings they had never known (or admitted) before, and that seemed quite inappropriate for "a man of the cloth." On one such occasion, walking through an area of recent combat, James Johnson saw an elderly Vietnamese couple sobbing and pointing at the remains of two bodies among the ashes of a hooch that had been torched by the VC. He wrote that he found himself "feeling an inner rage unlike any I have felt before. . . . I'd like to find a VC, any VC, and beat him senseless." Then, continuing his walk, he noticed that some of his troops were "almost gleeful" at the sight of dead VC soldiers lying on the ground. He admitted that he, too, felt "glad to see these dead VC," though "as a Christian I'm not proud of my feelings." Including himself among the troops, he wrote, "After what we have been through the past several days, it's hard to see these dead VC as human beings; we see them as good riddance. I'm stunned and sorry about the feeling."[72]

Chaplain Samuel W. Hopkins Jr. reported in his memoir of Vietnam that "the surprising self-discovery was about my hidden aggressions." In the midst of combat, he found that "you quickly lose objectivity" and that "it took great self-restraint to remain a noncombatant as my chaplain's role required." Had he borne arms in a different capacity, he realized how "enthusiastically" he would have used them.[73]

Just as chaplains' disillusionment with the war helped them to appreciate the troops' low morale, so did experiencing their own feelings of anger, distress, and aggressiveness enable them to understand why some soldiers might engage in acts of brutality. Chaplain Dulany was quite unprepared for the "extreme" feelings that overwhelmed him when he heard of the death of a first lieutenant he admired. "I was terribly upset. I was outraged," he wrote. "At the height of my emotional distress, I believe that I could have killed an enemy soldier with my bare hands if one had been within my reach." Recalling that experience in his memoir, he wrote that it helped him to "understand something of the anger that apparently leads to a person's loss of control and results in extreme action." He believed that he had "experienced emotions approaching the level of frustration and anger" similar to those of Charlie Company, Americal Division, "when they lost control and massacred civilians at My Lai." He hastened to add that "experiencing emotional pain and distress at this level does not justify or excuse extreme actions, but partially helps one understand situations and levels of intensity of feelings, which might engender them."[74]

Chaplains discovered how the experience of combat could numb emotions and moral sensitivity. We saw this earlier, in James Johnson's comment about not seeing the VC "as people." In another journal entry he described the day after a fierce firefight in which seven soldiers were killed and forty wounded. Johnson was helping load the dead bodies onto the helicopter. "We have to stack them aboard like logs," Johnson wrote. "It just doesn't seem right, but we have no choice." There was no time to grieve; he and the men had to move out, to get back to their base. In leaving the scene of battle, they passed through the positions the VC had held during the battle. The men paused at a VC machine-gun position. There, Johnson wrote,

> two dead VC are sprawled on their backs. Each dead VC must have at least twenty holes in his face and other exposed parts of their bodies.

> Someone in the column takes his white plastic spoon from his C-rations and sticks the handle into a wound made by the fragments. The next soldier in line does likewise. By the time I pass the position, each dead VC has a dozen or more plastic spoons protruding from his body.

Johnson observed that there was "no shouting or cheering as we pass the bodies." He concluded his description of the scene by writing, "The defiant symbolism of 'sticking it to them' may be desecration of the dead, but this is a way of emotionally coping with what's happening each day of this god-awful war."[75]

The Evolution of the "Ministry of Presence"

For the combat chaplains, recognizing and wrestling with the moral ambiguities of war was part and parcel of their experience in Vietnam. The chaplain manuals did not mention that as part of their assignment. Nor were they very specific about the nature of the chaplain's ministry on the battlefield. Thus the "ministry of presence" the chaplains developed in Vietnam was not so much a result of official Chaplain Corps guidance as it was a consequence of these chaplains' determination to be "where the action is," at the "forward edge of the battle area" with the troops who "hurt the most." It was a "work in progress" that also proved to be a significant learning experience for the chaplains. Just as their "life-sharing" ministry sensitized them to the soldiers' moods and needs, the experience of combat acquainted them with the moral ambiguities of warfare. In the process, many of the chaplains also acquired a high degree of self-understanding.[76]

As we shall see in a later chapter, in the mid- to late 1970s, when Chaplain Corps officials and Army leaders began formulating new battlefield doctrine for the coming decades, one of their primary concerns was the emotional, psychological, and spiritual needs of soldiers in combat. In developing a more "soldier-focused," "battle-focused" ministry for chaplains, they drew significantly on the experience of combat chaplains in Vietnam. What came to be known as "Religious Support Doctrine" incorporated several aspects of the ministry of presence they had developed in that earlier war.

5

Ministering to the Military Institution

In the late 1960s and early 1970s, while U.S. troops fought in the Vietnam quagmire and antiwar protest escalated in America, civilian and military commentators issued alarming reports of "an Army in anguish" struggling for "survival."[1] At no other time in its history had America's military experienced "troubles . . . in such general magnitude, acuteness, or concentrated focus as today," declared Col. Robert D. Heinl Jr. in the *Armed Forces Journal* in June 1971.[2] Well into the early 1980s, commentators continued to analyze the crisis, while military leaders went through a period of intense introspection, seeking to understand how and why it had developed and what should be done to resolve it.[3]

The unprecedented, multifaceted crisis involved all ranks—enlisted personnel, noncommissioned officers, and the officer corps, including general officers. It wreaked havoc in Vietnam and at military installations in Europe and the United States. According to the commentators, it stemmed from three basic problems: the decline of discipline and morale among enlisted personnel, incompetent leadership on the part of the officer corps, and the erosion of the professional military ethic. How military leaders and chaplains dealt with the military ethic problem will be discussed in the chapter following this one. In this chapter we shall consider their response to the first two problems.

The Decline of Discipline and Morale

The decline of discipline and morale was apparent in the wide range of serious behavioral problems that developed in Vietnam: drug abuse, racial conflict, refusing to obey orders (including combat avoidance and refusal), the "fragging" of unpopular officers, high absent-without-leave and desertion rates, and criminal activity. Desertion, drugs, and racial violence, including "large-scale riots," also troubled military installations in Europe and the United States. Colonel Heinl provided a detailed account of what appeared to be a general breakdown of authority. Among other things, he cited a Pentagon report that fraggings in 1970 had more than doubled the ninety-six of the previous year; he also mentioned an "authoritative estimate" of one a week in the Americal Division in 1971. Rates of drug addiction in the armed forces had reached "pandemic" levels in 1971, nearly double those of 1968. An estimated 10 to 15 percent of troops in Vietnam were using high-grade heroin. In 1970, the army reported 65,643 deserters, roughly equivalent to four infantry divisions.[4]

Antiwar and antimilitary sentiment and activity within the military was another symptom of the crisis. "Sedition . . . infests the Armed Services," Heinl declared. At least fourteen "GI dissent" organizations were operating more or less openly in the military community, and some forty-four "underground newspapers" were being published on or aimed at military bases in the United States and overseas.[5] Such sentiment and activity were part of what came to be known as the GI movement. It began in the mid-1960s with individual acts of protest against the Vietnam War. By 1968 it had developed a loose organizational network that published the underground newspapers and sponsored off-base coffeehouses. Besides opposing the war, the movement called for military reforms such as collective bargaining for GIs, elimination of racism in and democratization of the armed forces, and full constitutional rights for all enlisted personnel. Collective resistance soon replaced individual protests. Dissident GIs participated in civilian antiwar marches, teach-ins, and other demonstrations. They also organized their own protests: "pray-ins" and other demonstrations, as well as stockade rebellions in the United States, Germany, and Japan. At its peak, the movement represented a volatile compound of black nationalism, antiwar radicalism, and antimilitarism. The great majority of its participants were lower-ranking enlisted personnel.[6] Junior officers and former officers from all three service branches formed a parallel, less militant antiwar movement, acting through organizations such as the Concerned Officers Movement and the Concerned Academy Graduates, both founded in 1970.[7]

Heinl thought that the behavioral problems in the armed services mirrored the "social traumas of American society." Lt. Col. William L. Hauser, the author of *America's Army in Crisis,* attributed them to "a social revolution going on in America." Their view was common among other military as well as civilian commentators. Gen. Bruce Palmer Jr., army vice chief of staff, spoke of "an all-pervasive challenge to the traditional order of things: our basic value system and 'the establishment' which supports that system."[8] Like Heinl, military and civilian commentators regarded the behavioral problems exhibited by enlisted personnel as "spill-overs" or "inherited from" the civilian sector. The character of the young men and women entering the armed forces had been shaped by "a steady erosion of the authority and influence exercised by the church, the schools and the family." General Palmer declared that as a result of the assault on traditional values, "Our youth is being told that one's own personal values are all that count: do your own thing and ignore the traditional moral and ethical codes; the importance of the individual takes precedence over any collective structure."[9]

The Leadership Problem

In a 1971 interview published in the *Washington Post,* Gen. Michael S. Davison, commander of the U.S. Army in Europe, declared that the leadership problem was "the toughest problem we have" in the army.[10] Junior officers and noncommissioned officers incurred much of the blame, because of their ineffectiveness in maintaining discipline, unit cohesion, and morale among their troops.[11] They failed to carry out their basic leadership responsibility, which Davison defined as "getting your subordinates to work wholeheartedly toward the mission."[12] The solution to the problem, military leaders decided, was a new approach to leadership training. They believed that the armed forces had no choice but to recognize and deal with the results of the cultural revolution in American society, especially the nontraditional notions and values espoused by the "new breed" of young people coming into the military.[13] In November 1970, in the keynote address at an Army Commanders' Conference, Chief of Staff William C. Westmoreland pointed to American youth's concern for "personal freedom" and their aversion to submitting blindly to authority. "We simply have to recognize the fact that these changes have occurred and will continue. We cannot alter these trends ourselves, and we should forget about trying," he declared. In an interview in the *Baltimore Sun* that same year, Westmoreland insisted that "to ignore the social mores of this younger group is to blind ourselves to reality. Their values and attitudes need not necessarily be endorsed by Army leadership, yet we must recognize that they do exist."[14]

The new approach to leadership training involved a shift from authoritarianism to what Col. David R. Hughes, chief of staff of the Fourth Infantry Division at Fort Carson, referred to as "a participatory approach." Young Americans "want to know 'why,'" explained Lt. Gen. W. T. Kerwin Jr., army deputy chief of staff for personnel. They "feel an urgent need to communicate, to participate and to contribute," observed Col. Samuel H. Hays. Leaders should no longer demand or expect unquestioning obedience; they should explain why something should be done, not just how or that it should be done. They would have to "learn to promote discipline based on reason and conviction and never on fear or caste," wrote Lt. James H. Toner in *Army* magazine. In his 1970 keynote address to commanders, Westmoreland declared that leaders needed to discover ways to "engage the imagination and enthusiasm of the men," to engage in a "continuous dialogue" that is "sensitive to their needs and aspirations," to "enhance their self-respect," and to accept them "as partners in a mutual endeavor."[15]

As Westmoreland's use of the word 'dialogue' suggests, the new approach to leadership training put great emphasis on "communication" between superiors and subordinates as a means of establishing leaders' rapport with and respect for their subordinates.[16] In his 1971 *Washington Post* interview, General Davison observed that "many of our leadership failures" occurred because of the "inability on the part of the leader, whether he be a platoon, company, battalion, brigade or even a division commander, to communicate to his subordinates his appreciation of them as individuals."[17] A statement in a manual used in a leadership training course at the Army Infantry School underscored the new emphasis on recognizing the worth of the individual soldier: "Our style of leadership must respect the individual dignity of every man. There is no room in today's Army for the shouting, screaming, harassing style of leadership."[18]

Chaplains and the Crisis of the Armed Forces

The U.S. Army's Vietnam-era crisis recalls its post–World War II crisis. In the late 1940s, Chief of Chaplains Luther Miller had proposed a remedy for the VD epidemic that ultimately developed into a much broader program of moral, religious, and citizenship instruction. In the late 1960s and early 1970s, working with military leaders to resolve the army's behavioral and leadership problems, chaplains once again discovered new ways of ministering to the military organization.

Army chaplains were the first responders to the military's drug problem.[19] Indeed, they pointed out and tried to remedy growing drug abuse in Vietnam at a time when military leaders denied it existed. In a 1976 interview for the U.S. Army Military History Institute, Gerhardt W. Hyatt, the former chief of chaplains, recalled an occasion when Chief of Chaplains Francis L. Sampson had talked to a division commander in Vietnam about drug abuse. "And this guy looked at him and he said, 'What the hell are you talking about? We don't have a drug problem in this division.'" In fact, Hyatt remarked, "Frank had just come from a visit 20 minutes before where he saw half of the unit stoned."[20] Based on what he had seen in Vietnam, in 1969 Sampson took the initiative of making day-long workshops in drug abuse counseling part of chaplains' regular monthly training program throughout the army.[21]

Finally, in April 1970, the Department of Defense conceded the existence of a drug epidemic. In a directive issued that month, it admitted that since February 1968, "drug usage has increased at an alarming rate throughout our society and in the Armed Forces as well."[22] In 1971, a few months after President Richard M. Nixon announced a national campaign against drug abuse, the army created the Army Drug and Alcohol Prevention and Control Program (ADAPCP) to combat the abuse of alcohol and other drugs by American soldiers.[23] By that time, chaplains had been dealing with the drug crisis for several years. In the early 1970s, they continued to play an important role, serving with legal and medical officers (physicians and psychiatrists) on the various drug abuse councils, educational panels, and counseling teams created to combat the problem. The surgeon general strongly advised the inclusion of at least one chaplain on every counseling or "healing" team.[24]

Most chaplains agreed on the nature of the drug problem and how best to solve it. Because they recognized the similarities between drug abuse, which was mainly found among younger enlisted personnel, and alcohol abuse, which had long been a problem among older noncommissioned officers, they generally included alcohol abuse in their educational and counseling programs. Although they understood the medical and legal ramifications of substance abuse, they insisted that, fundamentally, it constituted a moral and spiritual welfare problem. Loneliness, boredom, stress, "problems about the meaning of life," "the collapse of social, moral and spiritual values"—these were the factors chaplains listed in discussing its causes.[25] In the 1970–71 "Historical Review," Chief of Chaplains Sampson observed that such factors were "outside the scope of any known medical treatment" but, he insisted, were "definitely . . . responsive to spiritual ministry" and to counseling "in the area of meaning, values, and goals."[26] Hyatt, his successor, made a point of defining the chaplain's primary role on the Prevention and Control Team as that of "a minister of religion," which entailed providing occasions for religious worship and study, as well as helping to plan and present educational programs and participating in "group therapy and rap sessions." Like Sampson, Hyatt also insisted on the greater effectiveness of the chaplain's ministrations as opposed to medical treatment. "The medical profession can treat an addict's symptoms—then treat them again and again and again," he observed in an interview published in the *Lutheran Witness* in 1971. "Only the Gospel of Jesus Christ can reach into the life and spirit of a man to get at the real reason he resorts to drugs or alcohol."[27]

Both Sampson and Hyatt aggressively promoted the special competency of chaplains in the areas of substance abuse prevention, treatment, and rehabilitation.[28] Their efforts were rewarded when army leaders sought chaplains' help in combatting the drug crisis.[29] The Office of the Deputy Chief of Staff for Personnel strongly supported chaplain training in drug abuse prevention and counseling and requested chaplains' advice in developing the army-wide drug and alcohol control program. Along with the surgeon general and many post commanders, DCSPER also advised including chaplains on antidrug teams organized at military installations. In the 1971–72 "Historical Review," Hyatt noted with some satisfaction that chaplains' "specifically religious leadership was welcomed as an integral part of the [antidrug] program."[30]

The Institutional Ministry

In responding to the armed forces drug problem, chaplains devised a new way of ministering in the military, what came to be known as "institutional ministry." According to Hyatt, the fact that chaplains saw drug abuse as "symptomatic of underlying social and spiritual problems" led them to focus attention on the military environment that generated them—and then to recognize other problems. "The 'drug problem' . . . forced us to become increasingly involved in other significant human problems—race relations, family problems, conflicting value systems and life styles," he explained. "We discovered . . . that we had other potentials in the chaplaincy that could serve the members of the Armed Forces." That, in turn, "forced a major reconsideration of the meaning of religious ministry." Chaplains began to think about how, in addition to playing their traditional priestly and pastoral roles, they could become "a humanizing influence in the culture."[31]

Certain words and phrases Hyatt used in describing the new institutional ministry give an idea of its varied objectives. As a "healing" ministry, it sought to remedy dysfunctional behavior, such as drug and alcohol abuse. As a "prophetic" ministry, it pointed out a new role for chaplains, facilitating change and reform in the military organization. As a "humanizing" ministry, it focused on human relations within the military community and sought to help military personnel live and work together "harmoniously" despite great variations in background, race, and rank. And, finally, as a "ministry of consultation" to commanders, it emphasized chaplains' role as advisors to the military leadership, not just helping to solve the "people-problems" in the military but also offering guidance regarding command plans, programs, policies, and decisions.[32]

Prophetic Ministry

The institutional ministry revolutionized the army chaplaincy, which was exactly the objective Chief of Chaplains Hyatt had in mind. While serving in OCCH in the late 1960s as director of personnel and then as deputy chief of chaplains, he had come to believe that the chaplaincy was not making the impact on the army that it had the potential to make. It needed to be brought "up to date" so as to confront the momentous changes taking place inside and outside the military. In addressing those changes, chaplains needed to focus their attention on the military organization—"its institutions, its policies, its practices, and its way of life."[33]

Hyatt also believed that chaplains needed new knowledge and skills, that one reason for their insufficient impact on the military organization was their lack of "expertise."[34] His predecessor had taken the first step in enhancing chaplains' professional development. In 1969, the same year he introduced chaplain workshops on drug abuse, Chief of Chaplains Sampson initiated a clinical pastoral education (CPE) program at Walter Reed Army Medical Center for selected chaplains.[35] The program originally was designed to develop counseling skills in a hospital setting. By 1970, increasing numbers of chaplains were enrolled in CPE at other army as well as civilian hospitals. The army chaplaincy adapted it to train chaplains in counseling on a variety of human relations issues, including drug and alcohol abuse, race relations, confinement, marriage and family problems, disaffection, and dissent.[36] Hyatt termed CPE "one of the greatest things that ever happened for the chaplaincy." Through it, he said, "the chaplaincy matured in quantum jumps."[37]

In the 1970s, chaplains also received training in a management strategy known as organizational development (OD). Based on behavioral science research and developed in the 1950s by industrial psychologists working in the civilian sector, OD offered a human-relations approach to management. The OD strategy emphasized such things as problem solving, goal setting, improving interpersonal and intergroup relationships, and conflict management. To have chaplains trained in OD, Hyatt contracted with the National Training Laboratories (NTL), an independent research and training organization that had been in the forefront of OD consultation since the late 1950s. Initially, NTL consultants instructed chaplains in interpersonal communication, conflict resolution, and decision-making skills, as well as chapel program development. After 1974, chaplains assumed the task of providing OD training.[38]

Hyatt anticipated that OD would "aid in the emergence of a proactive, rather than a reactive, chaplaincy, capable of providing an influential, prophetic ministry to the military institution, as well as a creative and innovative pastoral ministry to military personnel and their dependents within it." It would, he said, enable chaplains to discover where the system was hurting or hurting others and to propose remedies, and to recognize discord and bring about harmony. In this regard, he believed that OD harmonized with other chaplain programs that treated human relations issues and aimed at improving the quality of life in the military. "We are in business to help replace social injury with personal wholeness, schism with harmony and dysfunctional behaviors with positive life styles," he declared. "The OD program, CPE, alcohol and drug abuse prevention . . . all contribute to this ministry."[39]

Hyatt's concept of a "prophetic" institutional ministry was clearly influenced by the antiwar critics' harsh attack on chaplains. In articles in the *Chaplain* and the *Military Chaplains' Review,* he conceded that in the past chaplains had "neglected" or been "hesitant" to undertake their prophetic responsibility. Now, he asserted, "it is a task to which the situation in our country and in the Army calls us to meet with fresh determination." He spoke of chaplains becoming "agents of change" who pointed out problems and developed "innovative" programs to remedy them. Among the likely targets were policies that violated "the accepted ethics of our faith" or underwrote "racism or any of the other 'isms' subversive of human rights," as well as abuses growing out of standard military practices. However, he warned chaplains against adopting the "adversary posture" of the antiwar clergy. "Too many clergymen in recent years have accepted the easy work of eroding confidence in the institutions and criticizing those on whom the burdens and power of decision are placed and avoided the difficult task of building up something positive," Hyatt observed. He made the same point, in somewhat harsher language, in a keynote address to a Military Chaplains Association meeting. "It is easy to criticize," he observed. "To be negative takes no thought or wisdom at all. But to present helpful alternatives—to offer solutions, even partial solutions, takes intelligence and effort." In fulfilling their prophetic responsibility, chaplains should be unifiers, healers, and "instruments of reconciliation."[40]

Hyatt's concept of institutional ministry also reveals the influence of the secular theology that had gained credence among liberal mainline Protestants during the 1960s. Just as he urged chaplains to focus on the "structure" of the army organization and engage in "prophetic ministry," he also invoked the idea of a "theology of involvement," exhorting chaplains to "use the gospel of Jesus Christ in the way that He himself used it," by making it "actually work in the lives of people." To do that, they had to "go out into the world" and discover its needs. In the early 1970s that meant, among other things, addressing the army crisis, studying the "new culture" that had developed during the 1960s, and recognizing its influence on the young people coming into the armed forces. Hyatt also defended chaplains' use of secular skills taught by the behavioral sciences. We are not replacing religion with a behavioral science approach, he assured chaplains in a 1976 information letter. "We use these tools, when appropriate, but never as a substitute for God or His power, or in rejection of our high calling as ministers, priests and rabbis." In equipping themselves with "people skills," chaplains were extending their "basic clergy role" rather than developing a substitute for it.[41]

Humanizing the Military

In describing the purpose of institutional ministry, Chief of Chaplains Hyatt spoke of discovering where the military system was hurting or hurting others and proposing remedies as well as locating discord and bringing about harmony. Two army chaplains' programs, the Human Relations Ministry and Personal Effectiveness Training (PET), were established to accomplish these objectives.

The human relations program began in the early 1970s, with the implementation of an ad hoc race relations program for the U.S. Army in Vietnam. Chaplains established human relations councils "at all levels, down to and including company-sized units," and developed educational programs to assist personnel in "recognizing and changing discriminatory practices" within their units."[42] The OCCH organized the Human Relations Ministry in November 1973 for the stated purpose of improving race relations in the army by eliminating institutional as well as individual racism and by promoting equal opportunity and affirmative action.[43] In his March 1974 newsletter to chaplains, Hyatt told them,

> I stand behind your attempts to improve our ministries to racial, religious and cultural minorities. We must do all that we can to reduce the claim which prejudice and racism have on the lives of people we serve.

> No commander should be without our counsel on these matters and no EEO/RR staff officer should feel his work is complete without recourse to the spiritual dimension. All human beings must be freed from those personal or institutional abuses which rob life of meaning and fulfillment.

He concluded his statement by saying, "As clergymen, we are privileged to share in the humanization of organizations and in the process of social change that can only be achieved when people more perfectly understand the will of God."[44]

Hyatt ensured that chaplains played an integral role in the army-wide fight against racism and discrimination. He saw to it that chaplains participated in race relations conferences and served as staff and faculty in the Department of Defense Race Relations Institute. In 1973, he urged chaplains to "exert a strong moral leadership" by offering "positive input" to commanders who were drawing up Command Affirmative Actions Plans. The OCCH also offered advice and resource materials to assist chaplains in conducting race relations discussion groups and in organizing religious and educational programs to promote racial harmony. To equip chaplains with the skills they needed to improve race relations, OCCH contracted with the National Training Laboratories to provide sensitivity training in human relations to some nine hundred chaplains. Three hundred and fifty line officers also participated in the training.[45]

The OCCH Human Relations Ministry focused attention on race relations within the Chaplain Corps as well as in the larger military organization. In May 1971, at the request of OCCH, the Army Chaplain Board conducted a conference on "Ministry to Blacks" that considered such topics as black religion, preaching, worship and liturgy, and congregational participation—all designed to assist chaplains in understanding the religious needs of blacks in order to provide a "more viable ministry to them."[46] The conference marked the start of a long-term effort on the part of chaplains to incorporate elements of black religion and liturgy into their chapel programs.[47] In keeping with the army's promotion of equal opportunity and affirmative action, OCCH also implemented an elaborate plan for the recruitment of clergy from minority groups. It included making appeals to the General Commission on Chaplains and Armed Forces Personnel and to leaders of black denominations; sending "procurement chaplains" on visits to black theological seminaries and prospective candidates; and sponsoring a Conference on Recruitment of Minority Group Clergy to the Army Chaplaincy that brought together OCCH representatives, active-duty chaplains, the leaders of five black denominations, and officials of the Rockefeller Foundation and the Boston Theological Institute, among others.[48] Following the conference, OCCH intensified its recruitment efforts, setting a goal of 15 percent for minority-group chaplains and chaplains' assistants. In April 1973, such groups represented only about 5 percent of the Chaplain Corps. By 1985, however, the percentage had risen to 14.08.[49]

The OCCH developed the Personal Effectiveness Training program in response to a request from DCSPER in the summer of 1973 for assistance in solving the leadership problems of junior officers and noncommissioned officers (NCOs).[50] The program had a dual purpose: to increase officers' effectiveness in dealing with alcohol and drug abuse, racial conflict, and various disciplinary problems among unit soldiers and to improve relations and reduce tensions between unit leaders and their troops. A typical program consisted of sixteen hours of instruction. In weekly or sometimes daily sessions of two to four hours, chaplains taught small groups of twelve to twenty-four junior officers or NCOs using a combination of lectures, discussion, dramatization and role playing, and other participant-interaction exercises. Initial sessions encouraged self-awareness and openness on the part of the participants as a means of replacing indifference or hostility with a "caring" attitude toward the troops. Later sessions focused on the specific day-to-day problems the participants confronted in order to train them in communication, problem-solving, and counseling skills.[51] By the mid-1970s, as a result of a directive by Army Chief of Staff Gen. Bernard W. Rogers, PET was included in all Combat Army Primary and Basic NCO Courses.[52]

Assessing Institutional Ministry

Chaplains discussed and debated the institutional ministry throughout the 1970s and into the 1980s. It provoked a mixed response.[53] Most advocates depicted the new ministry as an outgrowth of profound changes in the civilian and military cultures that obliged chaplains to assume roles and responsibilities quite different from those customarily carried out by priests, ministers, and rabbis. Some chaplains justified the prophetic stance on biblical grounds, citing Old Testament prophets as well as John the Baptist, Jesus, and the Apostle Paul. Other chaplains, quoting Harvey Cox and Dietrich Bonhoeffer, invoked the secular theology in endorsing the new ministry. Some advocates had become disenchanted with worship-oriented ministry and frustrated by the limited appeal of chapel programs. They were excited by the prospect of engaging in a new kind of servanthood, learning and using new skills, improving the military institution, enhancing the lives of the individuals who worked in it, and becoming what one of them called a "vital force" in the military organization. They appreciated the way the institutional ministry enabled chaplains to be "proactive" rather than "reactive" or even inactive, as chaplains had tended to be in the past.[54]

Their excitement about engaging in the new ministry notwithstanding, chaplain advocates recognized its ambiguities and risks. As one recently retired chaplain pointed out, the chaplain who "set out to provide ways to live in a more human way in the system" might end up helping "to perpetuate that system." Or having started out, "as the prophets did, to 'speak for God,'" he might gradually substitute his will for God's. Taking a prophetic stance could encourage "personal self-righteousness and/or judgmental piety." And as another chaplain warned, "organizations always react angrily to an angry message of condemnation.[55]

Most of the chaplains who opposed or were unenthusiastic about the institutional ministry either preferred pastoral ministry or disapproved of the new ministry's secular orientation. A 1976 job satisfaction survey of army chaplains revealed that they regarded the traditional preacher and counselor roles as the most satisfying, whereas recently developed roles such as OD specialist, PET instructor, and Human Self-Development instructor were ranked least satisfying.[56]

Whether they advocated or disparaged the institutional ministry, most chaplains recognized that its programs helped them win the respect of military leaders—and at a crucial time, when they were buffeted by antiwar criticism and embroiled in the Character Guidance controversy. In 1973, speaking at the Annual Command Chaplains Conference, Chief of Chaplains Hyatt called attention to chaplains' enhanced reputation in the army. "Commanders are requesting chaplains to develop in their units and on their posts a more rewarding quality of life. This demand is testimony to the confidence they have in us," he said. "I am convinced that credence in our professional ability has never been greater, or more deserved."[57] As Hyatt recognized, the new appreciation for chaplains and their programs was a byproduct of the army crisis. "It was the easiest time for us to make our mark," he observed on another occasion. "The top leadership of the Army was looking everywhere. Who can help us? What can we do? They were so open to new things that had any possibility of success. We came in with programs that not only sounded good, but they had already been operational to some degree and had proven themselves to some degree, in models."[58]

Indeed, the programs chaplains developed to address the crisis not only promised but also produced tangible results in the form of reduced drug and alcohol abuse, less racial tension, lower AWOL and desertion rates, and more effective junior officer and NCO leadership. Confronted with what some commentators called the "disintegration" of the armed forces, military leaders were understandably impressed by programs that strengthened soldiers' readiness and cohesion and promoted combat effectiveness.[59]

Human Self-Development

The one stumbling block in the way of chaplains regaining the status and influence they had enjoyed in the 1950s and early 1960s was character education. In December 1971, Our Moral Heritage, which had been in operation less than two years, was replaced by a new program, Human Self-Development (HSD).[60]

The program was mandatory for soldiers in basic and advanced individual training, who received a total of six hours of instruction. All six hours for male personnel and two hours for female personnel were based on topics that had been taught in OMH. Four of the hours of instruction for female personnel were based on newly prepared topics: "The Military Woman," "Adjustments to the Military Service," "Marriage and the Military Woman," "Maturity and the Military Woman," and "Women's Service to the Nation."[61]

Like the army's earlier character education programs, HSD was a command program for which the chaplain had "primary staff responsibility" by virtue of his "training and interest." The OCCH was responsible to DCSPER for the preparation of instructional materials. Commanders exercised more authority than they had in earlier character education programs. They had the option of deciding whether their troops would undergo HSD training, since personnel other than those in BCT and AIT were to receive it only if prescribed by their commander. Commanders were also expected to be more involved in HSD than in earlier programs. They were encouraged to suggest topics the chaplain should cover in the HSD meetings and to lead some of the meetings themselves, using resource materials provided by OCCH. That would give them an opportunity to address problems specific to their installation. Indeed, chaplains were advised to tailor the subject matter of the training sessions "to the particular needs of the command."[62]

In certain respects, HSD was similar to earlier character education programs. The OCCH described it as a "system of value education" designed to encourage "personal moral growth."[63] The 1971 army regulation for HSD pointed out that the purpose of the program was "to encourage high standards of personal and social conduct among members of the Army" and "to strengthen in the individual the basic truths, principles, and attitudes that undergird our nation's heritage."[64] As the emphasis on "our nation's heritage" suggests, HSD continued OMH's focus on the historical, religious, and cultural foundation of America. The OCCH defended that approach by arguing that "as soldiers see themselves in relation to the fundamental values which undergird a free society, they are better able to realize their worth, to develop their full potential and to seek healthy goals for their lives."[65]

In other respects, HSD was a component of the institutional ministry. It was a secular program, it had a behavioral science orientation, and much of its subject matter focused attention on the army as an institution and its behavioral problems. The 1971 army regulation described the program as a means of assisting the commander in "today's challenging problems of racial tensions, drug abuse, poverty, dissent, and moral behavior." A few years later, the list was enlarged to include "questions about the morality of war, and poor communication between generations."[66] The titles of some of the HSD discussion topics reveal its institutional focus: "Cultural Awareness," "Alcohol and Alcoholism," "The Soldier and His World: As We Live and Breathe" (on "the ecological threat"), "The Soldier and His World: His Personal Response to Minorities," and "The Soldier and His World: Poverty."[67]

Human Self-Development used the same method of instruction employed in Our Moral Heritage. It was to be inductive, nondidactic, and carried out through dialogue and discussion, a mode of instruction thought to be attuned to the emotional needs and nontraditional values of the "new breed" of youth entering the military. Training pamphlets emphasized the great "flexibility" instructors had in choosing and developing topics for discussion. They were merely "resource materials" containing "suggestions for raising points and deepening the discussions." Instructors were expected to "enlarge" and "enliven" the materials with their own knowledge and experience.[68]

The OCCH did, however, insist on three rules. Instructors should "strive for dialogue and maximum participation by all members of the class," should adapt the discussion topic to the needs and/or problems of the local installation, and should "avoid any action which would tend to confuse [HSD] training with religious instruction." On this last matter, OCCH was adamant: "Specifically, the instructor will not, under any circumstances, use a scheduled training period to deliver a sermon, to sermonize parts of the topics or upbraid troops for nonparticipation in chapel programs, to show religious films or to expound his own personal theological views." The 1971 army regulation corroborated the third rule, declaring, "The instruction and the program [for HSD] will remain nontheological and nonsectarian. These sessions are separate and distinct from the voluntary religious program of the command which is the proper sphere of voluntary denominational religious activity."[69]

Our Moral Heritage had featured topics such as "Perseverance," "Charity," "Integrity," "Courage," "Honesty," and "Sacrifice."[70] By contrast, Human Self-Development did not *teach* values. Instead, it focused on the *process* of forming values. "There are no 'school solutions' to moral, morale, or value problems," OCCH observed in a comment on the program.[71]

This approach reflected the influence of behavioral science methodologies, in particular a type of moral education known as values clarification. Pioneered by Louis E. Rath, Merrill Harmin, Sidney B. Simon, and Howard Kirschenbaum, values clarification gained considerable attention in the mid-1960s and was widely used in civilian schools. In place of traditional methods of moral education such as moralizing (transferring a set of values from one person or group to another person or group) and modeling (living a set of values that others will emulate), the values-clarification approach recommended teaching a "process of valuing" whereby students learned how to discover and examine their values and ultimately arrive at a consistent set of values. The emphasis was on self-realization, self-worth, and personal growth. Values were broadly defined as anything an individual treated as right, desirable, or worthy and were considered to be personal and relative. Values clarification also emphasized that the values clarifying process was something that took place in a group, where individuals engaged in "social discourse" and "communication," sharing thoughts and feelings.[72]

The OCCH declared that one of the purposes of HSD was to "improve the soldier's self-image" by emphasizing "personal uniqueness and self-fulfillment."[73] Instructions for the discussion topic "Morality and the Conscience" emphasized that its purpose was not to promote any particular system of values but to make the soldiers aware of the existence of the individual conscience and of ways of "informing," "invigorating" and "validating" it. Another instructional pamphlet advised the instructor to establish rapport with the class by fostering "a spirit of permissiveness throughout the entire session in which expression of honest feelings and opinions would be encouraged." The instructor should also stress that there were no right or wrong answers and that every person would be treated with dignity and respect. Treating people that way would show "true democracy at work," the pamphlet declared.[74]

In a 1975 article in the *Military Chaplains' Review,* Chaplains Hugh J. Bickley and Ford F. G'Segner explained why they used values clarification in the HSD program at Redstone Arsenal, Alabama. They contended that the new participatory type of character education, with its emphasis on dialogue and discussion, was necessary because the soldiers attending HSD classes were different, culturally and psychologically, from earlier generations. Educational, management, sociological, psychological, and religious authorities all pointed out that "contemporary persons" are "*non-authoritarian, dialogical, interdependent, and participative.*" Thus the older, didactic, "moralizing" approach to character education was doomed to fail: "When a chaplain moralizes he imposes his values on others and, in so doing, he asks them to reject their operative values on the basis of blind submission to authority, or the power of propaganda, or peer pressure without allowing them an opportunity to measure and sort out their own beliefs." By contrast, values clarification was "a process which allows persons to build and to discover their own values systems," the two chaplains asserted.[75]

Assessing Human Self-Development and Values Clarification

Two articles published in the *Military Chaplains' Review* and the *Chaplain* give an idea of the arguments some chaplains made in favor of the new approach to character education embodied in HSD and other, similar programs. Chaplains Robert D. Crick and Douglas J. Groen cited "rumblings of a new Army," which indicated the need for "new and exciting models to facilitate constructive change." They urged chaplains to "accept the challenge of the times" rather than "compulsively" resisting "necessary change" by "patching up the old garment of authoritarian guidance and rejecting the new generation's demand for more colorful and perhaps more comfortable clothing." If the chaplain would give up the idea of being the "*only* authority" able to conduct the HSD program, and "share" that ministry with the commander and other staff members, he would come to be "valued and respected" by them. At the same time, in becoming a "group facilitator," he would develop a closer relationship with the troops. Abandoning his "self-image" as "an exclusive arbiter of morality," stepping down from "the pedestal required by the old stereotype," he would begin to "move among people to whom he belongs" and would discover "a grassroots existence of daring involvement."[76]

Former army chaplain Jack S. Boozer, a professor of religion at Emory University, focused his article on the OMH program, but his remarks also applied to HSD. Like Crick and Groen, he pointed to changes that chaplains needed to address: specifically, the 1960s theological revolution and the "serious and increasing criticism" aimed at military chaplains. He bluntly termed Character Guidance a failure. "In far too many instances, chaplains performed in this [program] as morale officers of the unit or as exponents of a 'parochial' faith and morality to all the men of the unit," he declared. By contrast, OMH "clarified" the proper roles of the chaplain. In effect, it prescribed two distinct roles, "that of 'secular teacher' in a command program on our moral heritage, and that of a convinced believer, a 'religious leader,' for those who voluntarily respond to his presence as priest, pastor or rabbi." Boozer believed that, given the religious pluralism of the military community, it was necessary for chaplains to continue to have a "religious presence," each serving as the "leader of a confessing community which voluntarily seeks and responds to his leadership." But chaplains also needed to project a "secular presence." He conceded that the word "secular" might seem threatening to some chaplains and that the idea of being a "secular teacher" might suggest "a compromise in one's calling." In Boozer's view, however, the new role held great promise, as he explained in the following statement:

> For the chaplain to appear in a "secular" role, without any religious or military authority which makes his views correct; for the chaplain to attempt in open honesty to communicate with and to another some meaning of human dignity and of citizenship in this nation—this, I suggest, is not a threat to a high calling but an opportunity of considerable preciousness.[77]

The newness of HSD, which so excited some of its advocates, provoked negative comment among rank-and-file chaplains. Some disliked the secular orientation, which seemed incompatible with the chaplain's religious calling. In the mid-1970s, some chaplains, commanders, and a few members of Congress objected to values clarification workshops for soldiers in the U.S. Army, Europe, as a "tool of secular humanists."[78]

Other chaplains found the nondidactic, participatory approach to character education daunting. Many of them had little, if any, training in the psychological and sociological theories on which the method of instruction was based. Feeling inadequate to the task of serving as facilitators in unstructured group discussions of values, they continued to use the deductive-didactic approach.[79] Chaplains also had trouble adopting the recommended "spirit of permissiveness." To tell soldiers, as OCCH advised, that there were "no right or wrong answers" to moral questions was especially problematic, in view of the fact that chaplains had long been considered—and considered themselves—authorities on morality and religion. Human Self-Development's emphasis on command involvement posed yet another problem. In earlier character education programs, chaplains had been able to work independently and had been regarded as the primary authority on moral education. Now they were expected to work closely with the commander and his staff in developing discussion topics and leading sessions. Thus their control over the program was significantly diminished.

As for commanders' assessment of HSD, it is clear that OCCH and the Army Chaplain Board designed the program to win command support. But while some of the components of chaplains' institutional ministry—drug abuse prevention and rehabilitation, the Human Relations Ministry, and PET—did earn command respect, HSD prompted little enthusiasm and even some resistance. Its much-touted flexibility made it easier for commanders to do without it. Some commanders disapproved of its secular orientation. Others doubted that the emphasis on individual self-esteem and personal growth contributed to military effectiveness. Commanders had long been skeptical of character education programs, and HSD did nothing to alter that view. Weakened by command resistance, the increasingly moribund program was discontinued in January 1977.[80]

Although commanders showed little enthusiasm for HSD, chaplains were successful in using other programs of the institutional ministry to prove their worth to the military organization. The substance abuse, race relations, and human relations programs not only won command respect but also advanced chaplains' long-term goal of playing an integral role in the military mission.

Chaplain Hyatt pointed out yet another way in which the institutional ministry enhanced chaplains' reputation. It "proved that we could minister to the power structure," he observed in a 1978 interview. "I don't say that the chaplaincy ushered in the kingdom of God by its efforts, but I do say that we were successful in demonstrating that the chaplaincy can be an agent of change."[81] The chapter that follows considers another component of the institutional ministry, the one developed to assist military leaders in remedying the erosion of the professional military ethic. It, too, expanded chaplains' moral and educational roles in significant ways.

6

"The Conscience of the Army"

The third component of the crisis the U.S. Army faced in the late 1960s and early 1970s was the erosion of the professional military ethic, what Edward L. King, in *The Death of the Army* (1972), called "pervasive ethical laxness at all levels."[1] Its most powerful symbol was the My Lai massacre and cover-up. During those decades, noncommissioned officers and members of the officer corps, including general officers, also engaged in other corrupt or illegal practices. The thriving black market in Vietnam involved officers as well as enlisted personnel. In Europe two general officers who headed the military post exchanges were implicated in and ultimately punished for corruption. The sergeant major of the army, along with a number of senior noncommissioned officers, became involved in a scandal involving the management of military messes and clubs.[2] Other officers engaged in minor cover-ups and the questionable practices that seemed endemic in Vietnam: "ticket-punching," inflating body counts, misusing officer-efficiency reports, and suppressing reports and recommendations that conflicted with current policies, strategies, or tactics.[3]

Analyzing the Problem

On March 18, 1970, when Lt. Gen. William R. Peers submitted the Peers Commission Report on My Lai to Army Chief of Staff William C. Westmoreland, he included a confidential memorandum. In it he concluded that, based on the more than four hundred interviews the commission had conducted with army officers, "there was something deeply and basically wrong with the moral and professional climate of the Army officer corps." According to Maj. Gen. Franklin M. Davis Jr., "The memo shook Westy to the core."[4] A few days after receiving it, Westmoreland directed the U.S. Army War College (AWC) to conduct "an analysis of the moral and professional climate" in the army.[5] He received its report, titled "Study on Military Professionalism," on June 30, 1970, and found it so unsettling that he ordered it "classified" and to be circulated only among the general officers of the army.[6]

In the preface to the report, the commandant of the Army War College wrote that it "deals with the heart and soul of the Officer Corps of the Army." Its subject matter included ethics, morality, professional competence, individual motivation, and personal value systems—all "inextricably related, interacting, and mutually reinforcing" and all "filled with emotional overtones." In accord with Westmoreland's directive, the investigators were mainly concerned with "the *perceptions* of the existing climate by members of the Officer Corps." Having interviewed 450 junior and senior officers representing "a broad spectrum of experience, grade, and branch," they summarized their main finding as follows:

> Officers of all grades perceive a significant difference between the ideal values and the actual or operative values of the Officer Corps. This perception is strong, clear, pervasive, and statistically and qualitatively independent of grade, branch, educational level, or source of commission. There is also concern among officers that the Army is not taking action to ensure that high ideals are practiced as well as preached.[7]

In the 1970s and 1980s a consensus developed among civilian and military commentators regarding the nature, cause, and remedy of the ethical crisis in the armed forces. They contended that the military's traditional professional ethic had been supplanted by a very different system of values drawn from the corporate world. Instead of "Duty, Honor, Country" and moral absolutes such as integrity, selflessness, loyalty, and truthfulness, the "managerial ethos" inculcated self-advancement, efficiency, and exploitation of enlisted personnel—values that motivated the so-called careerism of the Vietnam era. Although some commentators dated the erosion of the professional ethic from the post–World War II period, most thought it began in the 1960s. Some attributed it to the alleged "no-win" policy in Vietnam. Others blamed the 1960s cultural upheaval in civilian America, especially the transformation of religious and moral values.[8] However, most commentators agreed with the AWC *Study on Military Professionalism* that neither the Vietnam War nor political, cultural, and sociological "influences" from the civilian community were responsible for the erosion of the traditional military ethic. The *Study* emphatically declared that the erosion was "internally generated." And it called for "corrective measures"—to be "instituted from the top of the Army"—to "support adherence to traditional ethical behavior." The *Study,* as well as some commentators, urged the formulation of a written code or creed.[9] Although the idea prompted much discussion, no such code was ever promulgated.[10] However, over time, the army developed various programs of ethical and leadership instruction. Some were used in the training of officers and noncommissioned officers; others were instituted at the army service schools, something the *Study* had recommended.[11] Chaplains were in the forefront of that development.

"The Conscience of the Army"

In 1969, while serving as deputy chief of chaplains, Gerhardt W. Hyatt wrote letters to two friends, the commanders of the Armor School at Fort Knox and the Infantry Center and School at Fort Benning, proposing "an experiment." If they would put a chaplain on the faculty of their school to teach "moral leadership and moral responsibility," he would give them "the best chaplain we had in the corps to do the job." (He used the terms "moral leadership and moral responsibility" rather than "ethics" at that time, because, he said, everybody was aware of the "moral lapse" that had occurred at My Lai. "We used the word moral in order to sell the thing.") The overture worked. "Within days, I had a letter back—Send him!" Soon other commanders who heard of the experiment wanted to know why they had not been made the same offer. Hyatt had some difficulty keeping up with the requests—from the Artillery and Guided Missile School at Fort Bliss, and then from the Transportation, Quartermaster, and Medical Schools. By mid-1973, chaplains were teaching ethics at nine branch service schools, as well as the U.S. Military Academy. In 1974 the number rose to fourteen, and the following year, when Hyatt left his post as chief of chaplains, chaplains held instructor positions in twenty schools.[12]

Placing chaplain instructors in the service schools was one more component of the new institutional ministry Hyatt developed in response to the crisis in the army. Their placement also signaled the emergence of another new role for chaplains. Instead of serving as moral educators of enlisted personnel, as in the various character education programs, they were now teaching ethics and moral leadership to the officer corps. The new role led, logically, to the development of a new relationship vis-à-vis commanders and to the elaboration of the chaplain's responsibility for advising the commander and his staff on matters relating to religion and morality. Indeed, Hyatt anticipated this change. In his view, advising commanders on religion and morality was part of chaplains' prophetic responsibility—they were to serve as "the conscience of the Army." He thought that by taking up the role of teachers of ethics and moral leadership, chaplains would gain "a more influential role in the development of command policies and decision," which was one of his long-term goals as chief of chaplains.[13]

The men who succeeded Hyatt as chief of chaplains, Orris E. Kelly (1975–79) and Kermit D. Johnson (1979–82), continued to make the placement of chaplain instructors at the service schools a priority. During Kelly's administration, the U.S. Army War College instituted a twelve-session course in ethics as part of its regular curriculum. In the newly organized Sergeants Major Academy, a chaplain was responsible for more than half of the moral leadership coursework. During the 1980s, chaplain instructors, including two female chaplains, were teaching at West Point, the Command and General Staff College at Fort Leavenworth, and some twenty branch service schools throughout the United States. By the 1990s, according to Army Chaplain Corps historian John W. Brinsfield, ethics instruction had expanded to include "complete core and advanced courses in ethics at West Point, the US Army War College, the Command and General Staff College and the other 18 Army service schools, and the ROTC Cadet Command," much of it offered by chaplains. As of the spring of 1996, Brinsfield reported that "80 percent of the ethics taught in Army service school leadership departments was taught by chaplains."[14]

The knowledge chaplains brought to the teaching of ethics and moral leadership varied. Some had studied moral theology while pursuing graduate seminary degrees. Others had earned advanced university degrees in history, philosophy, literature, education, or the behavioral sciences. The U.S. Army Chaplain Center and School (USACHCS) played an important role in training chaplains to teach ethics and moral leadership. In the early 1970s, it became aligned with the newly created U.S. Army Training and Doctrine Command (TRADOC), which named it a "participating proponent" for military leadership training. As a result, USACHCS faculty and staff became involved in developing curricula for all the service schools, not just the Army Chaplain School.[15] They were able to contribute significantly to the development of informational materials and programs designed to infuse leadership training with ethical and moral principles. This, along with the placement of chaplain instructors in the service schools, proved especially gratifying to Chief of Chaplains Hyatt, for it signaled the extent to which chaplains were achieving their longstanding goal of playing an integral role in military training.[16]

The USACHCS also added new courses in ethics and moral leadership to the Army Chaplain School curriculum. By the early 1980s the chaplain officer basic course included instruction in "moral and ethics development," the code of conduct, and the Geneva and Hague Conventions. A publication describing the basic course described the scope of the "Leadership, Management and Ethics" component as follows: "The chaplain will acquire skills in utilization of a model for ethical decision making, problem-solving procedures, and techniques for effectively identifying and confronting unjust procedures and actions."[17] Another basic course publication pointed out the rationale for such instruction. "Society in general" and many religious denominations and congregations regard their clergy as "experts in the area of ethics," it observed. "Members of the Army Community expect the same of the chaplain," that its clergy "be knowledgeable about ethical matters." Indeed, the army was "presently experiencing a renaissance of concern with ethical behavior," which had been expressed "at the highest command levels" and was "reflected in training and command emphasis throughout the Army." Yet there were "substantial areas of disagreement . . . as to the exact nature of ethical behavior." As a result, "the chaplain is often called upon to be the one who confronts such conflict" and offers advice on how to resolve it. In the conclusion to the statement, the publication asserted that "the chaplain is a primary ethical player in the Army."[18]

Seminars, workshops, and conferences also provided chaplains with training in ethics instruction. The Military Chaplains Association and the Armed Forces Chaplains Board sponsored several one-day ethics workshops in the mid-1970s. During February 1977, for example, three such training sessions were held in Germany, at Nuremberg, Ramstein Air Force Base, and Frankfurt. The theme was "Ethical Issues of Military Leadership Today."[19] The Office of the Chief of Chaplains also organized various ethics seminars, workshops, and conferences during the 1970s and early 1980s, and it encouraged major command staff chaplains to plan workshops within their commands.[20] Some of the seminars and workshops included commanders and staff and line officers, along with chaplains, so that they could learn from each other what ethical issues needed to be explored.[21]

A five-day "Ethics Workshop" held at USACHCS in the fall of 1976 facilitated an interesting exchange of views among chaplains, military leaders, and civilian ethicists. According to Earl F. Stover, who reported on it for the *Christian Century,* twenty-eight of the participants were chaplains serving on the staff of major headquarters or the faculty of service schools. Another twenty-two were line officers, including two women, at the rank of colonel and lieutenant colonel. The workshop featured lectures by four civilian and three Army "resource leaders," including question-and-answer sessions afterward, two full-length motion pictures (*Billy Jack* and *A Clockwork Orange*), a panel discussion, and small-group discussions.

The four civilian "resource leaders" were Michael Novak, Mark H. Tanenbaum, William A. Jones, and George W. Webber, each of whom described his "ethical base" in the lectures. While Tanenbaum's and Webber's biblical perspectives seemed sound, Stover noted that Novak's presentation of what he called "the experience of nothingness" and Jones's "religious humanism" apparently "disturbed" some of the chaplains and line officers. The "relativism" of some of the civilian resource leaders also provoked alarm. The three military "resource leaders" were Maj. Gen. DeWitt C. Smith Jr., commandant of the U.S. Army War College; Chaplain (Col.) Kermit D. Johnson, a student at the AWC; and Chaplain (Col.) Joseph H. Beasley, an ethics instructor at the U.S. Military Academy. Some of the participants who had criticized the civilian resource leaders for being too "philosophical and theoretical" commended the military resource leaders for their "more practical approaches" to ethics. However Stover reported general agreement among the chaplains and line officers that chaplains had a responsibility to address and clarify ethical problems and that they should do so "regardless of the risk, potential conflict, and possible consequences."[22]

The Purpose and Content of Ethics Instruction

The purpose and content of the ethics instruction chaplains offered in the service schools varied and changed over time. In the early 1970s, as a member of the history faculty at the U.S. Military Academy (USMA), Chaplain Joseph Hodgin Beasley requested permission from his department head to develop and teach an elective course in ethics. In the proposal he wrote for the course, he argued that army leaders needed to know "something about moral philosophy" if, among other things, they were going to be able to talk to soldiers about following and applying the rules, traditions, and customs of the army. In particular, they would need to be able to explain why they believed army standards to be correct rather than simply insisting that "'that's the way we have always done it' or 'I know I am right and you had better agree.'" The leaders would need to "be capable of recognizing moral problems, of intelligently reasoning about the various solutions to those problems, and of justifying the decisions that they finally make."[23]

After some delay, Beasley's proposal was approved, and in the spring of 1976, he began offering elective course HI 378, "History of Western Ethics." It surveyed the history of ethical thought from Plato and Aristotle to modern thinkers and discussed the relevance of their theories to moral problems involving politics, war, and the use of weapons of mass destruction, sexuality, racism, and medicine. Beasley sought to help cadets recognize the complexity of ethical questions, to forbear "easy and pat answers," and to become "more tolerant of compromise and ambiguity."[24]

Although Beasley's course proved very popular among the cadets, USMA officials were less enthusiastic. In his University of North Carolina Ph.D. dissertation describing his teaching experience at the academy, Beasley lamented "the confusion about ethics that prevailed at West Point." He viewed (and taught) ethics as "an explorative method of inquiry, whereby teacher and student together seek to bring more insight into modern moral dilemmas where all possible answers are more or less unsatisfactory." His purpose in teaching ethics was to help cadets develop their powers of moral reasoning and to become "morally sensitive officers prepared to lead the military through the ethical problems of the next century." By contrast, the West Point faculty and administration wanted ethics taught in such a way as to bolster the professional military ethic and to make cadets appreciate and obey the honor code.[25]

In the early 1980s, when Chaplain John W. Brinsfield taught HI 378 at the USMA, his purpose was similar to Beasley's—to "help the cadets recognize ethical issues, develop analytical skills, assume personal responsibility for decisions, and stimulate their moral imaginations." Like Beasley, Brinsfield devoted the first half of the course to the history of the "major ethical theories of Western Civilization." But he expanded the section on the Judeo-Christian heritage and added a discussion of the ethical perspectives of Jesus, Mohammed, Machiavelli, Luther, Calvin, and William James. "I decided in the summer of 1980," he explained in his Drew University doctor of ministry thesis, "that the History of Western Ethics would be inadequate in scope without a fuller treatment of the historical quest for absolute Goodness (or Duty or Righteousness) as reflected not only in Plato's *Republic* and in Kant's *Metaphysics of Morals* but also in Jesus' Sermon on the Mount." Other moral values posited by Christianity, "such as the dignity of the individual life and the love of one's fellow man" seemed to him to "have had at least as great a social impact on Western culture as Kant's concept of duty as a moral imperative." Brinsfield devoted the second half of the course to analyzing contemporary ethical problems. He presented case studies involving "moral dilemmas," which the cadets were to analyze from various perspectives, such as epicurean, Thomistic, Kantian, utilitarian, and situational. He wrote that he was less concerned with "solving the dilemma" than with "uncovering the values which seemed to control the manner in which the problem was approached."[26]

Both Beasley and Brinsfield offered formal, academic instruction in the ethics of warfare at the U.S. Military Academy in the 1970s and 1980s. Beasley included lectures on "Ethics and Warfare," "Ethics and the Use of Weapons of Mass Destruction," and "Killing in War" in HI 378. Later, when Chaplain Brinsfield taught that course in the 1980s, he, too, discussed such topics. Brinsfield also developed a course titled "History of the Ethics of Warfare," which focused on the ethical theories undergirding the holy war, the just war, the crusade, the limited war, the total war, and the cold war. He invited guest lecturers, including a former battalion commander in Vietnam, to talk with the cadets, and he used case studies in analyzing the different ethical theories. He said that his "goal was not to pass judgment on different wars and war leaders from a modern perspective but to see if selected generals in history consistently followed the particular rules of war or military ethical systems known to them at the time."[27]

Teaching Moral Leadership

The chaplains who taught "moral leadership" in the service schools adopted a different approach from the one Beasley and Brinsfield took in teaching ethics or moral philosophy. In the early 1970s, when Hyatt coined the term "moral leadership," it connoted a broad range of instruction. In the 1972–73 "Historical Review," Hyatt endorsed the goals set forth by a Moral Leadership Curriculum Workshop held at the Army Chaplain School in June 1972. Chaplain instructors were to accomplish four objectives:

> (1) Cause faculty and students of the schools to question creatively their understanding of the moral dimensions of responsibility, leadership, and human relations. (2) Help them devise and justify acceptable alternatives. (3) Help interpret to the student and the staff and faculty the spirited aspect of the whole person, and (4) Identify and define the chaplain and his role and functions to young leaders (Officers and NCOs).[28]

Very soon, those broad goals were pared away. The main purpose of most chaplain instructors in the service schools was not to teach moral decision making, nor to enhance their students' self-understanding and concern for others, but to impress upon them the importance of and justification for the moral virtues deemed essential to the military profession—loyalty, selfless service, integrity, courage, and the like.[29] These chaplains taught professional military ethics. The army's *Historical Summary* for 1975 reported that chaplains on the faculties of the eight army schools and the Sergeants Major Academy had taught courses that "emphasized integrity, self-discipline, moral courage, and loyalty as ingredients of military professionalism."[30]

The chaplains' focus on professional military ethics suited the thinking and requirements of army leaders. Their approach was "practical," in contrast to the "abstract" approach of formal, academic ethics courses, and as we saw earlier in this chapter, that was the approach many line officers and chaplains preferred.[31] More to the point, this "practical" approach, like the term "moral leadership," spoke directly to the army's leadership crisis. And it synchronized with the "leadership doctrine" developed later, in the mid-1980s, which regarded ethics as one of the main components of leadership training. Maj. Robert A. Fitton, writing in the *Military Review* on the current status of army leadership doctrine and training, called ethics "the underpinning of leadership." Ethics instruction in the army "must address the body of principles and unique Army values essential to defining the military profession," he declared. Its objective must be "to provide a thorough foundation in the values, norms, and mores of the military profession." Its focus should be "directed toward instilling in leaders what it means to be a professional—the uniqueness, history and tradition associated with the military profession. Leaders need to continually reflect on the cornerstone values of our profession."[32]

The Nuclear Debate

The knowledge and confidence chaplains gained in exercising their new roles as teachers of ethics and moral leadership strengthened their commitment to prophetic responsibility. Chief of Chaplains Orris E. Kelly continued the emphasis his precedessor, Gerhardt Hyatt, had placed on it. The OCCH "Chaplain Professional Development Plan" of 1979 included the following declaration:

> Army chaplains demonstrate a prophetic presence. They are so in touch with their own value system and those of their churches that they boldly confront both the Army as an institution and individuals within it with the consequences of their actions. . . . They are knowledgeable, able and willing to confront both individuals and the Army with the ethical

> aspects of decision-making, policies and leadership, and the extent to which these, in both war and peace, reflect on basic Judeo-Christian ethical framework [*sic*]. They are prepared adequately to "stand up and be counted."[33]

In the 1980s, momentous changes in international relations and U.S. defense policy tested chaplains' commitment to prophetic ministry. During the first administration of President Ronald Reagan, tension between the United States and the Soviet Union increased markedly. Reagan initiated a massive buildup of military force, including the development of new, more lethal weapons systems, such as nuclear warheads.[34] Initially, he refused to engage in arms-control talks with the Soviets. In 1982, in an effort to undercut a burgeoning nuclear freeze movement in the United States, he did begin Strategic Arms Reduction Talks with them, in Geneva, but he continued to engage in bellicose rhetoric, culminating in a speech to the National Association of Evangelicals in March 1983 in which he called the Soviet Union an "evil empire" and "the focus of evil in the modern world."[35] In the midst of the military buildup, Reagan and members of his administration talked of limited nuclear war in Europe, of a winnable nuclear war with the Soviets, of not only using nuclear weapons but using them first, if necessary. News commentators speculated that the military program as well as the public statements of the Reagan administration signaled a dramatic policy shift—away from deterring nuclear war to developing the capability to fight and prevail in such a war.[36]

The new defense program provoked intense civilian debate over issues of "peace through strength" (Reagan's avowed policy), "nuclear deterrence," "mutual assured destruction," "nuclear freeze," and the "nuclear peril." Nuclear disarmament, declared Rev. Miles O'Brien Riley of the Catholic Archdiocese of San Francisco, "is the moral issue of the day. Everything else is a footnote."[37] The Reagan administration's massive arms buildup and its reluctance to negotiate arms limitation with the Soviet Union infused the American peace movement with new vigor. In addition, many of the mainline churches moved to the forefront of the antinuclear campaign during the 1980s.[38] By the mid-1980s the National Conference of Catholic Bishops (NCCB) and several major Protestant denominations had issued official pronouncements opposing many Reagan administration defense policies. In *Nuclear Weapons and the American Churches: Ethical Positions on Modern Warfare* (1983), comparing the views of the Catholic bishops and those of twelve Protestant denominations on nuclear weapons issues, Chaplain Donald L. Davidson found "striking similarity." None advocated nuclear superiority of the United States. All "registered significant concern for nuclear arms control and limitation." None advocated unilateral nuclear disarmament, but all supported some concept of deterrence based on goals of "sufficiency," "parity," "balance," or "rough equivalency" with the Soviet Union.[39]

Although neither the Catholic bishops nor the Protestant groups advocated pacifism, some of their leaders seemed to be moving in that direction. In May 1985, for example, some of them signed a statement, prepared and circulated by the American Friends Service Committee, which called upon Congress to refuse funding for the "Star Wars" defense system and rejected "any system of security based on fear and intimidation."[40] With the important exception of Billy Graham, conservative evangelicals and their main organization, the National Association of Evangelicals, generally supported the Reagan administration's defense policies. Radical evangelicals, however, favored complete nuclear disarmament.[41]

Besides official pronouncements, members of the civilian religious community engaged in other efforts to halt the testing, production, and further deployment of nuclear weapons. The so-called nuclear freeze movement generated a host of ad hoc committees, forums, workshops, and demonstrations, including one of the largest ever in New York City on June 12, 1982. In a referendum held during the fall elections of that year, twelve million Americans voted for a freeze on nuclear weapons.[42]

More than a few military leaders, in all of the service branches, recognized that the Reagan administration's defense policies raised significant moral and ethical questions.[43] Just war doctrine, the morality of strategic and tactical nuclear war, nuclear deterrence, and the nuclear freeze received considerable attention in various military journals, including *Parameters* and *Armed Forces and Society*.[44] Chief of Chaplains Kermit Johnson observed that Army Chief of Staff Edward C. Meyer and other leaders were discussing national defense policy and nuclear weapons, although it was being "done quietly" and therefore attracted little media notice.[45]

As the teachers and proclaimers of ethics in the military, and having specific competency in the ethics of warfare, chaplains were inevitably drawn into the debate over national defense policy and nuclear war. The upsurge of antimilitarist and pacifist sentiment sparked criticism of military chaplains. In 1985, for example, the Pax Christi USA Center for Conscience and War, a Catholic peace organization, voiced opposition to proposed changes in the Roman Catholic Military Vicariate that it feared would make the chaplaincy "an even more integral part of the military establishment" and compromise chaplains' ability to offer "detached moral guidance" to armed forces personnel.[46]

In 1986, members of the Lutheran Peace Fellowship (LPF) issued "An Open Letter Concerning the Military Chaplaincy" in which they urged Lutheran church bodies to "remove the uniforms, the rank, the accountability to military command, and the paycheck from the Department of Defense—all of which tend to compromise the military chaplaincy." Such changes, according to the LPF, would shape a ministry to military personnel "which is accountable only to the church." The letter also addressed individual chaplains, exhorting them to "strive for a civilian chaplaincy to the military" and to "minister with prophetic integrity—especially on military bases or naval vessels which deploy weapons of mass destruction—even if it would mean a refusal to serve in those places." The emphasis on chaplains' prophetic responsibility also appeared in a section of the letter that questioned the "apparent silence of the military chaplaincy" on the morality of nuclear weapons and nuclear warfare. The 150 signers of the letter challenged Lutheran chaplains to join them in "a conscientious outcry against weapons of mass destruction."[47]

The letter received a great deal of publicity, and a number of chaplains became concerned about its content and the fact that bishops and other church leaders had signed it. They wanted the American Lutheran Church (ALC) to disavow it. The ALC held a "consultation" to discuss the issue, which was attended by seven signers of the letter (including a bishop associated with the LPF), several high-ranking chaplains, and leaders of the ALC. According to minutes taken by one of the LPF participants, discussion at the two-hour session revealed that there were a few chaplains who spoke out "with some regularity" against the evils of war, but even when they spoke "quietly," their words were "not usually received with appreciation" from the military. During the consultation, "it became increasingly clear that *there are barriers within the present system* that make it more difficult for chaplains to speak with a prophetic voice about ethical issues raised by modern warfare." Nevertheless, LPF members continued to believe that "dialogue" with military chaplains was not only necessary but possible. "Our church . . . has spoken," an LPF participant observed; its position was that "weapons of mass destruction are immoral." The question LPF asked chaplains was, "Are you able from your position of power to speak against weapons of mass destruction?"[48]

Chaplains also heard from sociology professor Gordon C. Zahn, their nemesis during the Vietnam War. By 1982, when he published an article on chaplains in *America,* the Catholic pacifist had moderated his criticism of military chaplains and, to a lesser extent, the language in which he expressed it. He still favored demilitarization of the chaplaincy, but he also conceded that "justifications advanced for the present system" had some merit. His main concern was whether and to what extent chaplains were prepared and willing to "make and assist others in making moral judgments and decisions in the military milieu." The fact that the armed forces were being "expanded and equipped with new weapons of increasingly dubious morality" only intensified that concern.

Zahn pointed out that one of the duties of chaplains, according to army regulations, was to provide "an evaluation of the ethical and humanitarian depth of command policies." He wrote that he had seen nothing in the chaplains' training programs that seemed designed to help them carry out that duty. Moreover, he believed that a chaplain whose actions served to "trouble consciences" would "almost certainly invite sharp criticism and, quite possibly, punitive action on the part of the military authorities (including his own chaplaincy superiors)." And he noted that if any soldiers listened to him and followed "his moral lead in refusing to give or obey immoral orders," they would be "even more vulnerable to serious sanctions." However, neither of those possibilities would "be enough to free the chaplain from his obligation to provide moral direction." In a statement reminiscent of his articles of the late 1960s, Zahn declared, "Any chaplain who . . . would insist that the morality of a war, a given campaign in that war, or a specific action in that campaign is none of his concern is unworthy of his status as priest, as 'another Christ.'"[49]

Pacifist criticism played a minor role in prodding chaplains to respond to the debate over the ethics of nuclear war and weaponry. The main impetus came from the "Pastoral Letter on War and Peace" written by a committee of the NCCB and issued in early May 1983, as well as from other, similar pronouncements from mainline Protestant denominations.[50]

The NCCB committee discussed and revised the pastoral letter over a period of two and a half years, producing three drafts, which were publicly circulated. While the bishops deliberated, Roman Catholic military personnel, as well as Catholic chaplains, found themselves in a kind of moral limbo.[51] Chaplains felt pressed to try to explain the church's teaching regarding nuclear warfare to Catholic military personnel, but could not be absolutely sure what it was or was going to be and what it might say about the morality of military service. Terence Cardinal Cooke, the military vicar for Roman Catholic military personnel, sensed the foreboding among his parishioners. During the time the pastoral letter was being written, he discussed it in his annual letters to Catholics in the military, including chaplains, reassuring them that the church continued to "recognize and appreciate" their contributions to "the cause of peace with justice."[52] In April 1983, after the third draft of the pastoral letter had been released, some Catholic chaplains and Defense Department officials expressed relief that it was "softer" than the earlier ones had been with regard to U.S. nuclear policy and more supportive of Catholic military personnel. A State Department press release, which the Defense Department supported, conceded that the third draft was an improvement over previous ones but said that the Reagan administration still disagreed with portions of it.[53]

Less than a month later, on May 2–3, 1983, when the bishops formally approved the final text of the pastoral letter and released it to the public, it contained stronger criticism of U.S. nuclear policy, a result of last-minute amendments. It broadly condemned the use of nuclear weapons, based on just war principles. The bishops denounced the first use of such weapons, as well as their employment against civilian populations, as "morally unjustified." They also opposed "a nuclear response to either conventional or nuclear attack" on the ground that such a response could cause destruction far beyond "legitimate defense" and therefore "would not be justified." They expressed "a strictly conditioned moral acceptance of nuclear deterrence policy." Such a policy was allowable only as "a step on the way to progressive disarmament," in which the United States must seriously engage; the bishops did not consider it "adequate as a long-term basis for peace." And, finally, the bishops endorsed a nuclear freeze.[54]

The bishops' final document also contained a special section addressed to "The Men and Women in Military Service." In a letter to Roman Catholic chaplains, written in June 1983, Cardinal Cooke discussed its main points. He began by reiterating his previous assurances that the church appreciated Catholics' "sacrifices and service as peacemakers." Then, paraphrasing and quoting the pastoral letter, he observed that the bishops had not intended "to create problems for Catholics in the military." They did, however, believe that the letter's "teaching on war and peace . . . poses a special challenge and opportunity to those in the military profession." Cooke himself believed that the letter did question "the morality of some strategies of defense, and requires us to examine our individual consciences and our 'national conscience.'" In conclusion, he reminded the chaplains that the letter made a distinction between the moral principles of the church, which were binding, and "purely human judgments, which are not binding." Cooke also urged chaplains to join a "dialogue for peace" by sharing his message to them with their congregations and by airing their own views on the use of nuclear weapons. The bishops "want your opinions on such matters, particularly those of you who are highly skilled professionals," he explained. "But," he added, "the bishops do want to make it clear that matters of war and peace are not *merely* technical matters; they have crucial moral dimensions which must be given priority."[55]

Addressing Prophetic Responsibility

The "Pastoral Letter on Peace and War" had a profound effect on chaplains of all faiths, in all of the service branches—an effect compounded by the debate already going on in the civilian religious community.[56] In an article published in both *Parameters* and the *Military Chaplains' Review,* Chaplain Donald L. Davidson, a faculty member at the Army War College, where he taught ethics, European and Soviet studies, and military history, warned chaplains not to ignore "the nuclear dilemma," pointing out that "since 1980, churches and synagogues representing more than 100 million Americans have issued official statements that criticize nuclear weapons and U.S. deterrence policy." Among the groups Davidson cited as taking positions close to that of the pastoral letter were the National Council of Churches, the American Lutheran Church, the United Methodist Council of Bishops, the American Baptist Church, the Christian Church (Disciples of Christ), and the United Presbyterian Church.

Davidson presented several reasons why military leaders and chaplains needed to be knowledgeable about the debate taking place in the civilian religious community. Religious bodies exerted considerable influence in "the formation of public consciousness." Indeed, Davidson noted, one of the lessons learned from Vietnam was that national security policy required public support, and if it conflicted with "the public will," it was "in grave danger of failure." Davidson also pointed out the influence religious bodies had in "shaping individual conscience." The current debate among the churches had already caused many of their members to rethink their own positions on nuclear issues. And finally, speaking as a member of the armed forces and writing in a military journal, Davidson declared, "We should care what the churches say because as individuals we should be concerned to find moral truth for ourselves. We should also care because of the moral influence of religious teachings on personnel who implement national policy."[57]

There was some discussion, in all three military chaplaincies, as to whether each of the chaplain organizations should take up the prophetic role and, as a corporate entity, define its position in the nuclear debate. None of the chiefs of chaplains supported this idea. Chief of Chaplains Kermit Johnson had articulated his objections to such an approach as early as 1980. He had contended that "the Chief of Chaplains is not authorized to speak in terms of theological doctrine for the whole Chaplaincy and the churches it represents. Chaplains must be prophetic within the tenets of their own churches."[58]

There were other reasons why Johnson and the other chiefs of chaplains thought it unwise for the chaplaincies to take a position in the nuclear debate. Such action would no doubt increase the criticism already being directed at individual chaplains. It would likely be regarded as an unjustified, possibly unconstitutional, intervention in church and faith group policy making—which would harm chaplains' relationship with the civilian religious community. And even if the chaplaincies had wanted to express their corporate positions, they would have been hard pressed to accommodate the varying opinions of individual chaplains.

Rather than playing the prophetic role, chaplains for the most part addressed the "nuclear dilemma" in their role as teachers of ethics and moral leadership. In Chief of Chaplains Patrick J. Hessian's words, they sought to "educate and inform the consciences" of commanders and other military personnel by providing information on the moral and ethical issues raised by the debate.[59]

Hessian, himself, personally briefed high-ranking army leaders on the NCCB pastoral letter and Cardinal Cooke's commentary. He also sent a synopsis of his remarks to all senior army officers.[60] Chaplain Davidson briefed the Department of the Army Staff Council on the contents of the bishops' letter.[61] In July 1983, at Hessian's invitation, Cooke addressed the Army Command Chaplains Conference on the "Moral Responsibility of Command Leadership."[62]

Hessian's predecessor, Kermit Johnson, had helped prepare chaplains to deal with the bishops' letter and the lengthening debate over nuclear policy. In May 1982, he sent a letter to all U.S. Army chaplains suggesting that antinuclear sentiment and longings for peace in the United States and Europe were "a warning and a signal" from God, "possibly a life-and-death 'last chance' for human civilization." He said he welcomed "this widening awareness" and seemed to be urging other chaplains to welcome and encourage it, too. It was Johnson who had requested Chaplain Davidson to prepare a research paper on the just war tradition and the positions taken by religious ethicists, denominational leaders, and religious bodies concerning nuclear weapons and policies. This project resulted in the publication of Davidson's book, *Nuclear Weapons and the American Churches,* mentioned earlier in this chapter, which was used in ethics courses at the Army War College and the Army Command and General Staff College.[63]

The spring 1982 issue of the *Military Chaplains' Review* featured a statement by Johnson announcing a "new stage" in chaplains' contribution to ethics in the army. It would involve "surfacing issues beyond 'in-house' ethical issues," that is, issues having to do with the ethical values and climate of the army. Johnson proposed that chaplains focus on "the views of religious ethicists, leaders, and groups as they address the ethical implications of nuclear warfare." He told chaplains, "Provided we do our homework well, this research can be of utmost importance not only to us, but to the leaders of our Army and nation. It should be an important contribution to understanding and dialogue." As a result of Johnson's initiative, the fall 1982 issue of the *Military Chaplains' Review* was devoted entirely to the ethics of war and peace; all of the articles, by civilians as well as chaplains, focused on the just war tradition, the basis for much of the antinuclear protest in the civilian religious community.[64]

While he was urging chaplains to address nuclear issues, Johnson himself was nearing the end of a lengthy period of "wrestling" with those same issues. Taking into account "the exigencies of the nuclear age," he had arrived at "a new realism" that questioned "the traditional political 'realism' that assumes war is a regrettable but normal way of settling affairs between nation-states." In a 1983 interview with *Sojourners,* he said he had concluded that, according to just war criteria, the use of nuclear weapons was immoral. He also believed that it was "immoral to threaten to use . . . nuclear weapons," which, as he pointed out, was the basis of the policy of nuclear deterrence. He had not gone so far as to embrace pacifism. Neither did he endorse unilateral disarmament, because he believed "the defense of a nation" to be "a moral cause." Professing what he called "the Niebuhrian, and . . . biblical, understanding of being inevitably a part of this sinful world," he supported what he regarded as the least immoral option: bilateral phased reductions of armaments.[65]

Johnson's thinking about nuclear and other defense policy issues had been tested in 1982, when Army Chief of Staff Gen. Edward C. Meyer asked his staff for a paper on the moral aspects of two issues that were then being debated in civilian society: the U.S. government's nuclear policy and its support of the government of El Salvador. The task of writing the paper fell to Johnson, presumably because of his responsibility as a chaplain to advise his commander on matters of morality. In his memorandum to Meyer, "Moral Issues of Nuclear War and of Conflict in El Salvador," Johnson used just war criteria to raise questions about the morality of limited nuclear war, the use of strategic nuclear weapons, and nuclear deterrence, as well as U.S. support (including military assistance in the form of equipment, advisors, and training) for what he described as "a regime whose military forces systematically and spontaneously engage in violence against its own people[.]" The memorandum clearly revealed his disagreement with the Reagan administration's policies regarding nuclear weapons and El Salvador.

The chief of staff circulated Johnson's memorandum to other military leaders. Johnson later recalled that when it reached the office of the deputy chief of staff for operations, Lt. Gen. William Richardson, "it angered his 'Iron Majors.'" On the question of El Salvador, he and the director of the army staff, Lt. Gen. James Lee, "went head to head." As John Brinsfield has observed, Johnson "was increasingly being perceived as 'out of step' with the Army and with many of the senior leaders in the Chaplaincy who totally supported the policies of President Reagan as the surest and strongest deterrence to 'the evil empire' of Communism." In May 1982, faced with the choice of "working 'inside the box'" (remaining within the army chaplaincy and subject to its constraints) or going "'outside the box'" to engage in the more broadly focused prophetic ministry he believed God required of him, Johnson announced that he would retire early, one year before the conclusion of his four-year term as chief of chaplains.[66]

Johnson recognized that his opposition to Reagan administration policies regarding nuclear weapons and El Salvador had placed him in an untenable position. He had lost credibility and authority with the chief of staff. Moreover, as an active-duty officer in the armed forces, he was not free to criticize publicly the policies of the U.S. government or of the president and commander in chief. At the same time, as Brinsfield points out, Johnson had become increasingly uncomfortable "remaining solely inside the pastoral role 'box,' when his calling to speak prophetically outside the 'box' was so strong." In a personal note to General Meyer he had declared, "Even though much of my work is inside the [pastoral] box and I consider it to be important . . . increasingly I am seeing the impossibility of taking such a compartmentalized view. I cannot content myself with only looking inside the 'box.'"[67]

Assessing Chaplains' Prophetic Ministry

When, in the latter part of the twentieth century, army chaplains established themselves as "the conscience of the Army," they might have used their new status as a springboard for engaging in a broadly focused prophetic ministry. But only a few chaplains took up the challenge. The fact that the military institution imposed certain "intrinsic" limitations on such a ministry—something chaplains had recognized since the Vietnam War—was one reason.

Another reason was suggested by Brinsfield and Johnson. They focused less on the "instrinsic," institutional limitations and more on the limitations chaplains imposed on themselves, whether consciously or unconsciously. In a 1995 article assessing chaplains' moral leadership role, Brinsfield quoted some observations Johnson had made in discussing how individuals come to recognize something as an ethical issue. "Ethics in the military" was not "something 'out there,' something 'objective' which everybody recognizes," Johnson pointed out. What one person may see as an important ethical issue, another may see as a non-issue. "Each of us has a prescription in our lenses which determines how we see the ethical landscape. Depending upon where we come from—our family upbringing, our training, our denominational identity, our theological viewpoints, our personal experiences in life with events and people—a host of things determine how we see ethical issues."

Brinsfield also quoted from another article Johnson had written many years before becoming chief of chaplains, in which he declared that he had "reluctantly and tentatively concluded that, as chaplains, we are mainly interested in being pastors and priests to individual persons and small groups. Realistically, we are not prophets to the institution, but at best, and only occasionally, do we engage in prophetic acts." Citing, among other things, the general failure of chaplains in Vietnam to raise "questions or objections about the 'body count' policy," Johnson wrote, "It appears we have observed the 'off limits' signs to controversial areas of ethics and have chosen instead to tinker with issues like whether using a government ballpoint pen during off-duty hours is unethical or not."

Elaborating on Johnson's views, Brinsfield observed,

> We are not cowards when it comes to confronting command over issues for which we have proponency or responsibility. When confronted with mega-issues, however, those involving the entire defense establishment or the nation at large—the morality of nuclear deterrence, for example—some chaplains have been overcome by what may be called the "infinite

> question syndrome.” How do I personally feel about this issue? How does my denomination address it? What does the law say? If I oppose command as a chaplain prophet, what effect will that have on my ministry as a chaplain pastor, if indeed there is any difference? And, at a higher level, how can I draft a policy which will reflect the moral views of more than 100 denominations?

Such questions, Brinsfield observed, “symbolized the struggle many chaplains have encountered in addressing the complicated issues of a pluralistic, multicultural and largely secular world unlike that which existed just 50 years ago.”[68]

One other reason for the dearth of chaplain-prophets comes to mind. As we saw earlier in this chapter, when chaplains developed their new role as teachers of ethics in the military, their objective was not to teach particular ethical judgments but to help their students develop the analytical and moral reasoning skills necessary to good decision making. This instructional approach may have inclined chaplains to avoid making the kind of ethical and moral judgments necessary to engaging in prophetic ministry.

The distinction Brinsfield and Johnson made between “mega-issues” and those for which chaplains have “proponency or responsibility” is important. Army chaplains did not totally avoid prophetic responsibility, as their institutional ministry demonstrated. Beginning in the late 1970s, as we shall see in the next chapter, they were assigned, and accepted, much greater responsibility for evaluating and advising commanders on the “moral and ethical climate” of the command than they had ever had before. Significantly, and in keeping with Johnson’s observation, they combined that prophetic responsibility with an enlarged pastoral role on the battlefield.

7

Ministering on the Battlefield

In the latter part of the twentieth and the first decade of the twenty-first century, chaplains' role as "the conscience of the Army" expanded beyond the teaching of ethics and moral leadership in the service schools to advising and assisting commanders in meeting the challenges and requirements of the combat environment. In the process, chaplains continued or expanded traditional roles and developed new ones. By the end of the first decade of the twenty-first century, chaplains' battlefield ministry reflected both continuity and change.

The New Combat Environment

In the late 1970s and early 1980s, Army leaders began preparing for future wars by developing what became known as the AirLand Battle doctrine. Military strategists envisioned both nuclear and conventional war in Europe, as well as high- and low-intensity engagements in other parts of the world. The immediate concern was the possibility of a conventional war between NATO and the Soviet Union.[1] Chaplain field manuals presented graphic descriptions of the AirLand combat environment and its devastating effect on the soldiers fighting in it. The 1984 manual described the battlefield as "characterized by high mobility, lethality, and intense around-the-clock combat lasting several days at a time." It observed that "Soviet doctrine emphasizes striking at every echelon simultaneously." It also noted the "high probability of NBC [nuclear, biological, chemical] employment early in the battle." The "wide dispersion of troops" produced by large task-force areas of operation would result in "isolation—one of the greatest enemies of soldiers in combat." The manual predicted that "casualty rates, chaos, shock, and confusion will be greater than any this country has ever experienced." It warned that "only well-trained soldiers with high reserves of emotional, physical, and spiritual stamina will be able to function effectively in this environment."[2]

The 1989 chaplain field manual recognized the different strategic environment that had developed by the late 1980s. It predicted that "low-intensity conflicts" (including those that are terrorist inspired) would be "the predominant type of conflict in the near future, with the constant potential for escalation to mid-intensity conflict." It also took note of the threat posed by the "extreme religious and political beliefs" that prompted terrorist activity. In addition to warning of adversaries' use of chemical and/or nuclear weapons, it anticipated "artillery bombardment in the division area of operation and missile and air strikes as far back as the theater rear area."[3]

The collapse of the Soviet Union produced an even more dramatic change in the "threat environment." The 1995 and 2003 chaplain manuals pointed out that U.S. forces now faced "many and varied . . . potential threats," ranging from "a local populace with small arms to a conventional force possessing weapons of mass destruction." Religious or ethnic conflict would continue to produce regional instability. Military operations other than war (MOOTW) would generally focus on "stability" and "support" operations, but they might also involve offensive or defensive operations and adversaries' use of chemical or nuclear weapons. Manuals of the twenty-first century continued to emphasize the "lethality and intensity" of the combat environment and the "shock, fatigue, isolation, fear, and death" soldiers would undoubtedly experience.[4] In such circumstances, the religious support chaplains provided became extremely important—to the military operation as well as to soldiers.

Developing a "Battle-Focused Ministry"

When the Vietnam War ended, the army chaplaincy adapted to a peacetime role, focusing on religious support for personnel and family members on army installations. In the early 1980s, in response to changes in international relations and the development of AirLand Battle doctrine, the Office of the Chief of Chaplains turned its attention to formulating Chaplain Corps doctrine for a new, "battle-focused ministry" in the combat environment.[5]

The chaplain field manuals of the 1980s and later decades continued the longstanding regulation defining chaplains as noncombatants and stating that they "will not bear arms" and were not to receive training in their use. They presented the traditional rationale, pointing out that under the Geneva and Hague Conventions, chaplains have "protected status" and, when detained, are not considered prisoners of war. "Violation of this non-combatant principle by a chaplain would endanger the protected status of all chaplains captured by the enemy," the 1984 manual observed.[6]

Two new doctrines were Forward Thrust, approved in 1980, and the Unit Ministry Team (UMT), announced in 1984. In Vietnam, chaplains had been assigned at the brigade level. Forward Thrust provided for the assignment of the chaplain and his assistant "in direct support" of battalions.[7] The new doctrine positioned religious support "forward to the smallest most advanced elements of the battlefield." When "operationally feasible," the chaplain and his assistant could be assigned to "integral elements of the battalion," a company, platoon, team or squad. This enabled them to accompany soldiers on the battlefield, ministering to them "before, during, and after contact with the enemy" and "where ministry needs are greatest." Another advantage of Forward Thrust was that it brought the chaplain in closer contact with and improved communication between him and the battalion commander, which in turn facilitated the exercise of his advisory role to the commander.[8]

The Unit Ministry Team, comprised of the chaplain and his assistant, operated in the midst of battle. Chaplain Wayne E. Kuehne, who played a major role in the development of Forward Thrust and UMT doctrine, described UMT as "the vehicle, a force structural entity, to implement Forward Thrust doctrine."[9] The "primary mission" of the UMT, as defined in the 1984 field manual, was "to provide religious support to soldiers in combat." To do so, the UMT had to know how to move about, minister and survive on the battlefield, and how to adapt religious support to all of its phases and complexities. This entailed, among other things, being proficient in what the 1984 field manual called "common soldier skills, particularly communication, navigation, cover and concealment, escape and evasion, and NBC survival."[10] Thus the new Chaplain Corps doctrine, combined with the fact that less than 50 percent of active-duty chaplains had experience ministering in a combat environment, underscored the need for new battlefield training of chaplains. In 1985–1986, the Army Chaplain School integrated field exercises into all training for chaplains and their assistants. This involved what was called "realistic training," designed to simulate battlefield conditions, at the National Training Center at Fort Irwin and in army-wide battlefield simulation exercises at Fort Leavenworth, Kansas. Somewhat later, chaplains and their assistants trained at the Joint Readiness Training Center at Fort Polk and the Combat Maneuver Training Center at Hohenfels, Germany.[11]

Providing Religious Support to Soldiers

Beginning in the 1980s, chaplain field manuals offered detailed guidance on the chaplain's responsibility to provide soldiers in the combat environment with what was generally called "religious support," "pastoral ministry" or "pastoral care." The 1984 manual described it as "personal support to soldiers as they experience the stress and brutality of the modern battlefield," which involved "counseling, spiritual reassurance and encouragement." It was declared to be the chaplain's "primary mission" in the combat zone.[12]

The 1984 manual offered a detailed discussion of how the chaplain should provide religious support, geared to changing battlefield conditions and priorities on the one hand and soldiers' emotional and psychological state on the other. During the pre-battle phase, when the units were entering staging or assembly areas, it would be possible to provide large and small group worship services. However, once the unit began to move forward to battle position, troops would be dispersed and unable to come to the unit ministry team. At that point, the manual observed, "troop anxiety is extremely high as is the soldier's perceived need for counseling, spiritual reassurance, and encouragement." In that context, "denominational ministry" must give way to "unit ministry." The chaplain should visit, counsel, and pray with those soldiers who can be reached without compromising the unit or delaying the march forward.

Once the battle was underway, with troops "fighting or anticipating combat," they would be very anxious. "Adrenalin-induced surges" would be followed by great stress, and energy and stamina would be rapidly depleted. In the battle phase, the chaplain's priority was to provide religious support to casualties. But moving with the combat trains, the chaplain should also take advantage of halts and delays to minister to those soldiers he could reach, providing "individual and small group pastoral care." In the post-battle phase, according to the manual, stress remained very high, stamina very low. "Stress casualties and trauma casualties" were the chaplain's priority. Providing sacraments and rites to individuals, groups, and casualties was essential. Honoring the dead by means of memorial ceremonies and religious services might be possible. During the "reconstitution period," when the unit had withdrawn to a rear area, the troops needed "special religious support." Except for honoring the dead, the chaplain's "highest priority" was "rebuilding the emotional, psychological, and spiritual strength of the unit's surviving members and replacements." Worship services and "intense individual and group counseling" would be the means. The manual pointed out that "normally, the assigned ministry team stays with the unit through reconstitution because it has established an irreplaceable relationship (forged under fire) with the surviving unit's members."[13]

Manuals issued in the late 1980s and the early twenty-first century put special emphasis on the chaplain's responsibilities regarding "battle fatigue casualties." Battle fatigue was defined as "a broad group of physical, mental, and emotional reactions" caused by "the trauma, stress and chaos of battle," which depleted "essential physical, emotional and spiritual resources in the soldier." The 2005 UMT handbook asserted "the unique ability" of the chaplain and his assistant to assist the commander in its "prevention, treatment, and assessment." It pointed out the special training chaplains and their assistants received, as well as their ability to work with mental health providers and medical personnel in those areas. Equally important was the UMT's close relationship with the soldiers, which enabled the chaplain and chaplain assistant to identify potential stresses and intervene before battle fatigue set in.[14]

Advising and Assisting Commanders on the Battlefield

Just as their new battle-focused ministry thrust chaplains into the midst of combat and expanded the religious support they provided to soldiers in the field, it also enlarged their ethical, moral, and religious responsibilities vis-à-vis commanders and soldiers. In carrying out those responsibilities, chaplains drew on the expertise and authority they had acquired as teachers of ethics and moral leadership in the service schools.

Chiefs of chaplains played a significant role in gaining a more influential role for chaplains in regard to command policies and decisions. Chief of Chaplains Gerhardt Hyatt had taken the initiative in the early 1970s. His successors, Orris E. Kelly and Kermit D. Johnson, continued the effort. "Chaplains must be in the forefront of those who influence the ethical dimensions of military life and mission," Kelly declared. One of Johnson's "Chaplaincy Goals and Objectives" for 1983–84 was to "provide moral/ethical impact on decision-making and the system." Widely regarded as the "Army's Ethicist," he believed that the commander and the chaplain should act as an ethical team.[15]

The chiefs of chaplains had significant command support for their objective. In a 1977 article in the *Military Chaplains' Review,* Army Chief of Staff Gen.

Bernard W. Rogers noted that the officer corps had recently been advised that they were expected to be "the conscience of the Army." That was "doubly true" for the Chaplain Corps. "Through your words and your actions you must set the example," he told chaplains. "You must not temporize with wrongdoing, no matter how fashionable such attitudes may be. Soldiers expect chaplains to understand and sympathize with us as we are, but they also expect chaplains to lead us to where we should be."[16]

Beginning in the late 1970s, chaplain field manuals and army regulations assigned the chaplain a clearly defined moral advisory role that went beyond advising and consulting with the commander regarding religion, morals, and morale, to evaluating—and criticizing, when necessary—his decision making, policies, and leadership. The 1977 chaplain manual stated that in serving as an "advisor and consultant," the chaplain was to provide the commander with "periodic evaluation . . . on the moral and spiritual health of the command." The manual specified that the evaluation "will include ethical and humanitarian dimensions of command policies, leadership practices and management systems."[17] The 1984 and 1989 manuals assigned the chaplain a role as moral advocate for the soldiers. He was to inform the commander "personally of the impact of procedures or policies perceived by soldiers as unfair." This was part of his larger responsibility to provide the commander with "direct, personally verified information" concerning "the morale and moral climate of the command."[18]

The 1989, 1995, and 2003 manuals described the chaplain as the "conscience of the command," and the 1995 manual elaborated on the phrase, noting that "chaplains advise the commander on the moral and ethical nature of command policies, programs, and actions."[19] The late-twentieth- and early-twenty-first-century manuals listed a variety of specific moral advisory tasks. Assisting the commander with "ethical decision-making" appeared in several manuals.[20] The particular subjects on which the chaplain was to offer advice included the moral and ethical "climate" of the command and the commander's "policies, programs, initiatives, plans, and exercises." The latter included the rules of engagement as well as other operational decisions the commander and staff discussed during "mission analysis" and "course of action (COA) development."[21]

The chaplain's moral purview included violations of the laws of war. The 1984 manual directed him to assist the commander in preventing "immoral, even inhumane practices" provoked by the stress of combat, such as "dehumanizing treatment of friendly troops, enemy prisoners of war (EPW), or civilians; violations of codes of morality; illegal acts, desecration of sacred places, and disrespect for human life." The chaplain had the responsibility to inform the commander of "possible violations of the laws of war"; the manual even presented detailed instructions on proper reporting procedure. The 1989 field manual named the chaplain, along with other members of the commander's staff (the staff judge advocate and the inspector general) as one who had "a special duty" to receive soldiers' reports regarding acts they believed to be illegal under the laws of war and to ensure that those reports were properly transmitted to the commander (if they were not privileged) "as soon as possible."[22]

The Moral Leadership Training program, implemented in the mid- to late 1980s, was another means by which the chaplain assisted the commander in maintaining a positive "moral and ethical climate" in the command. When requested by the commander, the chaplain was to provide instruction and other activities to "promote soldiers' moral, ethical, and social development." This did not involve religious instruction. Indeed, chaplain instructors were cautioned to "avoid any action which would tend to confuse this training with religious instruction." Specifically, they were not, "under any circumstances," to "use scheduled training periods to deliver a sermon, to sermonize parts of the topic, to upbraid troops for nonparticipation in chapel programs, to show religious films or to expound their own personal theological views."[23]

The Moral Leadership Training program utilized concepts and skills chaplains had acquired in dealing with the Vietnam-era behavioral problems, as well as the experience and authority they had gained as ethics instructors in the army service schools. According to the 1984 chaplain field manual, the purpose of Moral Leadership Training was to "help solve critical problems such as racial tension, alcoholism, or drug abuse." It also involved teaching "such skills as counseling, communications, and team building" and "goal setting, decision making, and values clarification." By 2004, the program had acquired a somewhat different orientation. According to Army Regulation 165-1, it addressed "the full spectrum of moral concerns of the profession of arms and the conduct of war." The objectives of the program included the enhancement of "soldierly virtues and values within the members of the command" and the development of "common moral and ethical standards" to promote "section and unit cohesion." An appendix to the 2005 UMT *Religious Support Handbook* advised chaplains to address the subject of "ethics and the role of Army values" and "ethical problems and the ethical reasoning process."[24]

Ministering in Iraq and Afghanistan

Newspaper and magazine articles by and about army chaplains who served in Iraq and Afghanistan show how they were influenced by the soldier-focused doctrine that had evolved during the latter part of the twentieth century. Theirs was very much a "ministry of presence," similar to the one combat chaplains had developed decades earlier in Vietnam. Like that earlier generation, the chaplains who ministered in Iraq and Afghanistan wanted to be with the troops in the combat zone. Chaplain Ron Eastes, who served an Eighty-second Airborne unit, explained that "if I'm going to care for these guys, I need to be where they are." That meant, among other things, joining his soldiers in "a ritual of cigars and banter" or a walk through a Baghdad bazaar, as well as playing cards with troops visiting from a smaller outpost. For Eastes and other combat chaplains, it also meant accompanying the soldiers on patrol or riding in convoy.[25] Chaplain Dale Goetz of the Fourth Infantry Division was among five army personnel killed by an improvised bomb in Afghanistan in September 2010. He had hitched a ride on a resupply convoy. As a spokesman for the Office of the Chief of Chaplains observed, "Traveling in a war zone is very risky business," but "chaplains will continue to go where soldiers are on the battlefield to minister to their soldiers."[26]

Chaplains in Iraq and Afghanistan expressed the same sense of identification and "connectedness" with their troops that the Vietnam chaplains experienced. Chaplain Joseph Angotti said of his ministry to troops of the First Infantry Division, in Iraq, "The presence factor has made it very rewarding for me. I feel very useful and I feel needed." He felt "edified by the sacrifices that our soldiers make, that they are willing to make, that they are continuing to be willing to make, even when things aren't going well."[27]

Being "present" to their soldiers, physically as well as emotionally, enabled chaplains to fulfill two important responsibilities to soldiers: providing moral guidance and pastoral care in the combat environment. National Guard chaplain John Morris told an interviewer how he talked with soldiers about "killing another human being." "Not with great ease," he said. He told them,

> We are engaged in something that we believe, by our president and foreign policy, to be a needed task to combat a greater evil. We have people on the battlefield who want to take our lives. We have a great responsibility to discern who the enemy is and who isn't the enemy, and to use the proportionate amount of violence to address and kill our enemy, and to spare those who are innocent.

Morris also addressed the feelings soldiers might or would have after killing. "We cannot deny that when we take human life, it impacts us," he declared. "When we go in after killing people and search them and find their photographs of their families, we have killed human beings. We haven't killed things or people that we've managed to convince ourselves are robots. We've killed humans."[28]

Two reports of chaplains' intervention to prevent a possible war crime show how chaplains were in a position to offer immediate, on-the-spot ethical and moral advice. One involved an incident that occurred in July 2003 on the outskirts of Abu Ghraib in Iraq. According to Chaplain Glenn Palmer's description, a soldier had been killed when his convoy was hit by a car bomb, and his fellow soldiers were gathered around the body, "staring in shock and disbelief." He was the first casualty they had witnessed. Palmer said that he and another soldier "scooped up the [dead soldier's] remains" to prevent their being desecrated by "the locals." Then Palmer led the soldiers in prayer, committing the dead soldier to God's care "and asking God, in the midst of things we couldn't even begin to understand," to "comfort and care for" those who "grieved his loss." Meanwhile, a "crowd of locals" had gathered, and, seeing the dead soldier's "covered remains," began "cheering and singing and dancing." "They were openly celebrating the death of another human being, one of God's children," Palmer reported. "It was evil—in-your-face, raw and exposed." Then, he continued,

> As the intensity of the cheering increased, one of the soldiers beside me clicked the safety off his rifle, raised his weapon and pointed it at the crowd. I knew viscerally the rage and anger this soldier felt. I too had lost friends to horrible deaths in Iraq. Part of me wanted to say, "Go ahead. Do it." However, I am a pastor and an American soldier, and I know the "right thing to do." I put my hand on his M-16, gently lowering it, and said, "We can't become the thing we hate." It was one of the many times that God's presence guided me in being in the right place to help protect angry, tired, hot and edgy soldiers from their own worst selves.[29]

The other incident involved Chaplain Ron Eastes of the Eighty-second Airborne. As reported by Lee Lawrence, Eastes's soldiers had returned from a "routine patrol," which involved apprehending an "Iraqi sniper suspected of wounding a high-ranking U.S. officer." The "detainee" had been confined to a makeshift cell, and a soldier sat on a stool beside the door while two other soldiers stood nearby, watching. Then, Lawrence wrote,

> Well aware that a soldier's anger can flare at the sight of a man thought to have shot one of their own, Eastes strolls over with calm concern and pauses by the guard. From a distance it's impossible to hear what the chaplain and soldier are saying—but the words exchanged aren't what is important. What matters is that Eastes is getting the soldier to talk; if there is pent-up anger, he can spot and, he hopes, defuse it.

Lawrence described the incident as "a classic example of what chaplains call their 'ministry of presence'" and compared it to the effect of a guardrail on a mountain road: while no one could know how many accidents it might prevent, it was believed to make a difference. "In war," Lawrence explained, "when a sense of right and wrong can disappear in the fog of adrenaline and anger, the chaplain can act as a 'guardrail,' and officers who rely on them as such talk about the value of troops having a safe place to let off steam and regain equilibrium."[30]

The wars in Iraq and Afghanistan differed significantly from the Vietnam War, at least in their early stages, because they provoked much less debate in the civilian religious community. Chaplains in Iraq and Afghanistan noted that comparatively few soldiers raised questions about the morality of the U.S. interventions or the conduct of the wars—a difference they attributed to the soldiers being volunteers in a volunteer army rather than draftees. Instead, most soldiers sought chaplains' help in dealing with the emotional and psychological toll of combat and multiple deployments. They looked to the chaplains to make sense of the brutality and carnage of warfare, of buddies killed or horribly wounded, as well as to encourage them to keep on fighting. One chaplain wrote to a colleague, "I am very keenly aware of the responsibility that I and all chaplains in this theatre have to be symbols of sanity and normalcy in a world that often does not make sense."[31]

Chaplains described in some detail the occasions for and the kind of pastoral care they offered soldiers in Iraq and Afghanistan. Chaplain Angotti said that he told his soldiers "that it is important not to despair. It is important to keep hope. . . . that we have to pray, and that ultimately we are not in control of our destiny." When they asked for Scripture citations, he said he found himself "recommending the Book of Job quite a lot."[32] Chaplain David Sivret, who was wounded in a suicide bomber's attack on a mess hall, remained in Iraq for a few months, unwilling to leave "my soldiers." Like Angotti, he worked hard to encourage the soldiers, telling them that they were doing good work. "I was trying to give them perspective and hope," he said. "You have to build them up because they have to go back out there again."[33]

Chaplain Jeffrey A. Jencks, an Army National Guard chaplain, remembered the first time his unit was in a firefight. He was celebrating the Eucharist, "and the sergeant major came up to me and whispered in my ear, 'Chaplain, these people just killed.'" Jencks observed, "These were National Guard soldiers, and they were stunned. I had to sit down with them and tell them we are at war, tell them why we're at war, tell them they're protecting innocent people. But it's not easy. Still, God is there."[34]

Chaplains who led critical event debriefings also engaged in pastoral care and spiritual nurture. Glenn Palmer conducted more than one hundred of them, and he believed the training chaplains received in "mental health tactics, techniques and procedures" helped them counsel and support soldiers in combat. It helped, too, that the chaplain had "experienced the same horror of war as the soldier." He could offer "credibility and empathy and a safe, sacred, confidential space and environment in which the soldier is free to share his story." Palmer said, "We know that war can rob a person of meaning. We build on that knowledge by representing a God who is with us and for the soldiers, one who lets them know how much he cares, and one who carries their burden and pain for them, especially when that 'rucksack' is heavy."[35]

Chaplains and Indigenous Religions

Engaging in a ministry of presence and providing moral guidance and pastoral care in the combat environment were assignments army chaplains had carried out during much of the latter part of the twentieth century. They were basically soldier focused and keyed to chaplains' training and expertise as religious leaders and teachers of ethics. Another responsibility chaplains accepted in Iraq and Afghanistan, which also had a long history, involved advising the command and staff on the indigenous religions (IR) in the area of operations. The reasons for giving this advice and the purposes for which it was to be used changed over time. In Vietnam, for example, the chaplain's IR role was more soldier than battle focused, and the emphasis was on religious and cultural understanding rather than military operations. The 1964 and 1967 chaplain field manuals referred to the role as a "staff function." Among other things, the chaplain was to advise the civil affairs officer "on the theological considerations of the traditions and practices of local religious cultures and the degree of sanctity accorded their shrines, temples, and religious symbols." He was also charged with "encouraging military personnel in respectful attitudes toward other faiths, sacred rites, and places of worship."[36] In some cases chaplains actually instructed soldiers on the religion and customs of Vietnam. The advice chaplains provided the civil affairs officer and the guidance they offered the soldiers were part of the American effort to win the "hearts and minds" of the Vietnamese people.[37]

Teaching and Understanding World Religions

The 1967 chaplain field manual included a statement that had not appeared in the 1964 manual. It referred to the kind of advice the chaplain was supposed to give the civil affairs officer. The manual observed that "adequate execution of this function requires that the chaplain, through individual study or otherwise, be well versed in such matters."[38] This realization was one of many factors that resulted in chaplains receiving instruction in world religions. As early as 1966, Chaplain Albert Ledebuhr sent a report from the U.S. Army Combat Developments Command at Fort Belvoir, Virginia, to Chief of Chaplains Charles E. Brown recommending such training not just for chaplains but also for the army as a whole. Two years later, the basic course for chaplains at the U.S. Army Chaplain School included a course titled "World Religions" taught by Chaplain Delbert Gremmels. In the 1970s, chaplains then began teaching world religions in the army service schools.[39]

In the years immediately following the Vietnam War, the Army Chaplain School's interest in the teaching of world religions waned. But in 1987, recognition of increasing religious pluralism in the United States, as well as in the armed forces, spurred it to reintroduce such a course, which Chaplain John Brinsfield taught as part of the chaplain officer advanced course. By 2009, two graduate-level subjects required for appointment as an army chaplain were "religious ethics" and "world religions."[40]

Instruction in world religions also expanded in the service schools. Beginning in 1997, the U.S. Army War College offered "World Religions" as an elective for resident students.[41] During the first decade of the twenty-first century, instruction in world religions and related topics was being provided at a number of service schools, including the National War College, the Institute for Conflict Analysis and Resolution, the Army Command and General Staff College, Fort Leavenworth, and the John F. Kennedy Special Warfare Center and School, Fort Bragg.[42] Following 9/11, army units deploying to Iraq and Afghanistan also began receiving cultural and religious instruction that focused on the tactical, operational and strategic aspects of religion in those countries.[43]

In the late twentieth and early twenty-first century, the main impetus behind the army's increasing emphasis on religious and cultural instruction, for all ranks, was the recognition on the part of government and military leaders, as well as chaplains, of the pivotal role religion played in creating or exacerbating the ethnic rivalries, regional conflicts, and wars in which American military forces were becoming involved. The publication *Religious Support in Joint Operations,* issued in 2005, noted that "history and current US involvement around the world reveal religion's significant influence on nations, conflict initiation, resolution and reconciliation, and on post-conflict reconstruction efforts," a fact that needed to be taken into account in "all types, levels, and phases of planning and decision making."[44]

Developing a Battle-Focused IR Role

In the latter part of the twentieth century, chaplains continued to assist in training soldiers to understand and respect the beliefs, practices, and facilities of indigenous religions and to caution them against the desecration of consecrated places.[45] But beginning in the late 1970s, the chaplain manuals also assigned them two other responsibilities: advising the commander and staff regarding the "impact" of indigenous religions on the unit's "mission" and command "operations" and establishing "liaison" with local religious leaders. These two responsibilities significantly changed the orientation and purpose of the IR role. It became more battle than soldier focused.

As the advisor on the "impact" of indigenous religions, the chaplain was to present the commander and staff with a comprehensive, formal analysis or assessment of the faith and religious practices of indigenous religious groups in the area of operations.[46] Its purpose was not to promote cultural and religious "understanding." One of the main factors to be considered was "how religion plays a part in the enemy's capacity or inclination to fight." This assignment involved the chaplain in mission analysis and course-of-action development, offering the commander and staff not just ethical and moral counsel but also information on the "religious dynamics of the indigenous population in the operational area." According to the 2004 Joint Chiefs of Staff *Religious Support* publication, the command staff needed this information in "developing schemes of maneuver and rules of engagement or [in] planning civil-military operations, psychological operations, information operations, and public affairs activities." The handbook also noted that by recognizing the impact of indigenous religions, commanders "may avoid unintentionally alienating friendly military forces or civilian populations that could hamper military operations."[47]

Chaplains and Religious Leader Engagement

The third component of the chaplain's IR role, establishing liaison with local religious leaders, also became known as "religious leader engagement" and, later, officially, "religious leader liaison" (RLL). It received brief mention in the chaplain manuals during the 1960s, 1970s, and 1980s.[48] In the 1990s, it became a significant and increasingly well publicized activity in the Balkans, and then, even more so, in Iraq and Afghanistan in the early twenty-first century. Unlike the other two components of the IR role, this one proved to be somewhat problematic for chaplains and the army and thus occasioned more discussion and debate.

As Chaplain Kenneth E. Lawson has shown in his history of the army chaplaincy in the Balkans, 1995–2005, religious leader engagement constituted a significant element of the American intervention in that region. The context in which chaplains worked seemed to invite commanders and chaplains to experiment with a role that, up to that time, had not been well developed. What began as a humanitarian peacekeeping mission ultimately involved soldiers in "peace enforcement" in a hostile war zone. In addition, the Balkans were known for having a long history of religious, ethnic, and cultural tension.[49] The region offered a perfect testing ground for practicing, refining, and expanding religious leader engagement.

Lawson describes numerous instances of chaplains in Bosnia and Kosovo "networking" with civilian clergy of various faiths. Such activity had several purposes: to show respect and concern for the indigenous religious population, their leaders, and their religious traditions; to foster a good relationship with the U.S. military; and to promote religious reconciliation among the Balkan people. "Who better than a chaplain to talk about healing and reconciliation in a peacekeeping environment?" declared a chaplain who organized meetings between Muslim, Orthodox, and Catholic clergy in Bosnia. An Eastern Orthodox chaplain frequently met with indigenous Orthodox clergy, attended their church services, and helped his unit's Information Officer with "religious and cultural issues." Another chaplain who reported that he "met with and formed friendships with key Muslim, Orthodox, Catholic and Protestant religious leaders" in his sector said that he did not "participate in IO [Information Operations] targeting" but observed that his "relationship" with those religious leaders gave him "a better understanding of how to advise my commander on the influence of religion on our peace-enforcement mission and as to the religious perspective on the current situation." Lawson also quotes a chaplain who persuaded his commander to alter plans made for raiding a suspected al-Qaeda terrorist hideout, which turned out to house a family made up of an eighty-year-old woman, her thirty-five-year-old daughter, and her eleven-year-old grandchild. "I believe my role as chaplain gave me a different way of looking at these operations than the other officers," he wrote. "This assisted the Commander in making a prudent decision that showed compassion and restraint."[50]

As Lawson points out, chaplains' experience with religious leader engagement in the Balkans prepared them for similar involvement in Iraq and Afghanistan during the first decade of the twenty-first century.[51] According to one study of such activity, commanders in those two countries would sometimes direct their chaplains to engage with indigenous religious leaders. But often it was the chaplains who took the initiative, saying, "'I think I can help in this area.' And then commanders usually supported their involvement." The chaplain who wrote the study noted that "it was not difficult to convince commanders, but chaplains had to take the initiative and make the proposal."[52]

The first step in religious leader engagement, in Iraq and Afghanistan, as in the Balkans, was building a relationship with indigenous leaders. Some chaplains had monthly visits with imams, during which the two parties shared information about their respective cultures, customs, and religions. Other chaplains engaged in interfaith dialogue and worship with local clergy. In one case, a chaplain and a mullah met to study sacred texts together. In other meetings, chaplains explained the commander's mission to local religious leaders—partly to counteract misunderstanding and hostility, but also to learn about the problems the clergy and the civilian population faced. Commanders not only authorized and supported but also assisted their chaplains in the liaison activity, inviting the local clergy to conferences and prayer breakfasts or directing the chaplain to establish an interreligious council made up of indigenous religious leaders.[53]

As one chaplain observed, "building a relationship" was one aspect of religious leader engagement, but providing "something tangible," in the form of humanitarian or other assistance to the local population, was equally important for maintaining "credibility."[54] Some chaplains obtained funds to be used for improvements to or the reconstruction of mosques, or to build or renovate schools. Other chaplains cooperated with local clerics in addressing various "community concerns." In one example of such activity, a chaplain with the 101st Airborne Civilian Military Operations Center in Iraq acted as a liaison between a unit in Mosul and imams who complained of being treated disrespectfully by U.S. soldiers, and he made arrangements for the religious leaders to visit detainees being held by the 101st Division.[55] In another example, a Muslim chaplain, Capt. Mohammed M. Khan, also with the 101st Airborne Division, worked on several occasions to "calm tensions among the Muslim populace" and to "convince local leaders and imams not to launch attacks on coalition troops."[56]

Another aspect of religious leader engagement in Iraq and Afghanistan involved efforts to "advance reconciliation and healing across sectarian fault lines." According to Chaplain LaMar Griffin, who served as command chaplain for the Multi-National Forces–Iraq, "the focus of this effort was to bring religious groups together for discussions on reconciliation between Sunni, Shia, Kurdish, and non-Muslim leaders." In undertaking this mission, chaplains not only engaged with local religious leaders but also participated in "direct liaison with leaders of the Sunni Endowment, Shia Endowment, Non-Muslim Endowment, GOI officials who were influential in religious issues, Department of State representatives, and NGOs who effectively served as mediators between religious groups."[57]

Addressing Religious Leader Engagement

Although chaplains had been significantly involved in religious leader engagement since the 1990s, the chaplain field manuals and army regulations issued during the first decade of the twenty-first century continued to give the activity only brief mention, paying much more attention to the other two components of the chaplain's IR role. Thus the 2003 *Religious Support* manual devoted many pages to the chaplain's responsibility to advise the commander on the impact of the faith and practices of indigenous religious groups on the military mission and operations, but the only reference to religious leader engagement appeared in an appendix that listed the various ways chaplains advised the commander on "the interface between Religious Support Operations and Civil Military Operations (CMO)." The last item in the listing, the reference read as follows: "Relations with indigenous religious leaders when directed by the commander."[58] The reference in the May 2005 *Religious Support Handbook for the Unit Ministry Team* was equally brief: "providing liaison to indigenous religious leaders (with the G5[S5])."[59]

The paucity of official guidance regarding religious leader engagement contrasts sharply with the extended, public discussion it received among Army chaplains during the first decade of the twenty-first century. Most of the commentary appeared in articles written for various military and chaplain journals or in research papers completed at military service schools. In addition, some chaplains presented papers on the subject at the World Religions Summit held in 2007 at the U.S. Army Chaplain Center and School and in a special 2009 issue of *Army Chaplaincy* titled "World Religions: The Impact of Religion on Military Operations." Some of these chaplains had firsthand experience with religious leader engagement, while others did not. But the great majority were quite supportive of the activity. Indeed, some wrote and spoke as strong advocates, seeking to persuade commanders and their fellow chaplains of its utility. Other chaplains evinced a more skeptical attitude. Although they endorsed the activity, they called attention to several problems it posed.

The advocates of religious leader engagement did not ignore its humanitarian impact on the indigenous population, but they tended to focus more attention on pointing out its military relevance. Chaplain LaMar Griffin observed that religious leader engagement supported Department of Defense policy regarding "stability operations." Quoting from a DoD directive that defined the "long-term goal" of helping "develop indigenous capacity for securing essential services, a viable market economy, rule of law, democratic institutions, and a robust civil society," Griffin observed that "since indigenous religion and religious leaders are essential to a robust civil society and critical to our success, engaging religious leaders supports the DOD policy."[60]

Just as they argued that religious leader engagement was critical to mission accomplishment, the advocates also emphasized its importance in "force protection."[61] Chaplain Scottie Lloyd told of a brigade chaplain in Iraq who helped local religious leaders in the renovation of a mosque located on an American base, visited and talked with a group of Iraqi detainees whose arrest had provoked great concern among the indigenous population, and ultimately persuaded national-level Shia Muslim leaders to meet with a company commander and develop a working relationship. The results of that relationship included "opening lines of communication with Coalition Forces, prohibiting weapons in the mosque, [and] tempering their violent anti-coalition language" as well as getting them to "submit to monitoring of their Friday messages." But for the brigade chaplain, the most important result of the engagement process was that it "indirectly saved lives—the lives of our soldiers."[62] Two other chaplains in Iraq and Afghanistan also credited religious leader engagement with saving American lives. One had formed and met with an Inter-Religious Council of local clerics, which resulted in "a measurable decrease in the incidence of roadside bombings." The other, a Muslim chaplain, had worked successfully in several instances to "convince local leaders and imams not to launch attacks on coalition troops."[63] Chaplain William Sean Lee cited a statement by the coalition provisional authority chaplain, Col. Frank E. Wismer III, that "in areas of Iraq where chaplains and commanders engage the local religious leaders, coalition forces have had some success in decreasing anti-coalition actions."[64]

For a good many chaplains, the idea that religious leader engagement was integral to, even a critical factor in mission accomplishment was a point in its favor. Others, less sure, expressed concern about how the military used or planned to use information chaplains gained through such activity.[65] In an article in the *Military Intelligence Professional Bulletin,* Chaplain Kenneth Lawson expressed his own and other chaplains' thinking about the possible use of such information "for targeting or other offensive operations." He quoted one chaplain who observed that it "would be naive to think that the chaplaincy would be exempt from the pressure of a well-meaning but ill-advised commander" who demanded such intelligence gathering. Another chaplain warned his colleagues to beware of the temptation to seek "a deeper role" by adopting an IO role, at a time when commanders were putting increased emphasis on IO. Lawson, himself, declared that "chaplains, as religious advisors, must not allow themselves to drift too far into the realm of IO." He also cited chaplains who said they had felt pressured to provide "intelligence" to a commander or members of his staff and had refused to do so.[66] The main objection raised against chaplains' involvement in intelligence gathering was that it was an inappropriate role for chaplains, since, as Lawson pointed out, "IO is an element of combat power" and chaplains who engaged in it were in violation of their noncombatant status.[67]

The lack of official, specific guidance and training for religious leader engagement troubled a number of chaplains. In his 2005 *Army Chaplaincy* article, Chaplain William Sean Lee declared that allowing chaplains to "function unofficially as religious liaisons" invited significant risk. "The chaplain's role as religious liaison must be defined in official doctrine and delineated to conform to current standards and parameters of operation," he wrote. Chaplains should also be given formal training in the requisite "basic core competencies." Both the "doctrinal mandate" and "adequate training" were needed to "mitigate operational risk and protect the chaplain as a religious liaison." Other chaplains agreed on the need for more specific and comprehensive training for religious leader engagement in such things as language skills, cultural/religious understanding, and conflict analysis and negotiation.[68]

In 2003 and 2008, army chaplains did receive some specific guidance regarding limitations on religious leader liaison. The 2003 field manual declared, "Under Title X of the U.S. Code, Chaplains should not perform the following: Direct participation in negotiations or mediations as sole participant. Human intelligence (HUMINT) collection and/or target acquisition."[69]

Then, in 2008, Chief of Chaplains Douglas L. Carver issued a memorandum, "Religious Leader Liaison Policy Letter," dated September 30. Its declared purpose was "to establish both the use of the term Religious Leader Liaison (RLL) and the Chief of Chaplains' [*sic*] policy regarding UMT participation in the same." In a section titled "Background," Carver explained the reasons for issuing the policy letter. Citing the army's 2008 Field Manual 3-0, "Operations," he observed that "the simultaneous nature of full spectrum operations (offensive, defensive, civil support and stability operations) requires a functional understanding of the chaplain's role in RLL." In the current operational context, chaplains serving as "Religious Staff Advisors" to the commander "must understand how religion and religious tradition influence the center of gravity" and needed to "plan for liaisons with religious or secular leaders if ordered to do so by their command."[70] In meeting with such leaders, their purpose should be "to build relationships of mutual trust and respect, promote human rights, and deepen cultural understanding between unit personnel and host nation citizens."

In the "Policy" section, Carver declared that "Religious Leader Liaison" was now the "doctrinal term" the Chaplain Corps would use for such activity. He also emphasized that providing or performing religious support to "the Army Family" continued to be the "primary mission" of the Unit Ministry Team. By contrast, "RLL is a directed task to the UMT from the commander and does not represent a primary task; it is an operational capability the UMT can provide." The section on policy also listed several "principles" chaplains were to follow in engaging in RLL: conducting the mission "only at the order of the commander and in coordination with the supervisory chaplain"; demonstrating "a thorough understanding of the religious issues within the culture in which the liaison occurs"; fostering "an effective relationship between the command and a religious leader, emphasizing reconciliation (if appropriate), mutual respect, and relevant command messages or themes"; treating each leader "with dignity and respect"; and protecting "the non-combatant status of the chaplain." Prohibited activities included collecting or providing "information as a human intelligence (HUMINT) source," negotiating "as the sole or lead negotiator," and taking "any action that threatens [chaplains'] non-combatant status or violates the Law of Land Warfare."[71]

As for training and further guidance, the Office of the Chief of Chaplains looked to the Center for World Religions at the U.S. Army Chaplain Center and School as the main resource. In a message printed in the 2009 special issue of the *Army Chaplaincy* on "World Religions: The Impact of Religion on Military Operations," Chief of Chaplains Carver defined "a new mission" for the center:

> I envision that the Center for World Religions will be the Army's premier resource for educating and training the force on the impact of religion on Joint, Interagency, Intergovernmental and Multinational (JIIM) operations. It will foster the practice of religious analysis for strategic, operational, and tactical utilization; develop strategies and capabilities for conducting religious leader liaison; and provide a readily available resource to support the field. It will also develop and improve the quality, function and capability of both operational and institutional Chaplain Corps world religions subject matter experts by sponsoring regular world religions conferences; providing teaching and writing opportunities; building reference and resource capabilities; and assisting in the establishment of a world religions additional skill identifier to foster the continued development of this capability for the Army.[72]

Just as Carver's 2008 policy letter made RLL official Chaplain Corps doctrine, the 2009 Army Regulation 165-1 incorporated it into U.S. Army doctrine. In the "Chaplain as principle [*sic*] military religious advisor" section, the regulation listed "Religious leader liaison" as the first of several "religious support activities" the chaplain performed "for the commander."[73]

Some of the declarations Carver made in the policy letter and the message regarding the Center for World Religions addressed concerns chaplains had expressed in their extended public discussion of religious leader engagement—about the use of information chaplains acquired as a result of liaison activity, for example, and the need for specific guidance and training. But just as the chaplains' commentary contributed to the formulation of official RLL doctrine, it also provided an invaluable supplement to such doctrine—in the form of advice, encouragement, and reflection, often based on personal experience, addressed to their fellow chaplains.

The advice focused on the qualities and attitudes crucial to successful religious leader engagement. Chaplain advocates for RLL agreed that "not every chaplain can or should participate in RLL." More than "clergy credentials" were required. Factors such as a chaplain's personality, leadership skills, maturity, adaptability, and military experience, as well as his "theological education and world view," were equally or more important. The guidance chaplains offered regarding a chaplain's religious orientation was explicit. It was imperative that chaplains conducting RLL "reach out" to indigenous religious leaders in an attitude of "respectful engagement," treating them "with honor, dignity and respect." Therefore, those who were opposed to interfaith dialogue would not qualify. The same was true for chaplains who were "theologically and personally inclined to view those of other faiths as enemies."[74]

In making such frank judgments, chaplains were not suggesting that successful engagement required religious or theological neutrality. Indeed, the command chaplain for Multi-National Forces–Iraq, LaMar Griffin, stressed the importance of chaplains "retaining [their] own distinctive faith identity" while "building relationships with Iraqi religious leaders, cleric to cleric." His own experience with religious leader engagement had convinced him that maintaining his faith identity was "instrumental in building mutual trust and respect" because, among other things, it enabled both the American military chaplains and the Iraqi religious leaders to share "a mutual appreciation, each for the other, as representatives of an important, honored distinctive faith."[75]

Chaplain advocates recognized that some chaplains had no desire to perform RLL. They were not necessarily opposed to it. They simply preferred, and may have thought more important, the role of providing pastoral care to soldiers in combat. After all, it did remain, officially, the "primary mission" of the UMT. To encourage other chaplains to consider seeking an RLL assignment, the advocates pointed out its compatibility with military ministry. Just as the 2005 UMT handbook asserted chaplains' "unique ability" to deal with soldiers' battle fatigue, advocates of religious leader engagement contended that chaplains were "uniquely suited and positioned" to conduct RLL. Chaplain William Sean Lee cited their "skills, credentials and availability" for the role, Chaplain Kenneth Lawson their noncombatant status. Chaplain Thomas Vail observed that chaplains were already perceived as having special capabilities as "facilitators" of understanding, as "reconcilers" and "healers" of division, and as believers in "the notion of unity through the diversity and plurality of life"—all crucial to building bridges with indigenous religious leaders by developing a relationship based on respect and trust.[76]

While they encouraged chaplains to take up the RLL role, chaplain advocates did not minimize the limitations and problems they would face. Many of their colleagues lacked relevant language skills and training in conflict resolution and negotiation or "possessed only an elementary understanding of other religions and cultures." Some indigenous religious leaders were not interested in religious engagement with the Americans. It was difficult to conduct RLL in areas of intense combat and/or frequent anticoalition attacks. It worked better in a "fairly permissive environment." As Navy Chaplain George Adams observed, security was a "major factor in determining a chaplain's ability to interact with local leaders."[77]

Chaplains also cautioned their colleagues against unrealistic expectations of success. Quoting a 2006 study by Navy Chaplain George Adams, Chaplain Ira Houck pointed out the "limited, but useful, role" chaplains could play in peace building. "Chaplains are not positioned to take on such major conflict mediation tasks as healing historic wounds in ethnic and sectarian conflict," Adams had written. "The primary mediating focus of chaplains should be on establishing communications and building relationships with local religious leaders on the ground—not on attempting to negotiate the resolution of broad historical problems."[78] In his 2009 *Army Chaplaincy* article, Adams also pointed out that most chaplains were not likely to conduct RLL with national, high-level religious leaders but rather with mid-level or local leaders. However, they needed to recognize that local religious leaders, especially in Iraq and Afghanistan, exerted significant influence in their communities and that engagement with them had had a "positive impact," not just on those communities but also on American military forces.[79]

At the close of the first decade of the twenty-first century, RLL remained a work in progress. In a comparatively short period of time, it had become the most important—certainly the most publicized—component of the chaplain's indigenous religions role. It was now recognized in official U.S. Army, Army Chaplain Corps, and Joint Chiefs of Staff doctrine.[80] Some of its problems had been solved, but others awaited a remedy.[81] In the meantime, there was talk of expanding the chaplain's IR and RLL roles.[82] The military chaplaincies had begun to provide chaplains with instruction in "cross-cultural negotiations, interreligious dialogue and conflict resolution," and both the Army and the Air Force were sending chaplains to The School for Conflict Analysis & Resolution at George Mason University "for professional military training to enhance and expand [their] analytical and advisory capabilities."[83] Given army leaders' expectation that the United States "would continue to be engaged in an era of 'persistent conflict,'" religious leader engagement seemed destined to play a continuing and important role in military operations.[84]

The new soldier- and battle-focused doctrine the army and the Chaplain Corps formulated in the late 1970s and 1980s drew significantly on the wealth of experience and expertise chaplains had acquired during the Vietnam era, at the same time that it accommodated the new strategic environment and AirLand Battle doctrine. It codified the "ministry of presence," defined a new, prophetic, moral advisory role for chaplains vis-à-vis commanders, and expanded the chaplain's religious leader engagement assignment. Together with the institutional ministry and the teaching of ethics, world religions, and moral leadership, it invested chaplains with more authority and influence in the military organization than they had ever had before. Another battlefield role chaplains developed during this period also involved adapting—partly to the new military context but also to the changing religious context in which they ministered. But as we shall see in the next chapter, in comparison with the roles discussed in this chapter, chaplains' morale-building role proved to be much more controversial.

8

Building Soldier Morale

Building soldier morale has long been considered a military chaplain's responsibility, and military thinkers and leaders, as well as chaplains, have generally recognized religion as a chief component of morale.[1] During the six and a half decades following World War II, army leaders and chaplains adapted the chaplain's morale role to a changing military and religious context. Doing so involved pondering a number of questions regarding the purpose and methods of chaplain morale building in the American army: the nature and sources of morale, whether morale enhancement should focus on the needs of soldiers or on accomplishment of the military mission, and whether, in addressing soldiers' needs, chaplains should adopt an inclusivist or exclusivist approach.

Morale Building in the 1950s and 1960s

In 1950 the Army Chaplain School issued a 188-page pamphlet titled *The Chaplain and Military Morale.* The author defined morale as having "to do above all else with aims and purposes." Specifically, morale motivated a person to "give of himself beyond the call of duty." Thus, he wrote, the "sacrificial spirit is at the root of morale, and we find it of course only when a man knows what he believes in and what he stands for." Morale was "something to be created" and "systematically promoted." It was a command responsibility, but the chaplain, as a staff officer, was to consult with the commander in "all matters pertaining to the religious life, morals, morale, and related matters affecting the military personnel of the command." The wisdom of this assignment seemed obvious: "The chaplain is . . . a specialist in religion, which is inherently concerned with goals and purposes and the sacrificial spirit."[2]

The author of the pamphlet distinguished between two types of morale building. The one had a "positive" aspect, or result, inducing "fitness and readiness to act," "persistence, courage, energy and initiative," and a "spirit of aggressiveness." The other had a "*negative*" aspect, or result: "readiness to wait," "staying power," and especially "sacrifice and tenacity." The chaplain fulfilled his morale responsibility by promoting the "moral and spiritual factors of morale," which the author declared to be the chaplain's "specific province."[3] In effect, the author presented a rationale for a *religious* approach to morale building, the very kind for which chaplains had both training and aptitude.

Chaplain field manuals of the 1950s and 1960s provided similar but more extensive guidance regarding the chaplain's morale role and responsibilities. The 1952 and 1958 manuals defined the chaplain's morale role as a member of the commander's special staff and as a religious and spiritual leader. On the one hand, he was to advise and consult with the commander "in all matters related to religion, morals, and morale in the command," matters for which he was declared to have "specialized knowledge." On the other hand, he was to "stimulate and guide the growth of the spiritual and moral sense of obligation to enable the soldier to be . . . a devoted defender of the nation." In a chapter titled "The Chaplain in Combat Units and Organizations," discussing the chaplain's contribution to soldiers' code of conduct training, the 1958 manual observed that "a clear abiding faith in God and the spirituality of the soul provides [*sic*] a soldier with a purpose in life that will lead him to continue to faithfully serve his Home, his Church, and his Country whether in personal combat or in captivity." The chaplain was instructed to "use every opportunity" to "supplement" and reinforce code of conduct training by "leading [soldiers] to a deeper consecration to their God and their country." He should also make "every effort to conduct frequent appropriate religious services" in the combat environment. As the manual explained, "To lead men to God and to bring God to men is an even more basic requirement in combat than in garrison. Nothing reenforces the combatant more than to know that he is at peace with God. Spiritual stamina is the only dependable support for training, *esprit,* and morale."[4]

In the aftermath of the Korean war and the POW scandal, at a time when Cold War tensions were escalating, the OCCH felt pressured to defend chaplains' ability to enhance soldier morale and combat effectiveness. The 1961 Character Guidance manual described the program as "a moral management tool for the Army commander" that provided "a direct route to morale, efficiency, and discipline in the command."[5] Also in 1961, the OCCH, in reviewing a Continental Army Command training directive, urged military leaders to recognize that "religious and spiritual values" were "necessary motivational factors in human behavior," that they supported "effective leadership" and strengthened "the will to survive before and/or during capture," and that chaplains played a vital role in developing such values.[6] In 1962, responding to questions raised by the Senate Special Preparedness Investigating Committee, the OCCH asserted that CG was "one of the chief means utilized by the Army to maintain a high degree of morale, discipline, dedication, and principles in the current 'cold war' situation." It made its "most effective contribution" to high morale by "positively promoting a healthy response to American virtues, principles, and ideals."[7]

The 1964 and 1967 chaplain field manuals used during the Vietnam War repeated the earlier manuals' guidance about stimulating the soldier's sense of obligation to his country and supplementing code of conduct training. Discussing the chaplain's responsibilities in the combat zone, the two manuals expanded his advisory role to the commander and staff to include not just advice and information but also "plans" on matters of "religion, morals, and morale as affected by religion." The manuals specifically advised the chaplain to plan and lead soldiers in "group discussions in the positive aspects of moral courage, the spirit of sacrifice, sense of duty[,] and integrity." Like the manuals of the 1950s, they urged chaplains to hold frequent religious services in the field. "To the soldier in the front line, religion is extremely important," the manuals pointed out. "He looks forward to the chaplain's visits and the opportunity to pray, to partake of the sacraments, and to talk about his faith with someone who can assure him of Divine love, forgiveness, and comfort." Such activities were deemed especially necessary, given "the stress and violence of combat, [which] leaves men physically, emotionally, and spiritually exhausted."[8]

Clarifying the Morale Role

In 1970, U.S. Army leaders and the OCCH took two actions to clarify the chaplain's morale role. They were most likely a response to discussion in the Chaplain Corps and the civilian community, prompted by the Vietnam War, regarding the nature and purpose of chaplain's morale role. As we saw in chapter 4, chaplains were of two minds on the subject. Some engaged in the kind of morale building intended to promote combat effectiveness. Other chaplains, especially those serving in the combat zone, concentrated on equipping soldiers with "tenacity" and "staying-power." In the civilian community, antiwar critics focused much of their antichaplain critique on the morale role chaplains performed in Vietnam. The chaplains were charged with "sanctifying" war making, "using religion as a rationale for the righteousness of our cause," and "sacrificing faith to the fighting spirit." Their very presence on the battlefield was said to constitute an endorsement of an immoral, unjust war.[9]

The first clarification was announced in the May 1970 chaplain field manual: a revision of a statement in the December 1967 manual referring to the chaplain's responsibility regarding religion, morals, and morale. On page 3, the phrase "all matters of religion, morals, and morale" had been deleted, and in its place was the phrase "all matters of religion and morals, *and morale as affected by religion*." The change in punctuation and wording had been made throughout the 1970 manual, wherever the phrase "matters of religion, morals, and morale" appeared.[10] The manual provided no explanation for the revision. It would appear that the new formulation was intended to emphasize a religious, as opposed to what might be called a militaristic, orientation of the chaplain's morale role.

That was also the purpose of the other clarification, published in the 1970–71 "Historical Review." The OCCH reported that in October 1970 it had received a memorandum from the U.S. Army Combat Developments Command Chaplain Agency requesting comment on a document titled "The Role of the Chaplain in the Motivation of the Soldier."[11] OCCH did not paraphrase or quote at length from the document, but in recounting its response, it made its own view of the chaplain's morale role quite clear. It had pointedly rejected any characterization of the chaplain as a "psychological motivator" of soldiers. It had urged changes in the wording of the agency draft to avoid any suggestion that "the role of the chaplain is similar to that of a 'psychological' officer who instills faith to animate men in battle, thus improving military efficiency by his priestly incantations." For the chaplain, OCCH insisted, "religion is an end in itself. The chaplain is not an instrument for the conditioning process to make good fighting men. He is the spiritual leader to make better children of God out of men. The fact that men of great spiritual strength are usually good soldiers is a bonus effect the Army receives from support of religious programs which is [*sic*] its national duty to provide."[12]

Statements on chaplains and morale in two army publications of the early 1980s reflected the 1970 clarifications. A 1983 Army Training and Doctrine Command pamphlet on "Religious Support in Combat Areas" divided morale responsibility between the commander and the chaplain. "The commander creates the leadership climate that fosters the development of soldier bonding and unit cohesion," the pamphlet observed. "The chaplain, as a religious representative, points the soldier to a reality beyond himself. He assists the commander by facilitating *spiritual factors* that enable the soldier to strengthen his faith and achieve inner stability, calm, and peace."[13]

The 1984 chaplain field manual, *The Chaplain and Chaplain Assistant in Combat Operations,* echoed the qualms OCCH had expressed about chaplains being "an instrument for the conditioning process to make good fighting men." The manual acknowledged the chaplain's role as an advisor and consultant to the commander on morale. It also noted that "through pastoral counseling the chaplain lessens stress and enhances the soldier's morale and performance." But it stated the following qualification: "While the actions of the chaplain may improve the morale and fighting capability of the unit, the chaplain's main purpose is to assist the commander in providing for the soldier's religious needs in all phases and locations of the combat arena." Although "a strong religious program enhances soldier bonding and fosters unit cohesion," that was not the chaplain's primary mission. It was to provide soldiers with pastoral care in the form of "counseling, spiritual reassurance, and encouragement," thereby sustaining or "rebuilding" their "emotional, psychological, and spiritual strength."[14] Like the 1950 *Military Morale* pamphlet, both the TRADOC pamphlet and the 1984 chaplain manual emphasized the chaplain's reliance on "moral and spiritual factors" in enhancing soldier morale. They clearly endorsed a religious rather than a militaristic orientation for chaplain's morale building.

Addressing Chaplains' Morale Responsibilities

The 1983 TRADOC pamphlet and the 1984 chaplain manual signal the emergence of a new way of thinking about chaplains' morale responsibilities in the combat environment. The chaplains' Vietnam War experience and the new AirLand Battle doctrine made them keenly aware of the terrible strain combat placed on soldiers. The new approach to morale building envisioned the chaplain acting as a spiritual or pastoral counselor, attending to the emotional, psychological, religious, and spiritual welfare of the soldier rather than acting as a combat motivator. It defined a type of morale enhancement that was more soldier than mission focused and had a definite religious content. It was referred to as "spiritual nurture" and "spiritual support," but it was officially known as "religious support" (RS), a term that described both the providers and the nature of the support.

Two papers written by army chaplains in the mid- to late 1980s give an idea of the train of thought that inspired this new approach to morale building. The first, published in the *Military Chaplains' Review* in 1984, was titled "Preparation for Combat: Emotional and Spiritual." Pointing out the "bestiality and brutishness of battle," Chaplain Jay H. Ellens declared that the chaplain's "paramount" responsibility was to provide soldiers with a "moral universe of values and meanings," so that "*alone* men and women caught in the awful enigma of military fighting can continue to find the meaningfulness in their experience that keeps us human, humane, and filled with purposefulness." Ellens believed that emotional and spiritual preparation for combat was especially necessary because American soldiers, like American citizens in general, could no longer refer to "some general consensus" regarding "values and ethical-moral standards." That "ethical net" had been supplanted by a "radical," "alienated," and "self-centered individualism." In "the worst of the horrible times of war" soldiers needed the chaplain's ministrations to "provide and preserve rooted and durable meaning for life." Because, Ellens observed, "*meaningfulness* is, in the end, what stabilizes persons in trauma and makes endurance and triumph possible."[15]

The author of the second paper was Chaplain Wayne E. Kuehne, a major figure in the development of Army Chaplain Corps battlefield doctrine in the 1970s and 1980s.[16] In 1988, while enrolled in the U.S. Army War College's Military Studies Program, he wrote a research paper titled "Faith and the Soldier: Religious Support on the AirLand Battlefield." His thesis was that soldiers desperately needed faith on the AirLand battlefield. Both chaplains and commanders needed to recognize and understand that fact, because in the army of the 1980s, Kuehne observed, "personal experience of Vietnam and/or the combat zone is fading." An increasing number of future commanders and chaplains "will not have experienced fear on the battlefield, the trauma of the newly dead, or have heard the screams of pain and terror of the wounded and maimed." Thus it was vitally important that they recognize the lethality of current and future warfare and understand "the moral and spiritual capabilities, limits and needs of flesh-and-blood soldiers."[17]

To give an idea of the impact of combat on the soldier, Kuehne quoted David Marlowe of the Department of Military Psychiatry, Walter Reed Army Institute of Research. "The environment for the contemporary American soldier is the most stressful, threatening and alien that human beings can be subject to," Marlowe wrote. Indeed, the "power of the battlefield to break men can never be overstated. As intensity, lethality and the duration of time in which the troops are engaged in exchanging direct and indirect fire with the enemy increase, the potentiality for individual breakdown and unit disruption also increases." Marlowe cited World War II and the 1973 Arab-Israeli Yom Kippur War as "guidance" for "the worst case conventional wars of the future."[18]

To bolster his argument regarding soldiers' need of faith on the battlefield, Kuehne pointed out the considerable number of military theorists and leaders who had asserted the importance of "religious" and "spiritual" forces in battle. He named or quoted, in particular, Xenophon, Carl von Clausewitz, Maj. Gen. F. M. Richardson, Gen. Douglas MacArthur, Gen. George C. Marshall, Gen. John W. Vessey Jr., John Keegan, and Richard Holmes.[19] He even quoted sociologist Samuel A. Stouffer. Although Stouffer was best known for his emphasis on "group solidarity" as a "combat incentive" among World War II infantrymen, Kuehne noted that Stouffer also considered a class of "motivational factors . . . that did not impel the individual toward active combat but did serve the important function of increasing his resources for enduring . . . combat stress," one of which was faith in God and in the efficacy of prayer."[20]

Of all the commentators Kuehne quoted, the one most often cited by other chaplains in late-twentieth- and early-twenty-first-century discussions of morale was General Marshall's statement about "the soldier's spirit." (It was also ubiquitous in army field manuals of that period.) "I look upon the spiritual life of the soldier as even more important than his physical equipment," Marshall had declared.

> The soldier's heart, the soldier's spirit, the soldier's soul are everything. Unless the soldier's soul sustains him, he cannot be relied upon and will fail himself and his country in the end. . . . It's morale—and I mean spiritual morale—which wins the victory in the ultimate, and that type of morale can only come out of the religious nature of the soldier who knows God and who has the spirit of religious fervor in his soul.[21]

Whereas Marshall spoke of the soldier's "religious nature" and "religious fervor" as the source of "spiritual morale," Kuehne, in discussing soldier's need of "faith," differentiated it from religion. He used the word "religion" to denote "the cumulative traditions of the faith of a people in history," which included sacred writings, liturgies, and creeds and which bound people together in "a relatively stable and formalized structure of relationships." He defined faith as "a generic or universal feature of the human struggle for identity, community, and meaning." Quoting religious historian Wilfred Cantwell Smith, Kuehne pointed out that "faith" was "in some cases and to some degree" a product of religion, but it was "a quality of the person," not an outgrowth of religious tradition or doctrine. It was "an orientation of the personality, to oneself, to one's neighbors, to the universe; a total response; a way of seeing whatever one sees, and of handling whatever one handles; a capacity to live at more than a mundane level; to see, to feel, to act in terms of a transcendent dimension."[22]

Faith was "*one* of the major contributors to combat morale," Kuehne declared. He urged the chaplaincy to "continuously explore the ramifications of faith in the lives of soldiers and their family, particularly as they strengthen and sustain the soldier in combat," and to assign the Unit Ministry Team "functions and activities" that would enable it to be supportive of that faith.[23]

Religious Support in the Combat Environment

In the late 1980s and 1990s, chaplains' new thinking about morale building in the combat environment was recognized as Chaplain Corps doctrine. The 1989 chaplain field manual presented in boldface capital letters the following "Doctrinal Principle": "MEETING THE RELIGIOUS AND SPIRITUAL NEEDS OF THE SOLDIER IS THE PRIORITY FOR COMPREHENSIVE RELIGIOUS SUPPORT IN ALL COMBAT OPERATIONS."[24]

Both the 1989 and the 1995 field manuals pointed out the "religious focus" of the UMT, based on "the traditions of caring, compassion, and devotion to God." The 1989 manual explained that the UMT offered "a message of grace, hope, and forgiveness from God to the soldier." The 1995 manual observed that the chaplain and his assistant "provide encouragement, compassion, faith and hope to soldiers experiencing shock, isolation, fear, and death. In the chaos and uncertainty of conflict and war, the chaplain is a reminder of the presence of God. The chaplain serves 'to bring God to soldiers and soldiers to God.'"[25]

Both manuals provided detailed guidance regarding religious support on the battlefield, with particular attention to combat stress and battle fatigue. The 1995 manual pointed out that "stress is a reality of combat, and soldiers must deal with it to succeed." Combat stress could not be eliminated, but armies and their commanders, with the assistance of the unit ministry team, could "control" it or keep it "within an acceptable range." Indeed, according to the manual, "controlled combat stress gives soldiers the necessary alertness, strength, and endurance to accomplish their mission. It elicits loyalty, selflessness, and heroism." But "uncontrolled combat stress" could "cause behavior that interferes with the unit mission" and could even "lead to disaster and defeat."[26]

The 1989 and 1995 manuals asserted that the UMT was "a powerful asset to the commander" in dealing with combat stress and battle fatigue, mainly because of its "faith-based" relationship with soldiers and its "unique spiritual focus."[27] In building soldiers' morale, the UMT assumed that the need for "religion and spiritual values" was "inherent in human life" and that such values were the basis of the "inner resources" soldiers relied on in combat. Intense combat over an extended period of time caused the soldier to expend those resources "faster than they can be replenished" and thus led to the "state of exhaustion" known as battle fatigue. Some soldiers were able to "self-replenish," but others were not. The UMT was "the primary resource" available to soldiers needing or seeking to "refocus their spiritual values." Indeed, since the "spiritual dimension" was "an essential element" in the replenishment process, the religious support the UMT provided was indispensable.[28]

The Spiritual Fitness Program

In discussing the UMT's role in managing combat stress and preventing and treating battle fatigue, the 1995 chaplain field manual referred to a resource not mentioned in the 1989 manual—"spiritual fitness training." Such training bolstered the soldiers' "inner resources," helping them "to build spiritual strength" and enabling them to "draw upon faith and hope during intensive combat."[29]

Spiritual fitness training was the domain of a new army program established in 1987 called Spiritual Fitness. *Fit to Win: Spiritual Fitness,* the pamphlet that described the program, defined such training as a means of developing or strengthening "those personal qualities needed to sustain a person in times of stress, hardship, and tragedy." The qualities were said to come from "religious, philosophical, or human values" and to "form the basis for character, disposition, decisionmaking, and integrity." The particular "personal qualities" the fitness program enhanced were endurance, resilience, and emotional and moral stability.[30]

Spiritual Fitness was a command program, but chaplains assisted commanders in spiritual fitness training. The 1995 and later chaplain field manuals discussed such training as a component of the "preventive religious support" chaplains administered while soldiers were undergoing training or during deployment. According to the 1995 manual, it provided "a stabilizing influence" and helped them "strengthen or regain values." The activities or types of assistance the UMT offered soldiers during such training included worship or private and group prayer, religious literature, "Scripture readings with soldiers," "sacraments and ordinances" when possible, and "opportunities for soldiers to work through frustration, fear, anxiety, and anger."[31] The *Fit to Win* pamphlet pointed out other programs and activities in which chaplains could promote spiritual fitness, such as individual counseling, "chaplain-led study and meditation groups," "workshops/seminars on values and value clarification," "human self development activities," and courses in the professional army ethic and moral leadership.[32]

"Spiritual Readiness"

During the first decade of the twenty-first century, the chaplain's morale-building role continued to evolve. The most striking change was the absence of chaplains' earlier qualms about "conditioning" soldiers to be "good fighting men." The 9/11 attacks on the World Trade Center and the Pentagon, the advent of the Global War on Terror, and the involvement of American troops in counterinsurgency warfare in Iraq and Afghanistan turned chaplains' attention, once again, to combat motivation.

The chaplain field manuals of the 1980s and 1990s had viewed morale building as a form of pastoral care. Chaplains were to nurture, strengthen, and help soldiers replenish the "inner resources" that gave them the stability, resilience, and endurance they needed to deal with the stress of combat. Two manuals issued in 2003 and 2005 assigned the UMT a new morale role—helping soldiers develop the "personal readiness" they needed "to sustain them during combat." According to the 2003 manual, this effort began prior to the soldiers' deployment, at the installation, where chaplains conducted worship services for them, offered religious education, and provided spiritual fitness training and other activities. Once soldiers entered the "mobilization phase," the UMT worked to "maintain and improve" their "combat readiness posture." During the "Port of Embarkation Phase," chaplains were to initiate "worship, counseling, and other religious support tasks" in order to further "ensure the forces are spiritually prepared for deployment and combat."[33]

The 2005 *Religious Support Handbook for the Unit Ministry Team* devoted an entire appendix to "Spiritual Readiness," defined as "an indicator of the soldier's 'will to fight'" and of the "general well being" of personnel in the command. The chaplain was to use "spiritual leadership, professional skills, and personal insights and beliefs to define, interpret, and train spiritual readiness." Specifically, he was to "plan, execute, and evaluate spiritual fitness training" as a means of "sustaining spiritual readiness in soldiers and units."[34]

The 2005 *Handbook* was significant for two main reasons. On the one hand, it revived the idea of the chaplain as combat motivator. On the other, it presented a more detailed and explicit affirmation of the chaplain's morale-building role than chaplain manuals of the 1980s and 1990s had offered. In Appendix J, on "Spiritual Readiness," the *Handbook* not only assigned the chaplain the traditional role of *advising* the commander and staff on "matters of morale that may affect the mission." It also instructed him to *engage in morale building* by "planning and executing" spiritual fitness training. And it stressed the importance of that role by declaring that such training was not just an important factor in "unit morale" but also "an essential and critical component of the total soldier system in the Army." In Appendix C, on "Combat Stress Control," the *Handbook* asserted that in dealing with soldiers' combat stress and battle fatigue, the chaplain and his assistant helped soldiers "focus on their spiritual values" and "resolve doubts regarding combat morality." This, too, clearly qualified as morale building, because, as the *Handbook* observed, in doing so, the UMT contributed to "the positive mental health of soldiers, unit cohesion, and morale" as well as to "the overall combat effectiveness of the unit."[35]

In reinstating the chaplain's role as combat motivator, the 2005 *Handbook* pointed out how it made chaplains an integral part of the military mission. In Appendix J, it declared that "spiritual readiness . . . enables the soldier to live the Warrior Ethos and embody the Army Values." "The Soldier's Creed and Warrior Ethos," presented in another appendix, did indeed include a declaration of "readiness": "I stand ready to deploy, engage, and destroy the enemies of the United States of America in close combat." The soldier also pledged, "I will always place the mission first. I will never accept defeat. I will never quit. I will never leave a fallen comrade." And one of the official "Army Values," also listed in the *Handbook*, was "personal courage": "Face fear, danger, or adversity (physical or moral)."[36]

To show the coherence of chaplain and army values, the *Handbook* presented two cameos depicting the heroic actions of a chaplain and an army pilot in the Vietnam War. The one described how, during "one of the most ferocious battles of the war," the chaplain "continually exposed himself to enemy fire while running unarmed into the fray to pull fellow paratroopers to safety and administer physical and spiritual aid. . . . Even when the perimeter was established, this chaplain consistently ventured outside of it to retrieve the wounded." While engaging in "this selfless service" he was mortally wounded. The cameo pointed out how he embodied both Chaplain Corps and U.S. Army values. The chaplain exemplified the attributes of "spiritual leadership," the "core competency" of the Chaplain Corps, and "because he lived and breathed those characteristics, he naturally 'lived out' the Warrior Ethos. The impulse of his character was to put the mission first, never accept defeat, never quit, and never leave a fallen comrade behind."[37]

The other cameo, titled "The Fruit of Spiritual Fitness," told of an army pilot "flying through close-range enemy fire" and landing in enemy territory who "utilized three helicopters to evacuate 51 severely wounded soldiers" during the course of a single day. "In the years since that heroic day," the cameo pointed out, "he has consistently referred to the spiritual fitness that undergirded his actions." He believes that "because of his faith, he was able to do things that, for him, would have otherwise been impossible."[38] Just as the chaplain cameo revealed how chaplains exemplified army values, so did the army pilot cameo validate their conviction that soldiers needed religion to sustain and motivate them in the combat environment.

The Spiritual Awakening

At the beginning of the twenty-first century, chaplains' thinking about soldiers' needs on the battlefield had not changed much since Jay Ellens's and Wayne Kuehne's discussion in the 1980s. In articles published in 1998 and 2002, the Chaplain Corps' leading ethicist, John W. Brinsfield, declared that "all soldiers have human needs and most have spiritual needs broadly defined." Elaborating on the spiritual needs, Brinsfield echoed Ellens's and Kuehne's statements about soldiers' need for "meaningfulness" (Ellens) and to be able "to see, to feel, to act in terms of a transcendent dimension" (Kuehne). Brinsfield wrote of soldiers beset by questions about "guilt, pain, and death" and engaged in a quest for "ultimate meaning in life." He observed that some soldiers found it in religion, while others discovered it through "spirituality."[39]

Brinsfield used the word "spirituality" quite specifically, on the one hand to refer to "a broader and possibly less distinct category than institutional religion" and on the other to denote a spiritual awakening among the American people that was transforming the cultural-religious context in which chaplains conducted their ministry. As he explained in his 2002 essay, the spirituality that powered the awakening of the late twentieth and early twenty-first century was quite different from the traditional kind. Traditional spirituality involved the effort to achieve "union or connection with God or the divine" through "religious devotion, piety, and observance." In recent years, "spirituality" had acquired a "wider" meaning—as a form of religious expression other than that found in institutional or organized religion and most often exhibited in "private meditation and secular transcendent experiences including feelings of awe and oneness with nature." It signified "the individual quest for greater insight, enlightenment, wisdom, meaning, and experience with the numinous or divine." Brinsfield noted that "many scholars of world religions" agreed that the new type of spirituality "seems to fit the beliefs of many faith groups, even those with non-theistic views." He also speculated that since soldiers constituted "a microcosm of American society," many of those searching for "meaning in life" might well be caught up in the new spiritual awakening.[40]

Religious historian Charles H. Lippy offers further insight into the new spirituality. He observes that although it embraced great diversity of belief and practice, it did exhibit "some common elements." It offered "some sense of the Other, of a Higher Power, or something more than human," and most individuals who "embarked on spiritual journeys" did so because they believed that "there was more to life than empirical reality and their own life experience, as well as that of all humanity" and that such experience was endowed with "transcendent meaning." However, Lippy also emphasizes the "idiosyncratic and eclectic nature" of the new spirituality. Unlike traditional spirituality, it was not based in denominational or faith-group doctrine and guidance. For each individual seeker, "building a framework of meaning was very much a private, personal enterprise." It involved choosing from a vast array of beliefs and traditions whatever helped one to "understand and interpret [one's] own human experience." Lippy characterized the spiritual awakening as "an amalgamation of traditional religious thinking with everything from what was popularly called 'New Age' to a glorified form of nature mysticism."[41]

A more effusive overview of the "flowering of spirituality" in America appeared in an August 2005 *Newsweek* cover story, "In Search of the Spiritual." The authors found the awakening "everywhere we looked":

> In the hollering, swooning, foot-stomping services of the new wave of Pentecostals; in Catholic churches where worshipers pass the small hours of the night alone contemplating the eucharist, and among Jews who are seeking God in the mystical thickets of Kabbalah. Also, in the rebirth of Pagan religions that look for God in the wonders of the natural world; in Zen and innumerable other threads of Buddhism, whose followers seek enlightenment through meditation and prayer, and in the efforts of American Muslims to achieve a more God-centered Islam.

The authors observed that churchgoing and religious affiliation had declined in America, but spirituality was "thriving." A *Newsweek*/Beliefnet poll revealed that "more Americans, especially those younger than 60, described themselves as 'spiritual' (79 percent) than 'religious' (64 percent)."[42]

The *Newsweek* story focused on the spirituality of individuals who were apparently affiliated with some kind of institutional religion. But as Lippy points out, many individuals caught up in the spiritual awakening made their spiritual journeys alone, without relying on denominational or faith-group association or guidance. Recent surveys conducted by the Pew Forum on Religion & Public Life show that the new spirituality included the large number of Americans who professed a variety of Eastern or New Age beliefs or who responded affirmatively when asked if they had ever experienced "a moment of religious or spiritual awakening." And that number included some who identified themselves as atheists or agnostics.[43]

Chaplains and "Spiritual Leadership"

During the first decade of the twenty-first century, the Army Chaplain Corps experienced its own kind of spiritual leavening. The chiefs of chaplains made spirituality and "spiritual leadership" the "centerpiece" of chaplains' ministry to the military.

In 1999, the Chaplain Corps had formulated its six key values, declared to be "the bedrock and foundation on which we build our vision and mission." Heading the list was "Spirituality: Seek to know God and yourself at the deepest level." In an address to the Senior Leadership Training Conference in January 2000, Chief of Chaplains Gaylord T. Gunhus elaborated on the definition. "Spirituality depicts a Chaplaincy composed of Godly leaders who are actively engaged as people of faith, [and] committed to the devotional disciplines of their respective traditions, namely prayer, study of scripture and reverent attendance in worship," he explained. "This is the strength and hope that sustains us as we provide the message of strength and hope to the Army family." At another senior leadership conference in 2002, Gunhus offered his own definition of "spiritual"—"the conscious awareness of our God's presence in the moment." He said, "To me this means that God is a part of everything we do and to the degree we are conscious of the divine presence we are blest by it, and are able to bless those around us."[44]

In 2000, the *Army Chaplaincy* devoted its entire summer–fall issue to "Spiritual Leadership, an Army Tradition." In the introductory message, "From the Chief," Gunhus declared that "spiritual leadership in the Army is crucial because soldiering is a faith-filled calling." Soldiers have a "calling" to "serve the Lord." By the same token, Unit Ministry Teams, as spiritual leaders, "have the important responsibility of guiding, strengthening, protecting and challenging soldiers to find and live out their calling." They "focus on equipping soldiers by preparing them at every moment to live and perhaps die in the sight of the God whom they serve."[45]

In 2005, in another special *Army Chaplaincy* issue on spiritual leadership, the lead article, a "White Paper" for the Chaplain Corps, was titled "Taking Spiritual Leadership to the Next Level." In his introductory message, Chief of Chaplains David H. Hicks pointed out the paper's emphasis on "the pastoral identity of the Army chaplain" and asserted that "what we offer that is unique is the successful and personal delivery of religious support across the full spectrum of operations." In providing religious support, chaplains "must be spiritual leaders." This was "essential," for "if we are not spiritual leaders, we have no place on the battlefield." In providing religious support, chaplains "must offer a spiritual vibrancy and engaging example that draws soldiers to deepen their life of faith." Chaplain Maury Stout, the author of the white paper, began his discussion of spiritual leadership with the definition of spirituality Hicks had offered in an address to a recent Senior Leader Training Conference. He had termed it his "personal definition of spirituality" and said that he based it on "the ultimate definition from the Bible," found in Matthew 22:37, 38: "Love the Lord your God with all your heart and with all your soul and with all your mind. And . . . love your neighbor as yourself." For himself, Hicks explained, spirituality involved "a personal relationship with God through trust in Jesus Christ as Lord and Savior and it also includes others, my neighbors."[46]

Army chaplains' understanding of "the spiritual," of "personal spirituality," and of "spiritual leadership" was clearly based in the devotional guides, teachings, and doctrines of their endorsing denominations or faith groups. Theirs was the traditional kind of spirituality and they were well aware of how different it was from the spirituality of the spiritual awakening. They also recognized the challenge it presented for their ministry. Some, perhaps many, chaplains did not share the equanimity Chaplain Brinsfield displayed in writing about the impact of the awakening. Instead, they seemed uneasy, bewildered, and even hostile.

Two articles printed in the *Army Chaplaincy* issue on "Spiritual Leadership" illustrate this reaction. "Spirituality is a hot topic among Christians and unbelievers," observed Chaplain J. Gordon Harris. "Something is missing in today's world and spirituality fills the void." Chaplains and other members of the armed forces were "trying to understand" the new spirituality, but he himself expressed disappointment with the books that purported to explain it. Either they "define it so broadly that it loses any Christian distinction or so narrowly that it loses its relevance and biblical character," he complained. He described three "approaches" to spirituality being practiced by members of the armed forces. All of them produced "a sense of well-being and a measure of spiritual maturity," but in his opinion, none was based on "the full teachings of the Bible on spirituality." Another chaplain, Ronald Thomas, noting "a revival in how spirituality impacts our daily life," wrote that he had discovered during counseling sessions with soldiers that they were "concerned about spiritual matters." He wondered whether chaplains would be ready to respond to "this new spiritual hunger."[47]

It was a critical question. The spiritual awakening not only confronted chaplains with a myriad of beliefs and practices outside the fold of traditional, "organized" world religions but also had serious implications for chaplains' morale-building role. For decades, chaplains had insisted that soldiers needed religion to sustain or motivate them on the battlefield. Indeed, chaplain field manuals declared that the need for religion was "inherent in human life." *Religious* support, defined as "faith-based" and as having a "religious focus" and "a unique spiritual focus," was Chaplain Corps doctrine. In providing religious support, UMTs brought "a message of grace, hope, and forgiveness from God to the soldier." Initially, religious support involved chaplains offering pastoral care and counseling on the battlefield as well as traditional religious activities, including "rites, ceremonies sacraments, ordinances, services, . . . and religious education."[48] When Spiritual Fitness became a component of religious support, chaplains made worship, prayer, "Scripture readings with soldiers," and other religious activities part of such training.[49]

The problem chaplains faced in the early twenty-first century was that religious support was not (or no longer) inclusive. Rooted in the beliefs and practices of traditional, organized religion, it did not speak to the needs of the increasing number of soldiers who did not identify with traditional religion. These soldiers included those who embraced the new spirituality as well as those who, in religious preference surveys, identified themselves as "None," "No Religion," or "No Religious Preference," categories which often included nontraditional spirituality as well as atheism and agnosticism. In a 1985 Gallup Report on religious preferences of armed forces personnel, only 14.75 percent had identified themselves as "no religious preference" or "atheist." However, two surveys conducted by the Department of Defense Manpower Data Center in 2001 and 2009 showed 20 to 21 percent in that category.[50]

To continue playing their expanded morale role, developed over several decades, chaplains would have to adapt to the new cultural/religious context. To accommodate the needs of soldiers outside the fold of the traditional world religions, they would have to adopt a more inclusive approach to morale building. There was no need to formulate it anew. One already existed, elements of which appeared in several chaplain manuals and army regulations. As Chaplain Corps historian Brinsfield pointed out in his 2002 essay, the U.S. Army leadership had long recognized the existence of a "spiritual component" in the human personality, based mainly on research on human needs and motivation reported by psychologist Abraham Maslow. According to Brinsfield, Maslow did not define spiritual needs or "aspirations" in religious terms but viewed them as a "natural phenomenon" with a "naturalistic meaning"—they were, Maslow insisted, not "the exclusive possession of organized churches" and they did not need "supernatural concepts to validate them."[51]

Early evidence of a nonreligious approach to morale building appears in a 1962 statement written by Chaplain Lawrence K. Brady, assistant for plans, programs and policies in the Office of the Chief of Chaplains, to be sent to the Senate Special Preparedness Investigating Committee in response to questions raised about chaplains' morale-building efforts during the Cold War. Brady used the word "spiritual" not in a religious sense but to refer to "how a man thinks, what he considers important, what his scale of values are, and what he considers to be worthy of effort and sacrifice."[52]

During the 1960s and 1970s, most chaplain manuals used "spiritual" and related words in the religious sense or in such a way that they could be interpreted according to either a religious or nonreligious meaning. The breakthrough began in the 1980s. For example, in 1983, discussing the chaplain's responsibility for the "religious welfare" and "spiritual health" of soldiers, the U.S. Army Training and Doctrine Command assigned him the duty of advising and assisting the commander in "reinforcing" not just "spiritual, moral, [and] ethical values" but also "psychological dimensions." Another good example is the statement in the 1987 *Fit to Win* pamphlet describing spiritual fitness as grounded in "personal qualities" that came from "philosophical" or "human values," as well as religion.[53]

The identification of "spiritual" with "philosophical values" also appeared in the army's Well-Being Strategic Plan of 2001, formulated by the Office of the Deputy Chief of Staff for Personnel. It defined well-being as "the personal—physical, material, mental, and spiritual—state of soldiers, civilians, and their families that contributes to their preparedness to perform The Army's mission." The "spiritual state of well-being" was said to "center on a person's religious/ philosophical needs and may provide powerful support for values, morals, strength of character, and endurance in difficult and dangerous circumstances." Army Regulation 165-1 (March 2004) went further in separating the word "spiritual" from its religious connotation. It defined "spiritual" as "a term not limited to any specific religious definition of spirituality, but including qualities and values associated with the human spirit in general."[54]

These isolated statements in chaplain and army documents suggest that some chaplains and army leaders had begun to consider the need for, or had actually embarked on a process of defining and promulgating a nonreligious morale-building role for the Chaplain Corps. Observations Brinsfield made in his 1998 *Parameters* article and his 2002 essay reinforce that perception. In *Parameters* he pointed out that in the past three decades the army had "grown uneasy about publicly acknowledging religion as a support for soldiers" because it might prompt charges of an establishment of religion or violation of the right of free exercise of religion. He also made a point of noting the existence of an alternative to the "religious support" chaplains and their assistants provided. "For soldiers who do not choose to attend religious services or avail themselves of the support of chaplains, there are counselors, medical personnel, and other professionals who can provide some spiritual fitness support," he explained. "Commanders must of course be careful that in providing the resources to those who wish to worship freely they do not pressure others who may prefer their own private thoughts."[55]

In his 2002 essay, Brinsfield himself recognized the exclusivist character of chaplains' faith-based morale building. Having described chaplains' involvement in "ethical and spiritual preparation of units for Desert Storm," he wrote that "they helped *religious* soldiers find the bridge between their spiritual and professional values in a way no other staff officer was expected to do." Later, in concluding the essay, he raised the question of whether and "to what extent" the "profession of arms" (including chaplains) "should try to meet the spiritual needs of a military population becoming ever more ethnically, morally, and religiously diverse."[56]

Some statements in the December 2009 Army Regulation 165-1 may have been intended to alleviate concerns about religious support contravening the First Amendment or exemplifying an exclusivist approach to morale building. On the first page the regulation declared that "American Chaplains represent the unique commitment of the American social and religious culture that values freedom of conscience and *spiritual choice* as proclaimed in the founding documents." The same point was made in slightly different language in the next paragraph: "American Chaplains . . . demonstrate the values of religious freedom of conscience and *spiritual choice*." In chapter 2, "Religious Support in the Army," the regulation declared that "participation in religious activities is voluntary."[57]

"Spiritual Resilience"

As the first decade of the twenty-first century drew to a close, there was increasing evidence that army leaders and chaplains were developing a more inclusive morale-building role for chaplains, one that would accommodate all soldiers—those who acknowledged religious and spiritual needs as well as those who did not, those who identified with traditional "faith" as well as those who claimed "None." What did rank-and-file chaplains think of the effort?

A review of articles published in a 2009 special issue of the *Army Chaplaincy* may help answer that question. The topic was "Spiritual Resilience," a comparatively new locution for "spiritual fitness."[58] In some of the articles, chaplains and other contributors presented their own views of spiritual resilience, based on personal experience and/or the scriptures and history of their faith. In addition, some exhorted chaplains and chaplain assistants to assess and strengthen their own spiritual resiliency so as to provide spiritual leadership and religious support to soldiers and their families. Thus Chief of Chaplains Douglas L. Carver declared that "we do not maintain spiritual resiliency in and of ourselves, but through the 'inspiration' of the divine resources of our faith." Addressing chaplains who might sense a lack or waning of their own spiritual resiliency, he counseled them to seek the support of "faithful spiritual leaders and friends" as well as their endorsing agents, and to "call out to your Creator and ask him to '*renew a right spirit within you*.'"[59]

In addition to articles defining a Christian view of spiritual resilience, the *Army Chaplaincy* issue featured two that presented a Muslim and Jewish understanding of the term. In "TAQWAH and SABR: The Foundation of Spiritual Resiliency in Islam," Chaplain Abdul-Rasheed Muhammed described how "foundational components within Islam," mentioned in both the Qur'an and the Sunnah, functioned in the daily life of believers in Islam and contributed to their spiritual resilience. In "A Jewish Perspective on Spiritual Resiliency," Chaplain Henry C. Soussan discussed the "religious resources" he relied on in building his own "resilient spirit": "classical biblical and rabbinic sources" as well as "spiritual discipline, the Jewish experience and God's presence."[60]

Other articles addressed the question of how chaplains should or might assist soldiers in acquiring or strengthening their spiritual resilience. Chaplain Mike Dugal, director of the Center for Spiritual Leadership at the U.S. Army Chaplain Center and School, called for "present and future dialogue" within the chaplaincy regarding this matter. Perhaps to spark such dialogue, he presented two definitions of spiritual resilience. What he called "a liberal definition" viewed spiritual resilience as "the inner-life ability to respond to life's stressors, adversity or traumatic events and proceed in life without diminishment (chronic symptoms) to the soul (inner-life)." The other definition, which he himself professed, was based on what he called his "Christian world view." He contended that "this inner-life ability to respond to life's stressors is dependent on a relationship with the Living God as revealed through Jesus the Christ and is sustained by the imminent [*sic*] and abiding presence of the Holy Spirit." This "inner-life ability" was not "confined to the psyche or the physical." It was "part of man's metaphysical constitution" and "results from a passionate trust in a living God who promises never to abandon us."[61]

In fact, several articles in the 2009 *Army Chaplaincy* issue indicate that a dialogue was well underway. Chaplains and other contributors to the special issue mentioned several nonreligious definitions of "spiritual" and "spiritual resilience": Abraham Maslow's and the ones presented in the Army Well-Being Strategic Plan of 2001 and in the *Fit to Win* pamphlet. In various articles, the authors referred to spiritual resilience as "psychological fitness" and "moral courage," as a "learned behavior involving thoughts and actions," and as "an individual's journey to increase capacity to make meaning of life's events and then employ the learned lessons constructively."[62]

However, the main concern of these writers was not so much to develop a consensus in favor of a nonreligious definition of spiritual resilience but to point out the need for a more inclusive approach to such training and to describe how it might work out in practice. "How spiritual resilience can be nurtured depends on the individual's background and commitment," Chaplain Soussan observed. For decades military thinkers and leaders, psychologists and sociologists, as well as chaplains, had asserted that soldiers needed religion. Now, some of the writers in the special issue pointed out that in the new twenty-first-century cultural/religious context, an increasing number of soldiers did not profess or experience a need for religious faith. Chaplain William Scritchfield noted that chaplains were currently leading "spiritual renewal seminars and spiritual development programs through various chapel programs." The question he posed was, "How do we assist spiritual resiliency development in those who do not participate in such events?" One of the answers he supplied involved chaplains cooperating with and recognizing the contribution to be made by professionals other than chaplains. He recommended "a multi-disciplinary approach to spiritual resiliency," to be implemented by behavior health professionals, medical professionals, communities agencies, and law enforcement, in addition to commanders and chaplains. Scritchfield's suggestion recalls Brinsfield's observation, in his *Parameters* article, about soldiers who chose not to avail themselves of religious services or other forms of religious support. For them there were "counselors, medical personnel, and other professionals who can provide some spiritual fitness support." This statement was quoted, without attribution, in the 2005 *Religious Support Handbook*, yet another sign of chaplains' new thinking about morale building.[63]

Army chaplains' discussion of a more inclusive approach to morale building recalls an earlier time in Chaplain Corps history. In the 1980s, as we shall see in the next chapter, the issue was "religious accommodation"—ensuring that army and Chaplain Corps policies and practices did not violate the First Amendment guarantee of soldiers' right to the free exercise of religion. That issue provoked a protracted debate within the Chaplain Corps and a "culture war" in the civilian religious community. Whether chaplains would be able to avoid such controversy in dealing with the issue of "morale accommodation" in the twenty-first century remained to be seen.

9

Addressing Religious Pluralism

In 1979, two Harvard Law School seniors, Joel Katcoff and Allen M. Wieder, filed suit in the U.S. District Court for the Eastern District of New York against the Department of Defense, the Department of the Army, and the secretary of the army. They sought a judgment that "the Chaplaincy program [of the army] constitutes an establishment of religion in violation of the establishment clause of the First Amendment to the U.S. Constitution." They contended that the right of army personnel to the free exercise of religion should be served by a civilian chaplaincy similar to that provided by the Wisconsin Evangelical Lutheran Synod.

Litigation of the lawsuit that became known as *Katcoff v. Marsh* lasted for six years, during which time the district court decision, which ruled against the plaintiffs, was appealed to the U.S. Court of Appeals for the Second Circuit. In January 1985, the three Second Circuit judges affirmed the constitutionality of the army chaplaincy. In their decision they wrote that the primary function of military chaplains was to engage "in activities designed to meet the religious needs of a pluralistic military community." Army personnel "experience increased needs of religion as the result of being uprooted from their home environments, transported often thousands of miles to territories entirely strange to them, and confronted with new stresses that would not otherwise have been encountered if they had remained at home." Citing both the establishment and free exercise clauses of the First Amendment, the judges asserted that "unless the Army provided a chaplaincy it would deprive the soldier of his right under the Establishment clause not to have religion inhibited and of his right under the Free Exercise clause to practice his freely chosen religion."[1]

The filing of the lawsuit reportedly sent "shock waves" through the Army Chaplain Corps. Although the constitutionality of the chaplaincy was ultimately affirmed, the suit and its implications made a lasting impression. The Chaplain Corps felt impelled "to focus as never before on its responsibilities to provide for and to defend the rights of soldiers to free exercise of their faith (or lack thereof)."[2]

The Challenge of Religious Pluralism

In effect, *Katcoff v. Marsh* forced the army and the Chaplain Corps to address increasing religious pluralism in the nation and the armed forces. Two American Religious Identification Surveys (ARIS) taken in the late twentieth and early twenty-first century revealed a religious landscape markedly different from that of the 1950s and 1960s.[3] Some of the changes resulted from an acceleration of trends that began during the 1960s: the decline in the number or proportion of Roman Catholics, mainline Protestants, and Jews, and the increase in the number of evangelical Christians. In 1990, according to the surveys, Roman Catholics made up 26.2 percent of the total adult population in the United States; in 2001 they were 24.5 percent, and in 2008 they were 25.1 percent.[4] The ARIS surveys gave the following figures for the "Mainline Christian" category, including Methodists, Lutherans, Presbyterians, Episcopalians/Anglicans, and members of the United Church of Christ: 1990, 18.7 percent of the total population; 2001, 17.2 percent; and 2008, 12.9 percent.[5] The percentage of Americans who identified themselves as Jewish (referring to religion, not ethnicity) also declined: to 1.8 percent, in 1990, 1.4 percent in 2001, and 1.2 percent in 2008.[6]

While the mainline denominations experienced marked declines, the ARIS surveys showed an increase in the "Evangelical/Born Again" category.[7] In 2008, 34 percent of the total adult population identified themselves as a "Born Again" or "Evangelical Christian." Not all of these individuals were affiliated with fundamentalist, evangelical, or pentecostal churches or denominations; millions of mainline Protestants and Roman Catholics also identified themselves using those labels.[8]

In addition to continuing change in the Christian and Jewish communities, two newer trends appeared. These were mainly a product of the 1990s, "a period of significant shifts in the religious composition of the United States."[9] One was the growth in the percentage and numbers of Americans who claimed affiliation with religions other than Christianity. These "Other Religion Groups" included Judaism and the Unitarian Universalist Association, which had a long history in the United States, as well as newcomer faiths such as Islam, Buddhism, and Hinduism and cults and sects such as Eckankar, Baha'i, est, Transcendental Meditation, various New Age groups, spiritualist groups, paganism, the Hare Krishna movement, and Wicca. In 1990, individuals who identified themselves as belonging to "Other Religions" constituted 3.3 percent of the total adult American population; in 2001 they were 3.7 percent, and in 2008 they were 3.9 percent.[10]

The percentages seem small, but the ARIS 2008 survey pointed out that "in absolute numbers" the "Other Religion Groups" increased by 50 percent between 1990 and 2008.[11]

Even more remarkable was the growth of the "Nones"/"No Religion" category, comprised of Americans who identified themselves as "Atheist," "Agnostic," or "No Religion." Between 1990 and 2008, the "Nones" increased by almost 20 million people, to a total of 34,169,000, and their percentage of the total U.S. population rose from 8.2 percent to 15.[12] Although the growth rate of individuals who expressed no religious preference declined after 2001, according to the 2008 survey, the numbers of those who identified themselves as atheists or agnostics "rose markedly from over a million in 1990 to about 2 million in 2001 to about 3.6 million today."[13] Of all the groups named in the surveys, the "Nones"/"No Religion" category had the largest net increase in numbers, having grown by 138 percent during the period 1990–2008. (The total religious population had grown during that period by 30 percent.)[14]

In summarizing its results, the 2008 ARIS survey pointed out three significant religious trends of the late twentieth and early twenty-first century. First, Americans were becoming more evangelical and thus more conservative in belief and practices. Second, and somewhat paradoxically, the American population as a whole was "becoming less Christian" and "less religious."[15] The second trend was partly a consequence of Americans' increased identification with other world religions and new religious movements. But it also resulted from the third trend, the considerable growth in the number and percentage of the "Nones"/"No Religion" group, which revealed "a rejection of all forms of organized religion."[16] All three trends affected the ministry of military chaplains, partly because of their relationship with the civilian religious community but also because the trends were reflected in the military community.

Implementing Religious Accommodation

A May 1985 Gallup report gave a general idea of the extent to which the religious preferences of armed forces personnel mirrored the changes taking place in the civilian religious community. It showed that 54.7 percent of the military population favored specific Protestant faith groups; 26.19 percent, Roman Catholicism; 0.37 percent, Judaism; and 14.75 percent, "no religious preference" or "atheist."[17]

A 2001 survey of the religious preferences of the U.S. population and military personnel, issued by the Department of Defense Manpower Data Center, showed significant change in the Protestant, Catholic, and atheist categories. Thirty-five percent of military personnel indicated a preference for "Protestant" and another 11 percent for "Other Christian" (which included, among other groups, Latter-day Saints, Seventh-day Adventists, Jehovah's Witnesses, the Christian and Missionary Alliance, Church of God, and Assemblies of God). Twenty-two percent of military personnel identified themselves as "Catholic/orthodox." The percentage for "Atheist/no religion" was 21. (Among civilians, age twenty to thirty-nine, it was 3 percent.)[18]

A 2009 survey of 1,398,881 active-duty military personnel in the army, navy, Marine Corps, and air force, produced by the Defense Manpower Data Center, presented more detail. It reported that 49 percent listed themselves as members of various Protestant denominations; 20.8 percent, Roman Catholic Church/Catholic Churches; and 0.33 percent, Judaism/Jewish. The percentages of personnel who checked Islam and Buddhism were 0.25 and 0.37 percent, respectively. The percentage of Eastern Orthodox/Orthodox churches was lower, at 0.108 percent. Only 739 people checked Hinduism, while 260 checked Native American. "No Religious Preference" accounted for 281,710, or about 20 percent. The percentage who described themselves as atheist was 0.479 percent, and agnostic, 0.09 percent.[19]

To address the challenge of religious pluralism in the armed forces, military leaders implemented the system known as "religious accommodation." Beginning in the 1980s, the Department of Defense, the three service branches, and their chaplaincies issued a host of directives, regulations, and instructions, all aimed at ensuring that armed forces personnel enjoyed their First Amendment right of free exercise of religion within the military organization. One of the earliest and the most inclusive of these, covering all of the services, was Department of Defense Directive 1300.17, dated June 18, 1985. It began with a general policy statement, that "a basic principle of our nation is free exercise of religion." Then followed the directive that "requests for accommodation of religious practices should be approved by commanders when accommodation will not have an adverse impact on military readiness, unit cohesion, standards, or discipline."[20] The "religious practices" to be accommodated varied. Some had to do with military apparel and dietary, medical, and/or burial requirements; others related to the observance of holy days, opportunities for worship, and worship practices.[21]

Religious accommodation was a command responsibility. The commander made the decision to approve or deny requests for accommodation, in accord with specific, detailed guidelines and procedural rules.[22] However, he might choose to discuss the request with the unit chaplain or suggest that the individual seeking accommodation talk with a chaplain. DoD and service regulations emphasized the important role the chaplain played in the system of religious accommodation. The chaplain had a duty to "facilitate" and "ensure" the free exercise of religion for all personnel, "regardless of religious affiliation of either the chaplain or the unit member." The chaplain was also expected to advise and consult with the commander on matters of religious accommodation. Indeed, Army Regulation 165-1 (2000) urged a proactive approach: "Even though the chaplain is an ecclesiastically endorsed representative of his or her faith group, the chaplain has the responsibility to confront the command when the religious rights of any soldier are affected."[23] The commander made the final decision, but a chaplain might exert a significant influence on it or even become an advocate for an individual or group requesting accommodation.[24]

Several DoD directives specifically targeted chaplains and their endorsing agencies. A 1988 directive required endorsers to "be able to certify clergy who are qualified to provide directly or indirectly for the free exercise of religion by all members of the Military Services, their dependents, and other authorized persons."[25] A 2004 directive on the appointment of chaplains declared that the chaplaincies were established to advise and assist commanders in providing for "the free exercise of religion in the context of military service as guaranteed by the Constitution" and that chaplains "shall serve a religiously diverse population." Endorsers must recognize this "command imperative and express willingness for their [chaplains] to perform their professional duties . . . in cooperation with [chaplains] from other religious traditions."[26]

At the U.S. Army Chaplain School the officer basic course included instruction and guidance in matters of religious accommodation.[27] According to Army Chaplain Corps historian John W. Brinsfield, "Chaplain instructors would ask chaplains just coming on duty if they could help a soldier of a completely different religion practice his or her faith 'without qualms of conscience.' New chaplains were also asked if they could cooperate with chaplains of other faiths in implementing joint religious programs without compromising their beliefs."[28] The chaplain school was viewed as a place where newly accessioned chaplains could experience and come to accept religious diversity within the Chaplain Corps.

Chaplains also received instruction and assistance in religious accommodation in various programs, workshops and conferences, which treated such subjects as religious pluralism, racial and ethnic diversity, affirmative action, multiculturalism, inclusive language, and minority and ethnic worship services.[29] The *Military Chaplains' Review* published articles on matters relating to religious accommodation, including a special issue in the summer of 1992 on "Pluralism and Ministry Issues."

Religious Pluralism and Imbalance within the Chaplain Corps

Charged with the responsibility of facilitating and ensuring soldiers' free exercise of religion, the Chaplain Corps recognized the importance of increasing religious pluralism within the chaplaincy so as to better serve soldiers' diverse religious needs. Among other things, this required increasing the number of chaplains endorsed by underrepresented religious faiths as well as adding chaplains from faith groups that previously had had no representation. As we saw in chapter 5, the chaplaincy had begun a concerted effort to recruit chaplains from black denominations in the early 1970s. By the mid-1980s, the percentage of black chaplains had risen to 14.08, and the recruitment effort continued. Of the non-Judeo-Christian faith groups, Islam was the first to have representation in the Chaplain Corps. In 1994, Abdul Rasheed-Muhammad became the first Muslim chaplain in the army, endorsed by the Islamic Society of North America. The first Buddhist chaplain in the army was a former Southern Baptist pastor, Thomas Dyer. By 2011, there were six Muslim and two Buddhist chaplains, and the first Hindu chaplain had recently been appointed. In the early 1990s the Universal Fellowship of Metropolitan Community Churches, a denomination with a mostly homosexual membership, had requested that one of its clergy be appointed a military chaplain, but the application was turned down by the Armed Forces Chaplains Board. During the late twentieth and early twenty-first century, Wiccans, members of the Military Association of Atheists and Freethinkers (MAAF), and humanists requested the appointment of chaplains to serve their members in the military, without positive results.[30]

The other problem the chaplaincy faced was a faith-group imbalance. In the Army Chaplain Corps, the most important change in religious composition was not the increasing number of chaplains in the "other religions" and "none" categories, which was negligible. It was the considerable presence of chaplains from conservative evangelical, fundamentalist, and pentecostal churches and faith groups. They had registered a dramatic increase in numbers during the mid-

1960s.[31] Two surveys done in 1987 and 2009 showed that the trend continued. (Although the surveys comprehended chaplains in all of the service branches, they give some idea of the changing composition of the army chaplaincy.)

According to the 1987 "Study of Representation of Religious Faiths in the Armed Forces," the 3,488 chaplains of the armed forces came from 205 faith groups. The percentages for the largest faith groups and denominations were as follows: Roman Catholic, 19.84 percent of the total; Southern Baptist Convention, 12.96; United Methodist Church, 10.72; and Lutheran Churches, 7.92. Jewish chaplains represented 1.28 percent of the total.[32]

Twenty-two years later, the 2009 Defense Manpower Data Center survey of 2,930 active-duty chaplains in the army, navy, and air force revealed significant change. It showed a decline in the number of Roman Catholic chaplains to about 9 percent and a considerable increase in the number of Protestant chaplains, who represented almost 90 percent of the total. Within the Protestant fold, the number of chaplains from evangelical, fundamentalist, and pentecostal denominations and churches greatly exceeded those affiliated with mainline Protestant denominations. The three largest evangelical groups, the Southern Baptist Convention, Baptist Churches, and the Assemblies of God, had a combined total of 757 chaplains, compared to only 255 for the United Methodist Church, the Episcopal Church, the Evangelical Lutheran Church in America, the Presbyterian Church in the USA, and the United Church of Christ. A great many other, much smaller evangelical groups also endorsed chaplains for the military.[33]

Evangelical Sectarianism

In 1984, in a *Military Chaplains' Review* article, retired chaplain Bertram C. Gilbert observed that "the more inclusive the Army chaplaincy has become in recent years, the harder it has been to persuade chaplains that their task was not primarily the promoting of their own brands and styles of religion."[34] Increasing religious diversity within the Chaplain Corps seemed to intensify sectarian conflict among the chaplains, especially among the Protestants.[35] Mainline Protestant chaplains were certainly not immune to denominationalism and sectarianism, but evangelical sectarianism proved to be the more intractable.

It also had a long history in the army, dating back to the 1940s and 1950s, when chaplains and their allies were mainly concerned about the Unified Protestant Sunday School Curriculum and the General Protestant Service. By the late 1980s, those issues became moot, as a result of Defense Department and Army Chaplain Corps actions. The evangelicals had succeeded in getting the Department of Defense and the Army Office of the Chief of Chaplains to cease "mandating" the use of the Unified Curriculum. An evangelical chaplain, James A. Edgren, had helped rewrite the army regulations in the early 1980s.[36] The OCCH also changed its policies regarding the General Protestant Service. In January 1988, it published a statement in its chaplain newsletter that established new guidance regarding worship services on army installations. Traditionally, the GPS had been regarded as the primary religious service and denominational services as a "secondary responsibility," although chaplains were not prohibited from holding them.[37] The new guidance was clearly prompted by the concern for religious accommodation and the free exercise of religion. Formulated by Chaplain Edgren for the chief of chaplains, it declared that "to protect and enhance the free exercise rights of soldiers and family members, all 'distinctive faith groups' have equal claim upon Chaplaincy services." Religious groups that wished to worship separately, in denominational services, were considered to be "distinctive faith groups," and as such, according to the guidance, "should receive an equitable share of resources, including appropriate funds." The new policy effected a dramatic change in the Sunday morning schedule of installation worship services. The "General" or "Collective" Protestant service—sometimes referred to as simply the "Protestant worship service"—now shared placement with a great variety of "distinctive" worship services.[38] The evangelical chaplains who held such services for evangelical, pentecostal, and fundamentalist soldiers and their families were free to conduct them according to the tenets and practices of their denominations, faith groups, and/or endorsing agencies.

Evangelizing and Proselytizing

Although they were mollified by the new, more pluralistic policies regarding the Unified Protestant Sunday School Curriculum and the General Protestant Service, evangelical chaplains did not abandon their sectarian orientation in the 1980s and 1990s. The demise of Protestant Christianity as the "national religion," the diversification of religious faith in America and the armed forces, and the consequent increase in secularism alarmed evangelicals in both the military and civilian communities. In their view, such developments made evangelizing and proselytizing even more imperative. Another factor was the military's new emphasis, in the aftermath of *Katcoff v. Marsh,* on religious accommodation and free exercise of religion. The chiefs of chaplains and chaplain field manuals not only recognized the extent of religious pluralism but preached the value of diversity and soldiers' right to the free exercise of religion (including atheism or no religion), which seemed to laud the very developments evangelicals found so alarming. At the same time, however, they also acknowledged *chaplains'* right to free exercise, which evangelicals had invoked in challenging the requirements of the GPS and the Unified Curriculum. Now, enshrined in DoD directives and army regulations, the free exercise principle offered them and their civilian allies the perfect, seemingly indisputable rationale for a concerted effort to combat secularism and spread "true Christianity" throughout the army.[39]

Chaplains' evangelism and proselytizing had not been a major issue in the Chaplain Corps in previous decades, except on occasions when a supervisory chaplain raised questions about the use of evangelistic doctrines or practices in the General Protestant Service. Most liberal, mainline Protestants had abandoned evangelistic activity in the 1960s and shifted their energies to social activism, although some conservative mainline Protestants still believed in it.[40] During the 1960s and 1970s, OCCH and military authorities tolerated evangelical chaplains' soul-winning efforts, so long as they followed certain generally accepted, unwritten guidelines.

Traditionally, and by custom, evangelism was defined as "proclaiming" the gospel. Proselytizing involved a specific effort to persuade an individual to join or convert to a denomination or faith group. The unwritten guidelines allowed a chaplain to engage in the act of persuasion if the individual were "unchurched" (belonged to no church or faith group), voluntarily expressed interest in the chaplain's denomination or faith group, or attended a service or program led by the chaplain. But if the individual were not unchurched, had not expressed interest in the chaplain's faith or voluntarily met with him, then the chaplain was guilty of "sheep stealing" (persuading an individual to leave a religion to which he or she was actively committed and to join another), and this was generally frowned upon, albeit not officially prohibited.

This was the position OCCH took in 1969, in a statement published in its monthly newsletter to chaplains. It referred to the objectionable kind of proselytizing as "active proselytism" and declared it to be "impolitic within the military framework." It presented a guarded endorsement of the acceptable kind of proselytism, urging chaplains to exercise caution in dealing with individuals who had expressed an interest in the chaplain's faith or a desire to convert. The chaplain could counsel them, but he should "make every attempt to have the potential convert establish contact with a chaplain or civilian minister of his denomination before embarking on a course of instruction leading to conversion." If the would-be convert were under twenty-one or seemed "emotionally immature," the chaplain should inform his or her parents or next of kin of the desire to convert.[41]

Evangelical chaplains and their endorsing agencies were quite sensitive to any scrutiny of or directives issued regarding their evangelistic activity. Floyd Robertson, the executive secretary of the NAE Commission on Chaplains, sent OCCH a lengthy rejoinder regarding its statement on "proselytism." According to the summary presented in the 1968–69 *Historical Review,* "he granted that chaplains should refrain from unethical and objectionable types of proselytism, but insisted that a chaplain has complete freedom to present the claims of the Gospel to anyone who seeks spiritual help." Robertson made clear his dislike of official OCCH guidance on the matter. He conceded that there were times when a chaplain should follow the directive contained in the monthly newsletter, but he hoped that OCCH would generally rely on the chaplain's "discretion." If questions should arise as to the chaplain's judgment, his endorsing denomination should provide "appropriate guidelines." Such an arrangement, Robertson noted, "would obviate the sensitive church-and-state issue."[42]

During the 1970s, OCCH actually sanctioned evangelism, by encouraging chaplains to work with various evangelical parachurch groups, such as Youth for Christ and Campus Crusade for Christ, in planning religious activities on army bases. In 1973, Chief of Chaplains Gerhardt Hyatt joined with the navy and air force chiefs of chaplains in urging chaplains to play a "strong role" in Key 73, a year-long, ecumenical evangelistic campaign that involved Roman Catholics and mainline Protestants as well as evangelicals, fundamentalists, and pentecostals.[43]

But then, in the 1980s and 1990s, religious pluralism in the armed forces and the military's new focus on religious accommodation and the right of free exercise changed many military leaders' and chaplains' perception of evangelism and proselytizing. Activities once tolerated or encouraged became a matter of contention. NAE-endorsed chaplains began to complain that their advocacy of evangelism was being "distorted or violated" by "ecumenical or pluralistic" chaplaincy decisions. A fundamentalist chaplain reported that "Bible-believing chaplains" were experiencing "increased discrimination" and "pressure" on the part of "theologically liberal chaplains and compromising conservative chaplains" who objected to their evangelistic "witnessing." A Seventh-day Adventist chaplain pointed out that many Adventist chaplains felt that "military strictures" prevented them from engaging in "overt" forms of evangelism and, as a consequence, they were developing "covert methods."[44]

Evangelical chaplains' endorsing bodies continued to support evangelism and proselytizing. A 1983 Christian and Missionary Alliance chaplain manual seemed to agree with the customary prohibition against "sheep stealing." It declared that "Alliance chaplains will seek opportunities to minister to the *unchurched,* encouraging them to make a Christian commitment." But it was more strongly and absolutely opposed to OCCH directives than Floyd Robertson had been in 1969. "The chaplains' only religious authority is that given them by their church body," the C&MA insisted. "Denominational limitations or religious authority *cannot be removed or changed by military command or military necessity.*"[45]

In the 1980s and 1990s, religious pluralism, religious accommodation, and recognition of the right of free exercise not only raised questions about the propriety of evangelism and proselytizing, but also about the very meaning of such terms and formulations. Evangelical chaplains generally distinguished between evangelism and proselytizing, and between acceptable and objectionable proselytizing. Opponents of evangelical sectarianism tended to conflate evangelism and proselytizing, regarding both as manifestations of an unacceptable, exclusivist point of view. The disagreement was partly a result of longstanding theological differences between the two camps, but it was also a consequence of the new religious context in which chaplains ministered. Given increasing religious pluralism in the military community, what did it mean to say that chaplains should limit their proselytizing to the "unchurched"? Did that term apply to individuals who had a "preference" for, but were not members of, a religion or church? Did it include those who identified themselves as "none" or "no religion," or "atheist" or "agnostic"? The military's emphasis on religious accommodation and free exercise posed yet another question. Did a soldier's right of free exercise preclude any and all efforts to persuade him or her to adopt or convert to a particular religious faith or denomination, regardless of his or her current religious status? And, the most difficult question of all, did the chaplain's right to free exercise—as well as his loyalty to the tenets of his endorsing agency—trump the soldier's right to free exercise?

In the 1980s and 1990s, the *Military Chaplains' Review* published several articles that give an idea of whether and to what extent evangelical chaplains were adjusting their approach to evangelism and proselytizing to the new military and religious context.[46] Not surprisingly, none of the authors, or the other chaplains they quoted, endorsed objectionable proselytizing. A few even made a point of criticizing "coercive or repressive" proselytizing. In using the word "evangelism," some of them signaled a growing tendency on the part of evangelicals to conflate it with the acceptable kind of proselytizing. Thus Roy N. Mathis, a staff chaplain at TRADOC, defined evangelism as "seeking to share the gospel of Jesus Christ with soldiers and their families who do not have a faith, a church association or commitment to any religious organization." "Sharing" not only suggested more of a one-on-one relationship between the chaplain and a potential convert but also dispelled any sense of coercion or pressure on the part of the chaplain. Another synonym evangelical chaplains used for evangelism was "witnessing," which could mean both sharing and the traditional "proclaiming."[47]

In the *Military Chaplains' Review* articles, all the authors or the chaplains they quoted seemed intent upon making a case for "evangelism" in the new context of religious accommodation. But they were not all of one mind. Some advocated what might be called "old school" (pre-religious accommodation) evangelism and proselytizing, while others presented a "new school" rationale tailored to the directives and regulations promulgated in the 1980s and 1990s.

A National Guard chaplain and a command chaplain who had recently served in Saudi Arabia defended the old school approach. The National Guard chaplain declared, "Every time I preach, I ask people to respond, to make a public decision, a faith commitment, and to invite Christ into their lives. . . . I've closed every worship service I've led by giving an invitation, to exhort people to act on what they have heard and felt in mind and heart, . . . and I plan to continue." The command chaplain exhorted his fellow evangelical chaplains to be "true" to their "calling." "Those of us who come from an evangelical background need to be true to our God, true to our denominations, true to our church," he asserted, "and this means evangelism." They needed to "get over [their] timidity" and make their distinctive doctrines part of religious pluralism. They needed to "emphasize their personal traditions," just as "other faith groups" were doing in the army. Could they announce altar calls in General Protestant Services? "Yes, of course you can!" he replied, although he admitted that sometimes it was a "sensitive issue." He believed that supervisory and senior chaplains needed to "provide guidance, support and encouragement to younger and lower ranking chaplains for evangelism."[48]

Chaplains who favored the old school approach often used counseling sessions for evangelism. The National Guard chaplain pointed out that the soldiers who came to see him for counseling often articulated "a perceived need." Since he believed that "the greatest need is to have a personal faith, and a purpose and power for living through Jesus Christ," he sought to help the soldier by "sharing Jesus Christ and the Gospel" with him. Soldiers often had "a desire, a hunger to know God in a personal way," he noted. Indeed, he estimated that some 75 percent of the members of his unit did not have "a relationship with any religious organization, or church."[49] Chaplain Robert G. Leroe, endorsed by the Conservative Congregational Christian Conference, also observed that soldiers who sought counseling with a chaplain often had "little spiritual background or interest." He would "occasionally" say to them, "'Let's talk for a moment about the possible spiritual dimension of your concern." He justified this approach by saying, "If we believe that there is a word from the Lord for those who seek out our help, we are neglecting their needs if we keep silent. . . . we are obliged to share our faith. If we do not somehow communicate the love and truth of God, we are doing social work and not ministry."[50]

The new school approach to evangelism was less exclusivist and more pluralistic, revealing the influence of religious accommodation and the emphasis on free exercise. Chaplain Jerry E. Malone, another clergyman of the Conservative Congregational Christian Conference, provides a good example of such thinking. On the one hand, he identified himself as "mission oriented" and opposed to those who would "limit my freedom to express evangelism." But he also observed that the United States constituted a pluralistic, "religiously neutral" society. In his view, religious neutrality meant that the government could not "control . . . present and future forms of evangelistic outreach." Citing the "religious rights of soldiers and their families" and the chaplain's duty to "provide for the free exercise of religion of Army personnel," he contended that evangelism must be allowed. But he stated his opposition to "coercive or repressive" proselytizing and the "degrading" of individuals' "personal faith." Indeed, he claimed to be an "advocate" of "those who hold to the equal validity of many religions." He urged chaplains, "as the Army's defenders of faith and freedom," to show "high esteem and value (respecting what is right in the sight of all men, Romans 12:17) for all persons, particularly minority religious views and views which are offensive or contrary to our own."[51]

Chaplain Rick D. Mathis, a Foursquare Gospel minister endorsed by the NAE who had formerly served in the Judge Advocate General Corps, presented a variation on the new school approach. He argued that "constitutional guidelines" required all chaplains to exercise "constant vigilance in defending religious pluralism and advocating free exercise rights." He recognized that some evangelical chaplains might have a problem with that responsibility. "Can a chaplain evangelist still be true to the gospel, while ensuring that the needs of Muslims and Hindus, among others, are met?" he asked. "The question is answered by his or her office as a military chaplain. Constitutionally, the chaplain evangelist *must* ensure that the needs of all are met. If the chaplain evangelist believes that the gospel prohibits obeying this requirement, then he or she must seek employment outside the military."

Mathis himself clearly had no problem with the requirement. He did not believe that protecting and facilitating other faiths threatened his own religion. Indeed, he argued that by helping others to exercise their religious faith, he ensured his own freedom to practice and proclaim the gospel as he understood it. He believed that free exercise, by chaplains as well as other military personnel, was a "God-given liberty" as well as a First Amendment guarantee and that it included the right to evangelize. He thought it wrong to denigrate other religions ("seeking to win the lost by assaulting other religions"). However, unlike Malone, Mathis did not believe in the "equal validity of many religions" and he still harbored the sectarian spirit. In an interview published in the NAE Commission on Chaplains *Centurion,* he said that one of the reasons he supported pluralism was because he believed that "in a pluralistic religious setting, when freely presented in a ministry forum unrestricted by abuse or restraint, the gospel will triumph over any false gospel."[52]

In the *Military Chaplains' Review* articles, evangelical chaplains defended their free exercise right, which they believed included the right to evangelize, on both sacred and secular grounds—as "God-given," commanded by Jesus in the "Great Commission," and "legitimized" by the First Amendment to the Constitution. They saw no conflict between their right and soldiers' right to free exercise. Indeed, two of the chaplains stressed their reciprocal relationship. Chaplain Rick Mathis asserted that by enhancing soldiers' right to free exercise, he also ensured his own free exercise right. The TRADOC staff chaplain, Roy Mathis, contended that unless chaplains engaged in evangelism—by offering the "invitation" at a GPS service or encouraging worshippers to fill out a "decision card"—they would actually be denying free exercise to "a large segment of people."[53]

Debating Public Prayer

The implications of soldiers' and chaplains' right of free exercise in the military environment were central to another issue debated in the 1980s and 1990s—the content of the "public prayer" chaplains offered at joint, ecumenical, or interfaith religious services and military or patriotic ceremonies. Traditionally, the prayers were supposed to be nonsectarian, since the gathering would be composed of individuals of varying faiths, or no faith, and/or because attendance was mandatory. Chaplains who did not wish to participate in or offer prayers at such events, on the grounds that doing so would violate their religious beliefs, were granted an exemption. In the 1980s, the challenge of religious pluralism in the military and the new emphasis on religious accommodation and free exercise, combined with a spate of Supreme Court decisions involving the First Amendment, focused chaplains' attention on public prayer. There was much discussion of the issue, especially the matter of evangelical chaplains "praying in the name of Jesus."[54]

In 1987, the *Military Chaplains' Review* published two seminal articles on chaplains and public prayer, one by a Navy chaplain, Rabbi Arnold E. Resnicoff, and the other by a retired chaplain, Bertram Gilbert. The title and language of Resnicoff's article, "Prayers that Hurt: Public Prayer in Interfaith Settings," surely evoked surprise, perhaps even consternation, among the journal's readers. In personal, sometimes passionate language, Resnicoff described his feelings upon receiving or hearing some of the prayers of his "fellow chaplains." After his father had died, a chaplain had written to him "with the prayer that I would accept the resurrection of Jesus." Although "many other notes from chaplains brought me comfort during that time of grief, this letter brought me pain," Resnicoff wrote. He recalled being part of a group of navy chaplains whose leader exhorted them to work together as a team and invited them to join in a moment of prayer, which he ended in the name of the Trinity. "I could not add my Amen," Resnicoff observed. "Hadn't I been invited to pray with the group? I felt out of place. I wanted to be a part of this prayer, . . . and it hurt me that I could not." Resnicoff also told of chaplains who had let him know that they could not pray with him in interfaith gatherings, in effect saying, "we may work together, but we cannot face God together, as servants or as children, not even for a moment."

"Does not each of us have the right to pray as he or she pleases?" Resnicoff asked. In answering that question, he emphasized "the right of the listeners" as against "the right of the speaker." He believed that a chaplain who for reasons of conscience or faith could not offer "a 'general' prayer" had the right to decline an invitation to participate in a public ceremony or gathering. But, Resnicoff declared, "if we accept the invitation . . . we have a responsibility to understand that we have been asked to add a reminder of the holy and challenged to touch and inspire those present through a moment of shared prayer. We have a responsibility to our conscience and our faith, but we also have a responsibility to those before whom we stand. Neither can be ignored."[55]

Bertram Gilbert, who had served as a chaplain in both the army and navy, was a director of the chaplain endorsing agency of the Lutheran Council of the USA when he wrote his article, "On Prayers in Jesus' Name." He believed that "no one may or should tell a chaplain how to pray" and that chaplains had a "basic right to pray as conscience leads." But he urged chaplains, in composing their prayers, to consider the new religious and cultural context in which they worked. In the late 1980s, Gilbert pointed out, people were more sensitive to exclusivist language of all kinds—racial, sexual, and religious. "It will not suffice simply to deny guilt or claim that our way of praying is not intended to denigrate other faiths or to suggest superiority," he explained. However a chaplain defended his choice of words, "it will be possible and even legitimate for someone to claim, 'Your prayer offended me.'" So while the chaplain had a "right" to pray according to his conscience or faith, and no one could or should tell him or her how to pray, it was necessary for the chaplain to consider how the prayer would be understood by individuals who were likely to be sensitive to exclusivist language: "The chaplain must project the event in his or her imagination to ask what words will be helpful, soul enriching, inspiring, and what words will cause all the hearers present, no matter their ecclesiastical identity, to say 'Amen.'"[56]

Neither Resnicoff nor Gilbert directly addressed the matter of praying in Jesus's name, which was at the heart of the debate over public prayer. In 1992, an evangelical chaplain, Robert G. Leroe, spoke to that issue in "Public Prayer," a *Military Chaplains' Review* article. Like Resnicoff and Gilbert, Leroe emphasized the context of public prayer, which made it different from private prayer and the prayers offered during a worship service. "When praying in public, one is addressing the concerns of the corporate body" and "interceding for the congregation," he wrote. Unlike those two chaplains, he offered specific guidance on closing the prayer. "There is great beauty in a benediction or doxology," he observed. "Amen" was "certainly appropriate" but not required. "Neither is the phrase 'in Jesus name' mandatory to close a prayer," he asserted. "The spirit and content of prayer will determine whether we are praying in Christ's name."[57]

In the early 1990s, two of the largest chaplain endorsing agencies issued statements that had a significant bearing on chaplains' debates regarding evangelizing/proselytizing and public prayer. One was a "Code of Ethics" for chaplains promulgated by the National Conference on Ministry to the Armed Forces (NCMAF), formerly the General Commission on Chaplains and Armed Forces Personnel, whose membership embraced more than one hundred ecclesiastical bodies representing a very broad theological spectrum.[58] Several of the pledges in the code invoked the principles of religious accommodation and free exercise. Thus the chaplain acknowledged that he or she functioned "in a pluralistic environment with chaplains of other religious bodies to provide for ministry to all military personnel and their families." This meant, among other things, recognizing the chaplain's "obligation" to "provide for" their free exercise of religion; working "collegially with chaplains of religious bodies other than my own"; and respecting "the beliefs and traditions of my colleagues and those to whom I minister." Although it did not mention the General Protestant Service by name, the code pledged the chaplain to conduct a nondenominational service when leading worship for "persons of other than my religious body," based upon "those beliefs, principles, and practices that we have in common." The code spoke directly, albeit somewhat ambiguously, to the ongoing discussion of evangelism/proselytizing. It declared, "I will not proselytize from other religious bodies, but I retain the right to evangelize those who are not affiliated."[59]

In November 1992, the NAE Commission on Chaplains issued two policy statements regarding evangelism/proselytizing and public prayer, "Guidelines for Cooperation with Chaplains of Other Faiths" and "Freedom of Expression in Public Prayer: Guidelines for Evangelical Chaplains." In the "Guidelines for Cooperation," the commission warned its chaplains of "mounting pressures for a more ecumenical approach to chaplain ministry" and increasing stress on "a broader form of worship," which deemphasized evangelicals' distinctive worship practices. It noted that evangelical chaplains were being encouraged to engage in jointly led worship services with chaplains of different beliefs. As in the past, the endorsing agency assured evangelical chaplains that they were not required to compromise their "personal faith convictions" in the conduct of religious services and that they had the right "to determine the extent of [their] cooperation in jointly conducted services." While the commission advised its chaplains to lean, whenever possible, toward "cooperation rather than non-cooperation" and to "avoid entering into an adversarial relationship," it concluded the guidelines with an unequivocal offer of assistance to NAE chaplains who felt their rights had been violated by an "ecumenical or pluralistic military chaplaincy decision." They "can expect full support from the Commission in any case where the action taken is consistent with the Commission position," the commission declared.[60]

In its "Freedom of Expression in Public Prayer" statement, the NAE commission observed that in recent years, some evangelical chaplains had felt pressured to avoid mentioning Jesus's name on the grounds that some people in the gathering or audience might be offended. The commission was adamantly opposed to any effort by the chaplaincies to discourage particular forms of prayer. Such action would constitute a violation of a chaplain's free exercise right, which included his freedom "to pray in a manner consistent with his faith." The commission recognized the problem this position posed: how to reconcile the chaplain's right to offer a prayer "consistent with his faith" with the service member's right to be protected "from being forced to listen to such a prayer if he or she chooses not to do so." The solution the commission proposed was for chaplains to observe the distinction between mandatory formations and voluntary gatherings. In the case of a voluntary gathering, proscribing any form of prayer would be inappropriate, and "it would appear that it would be perfectly legitimate to pray 'in Jesus name,'" since those attending the gathering would or should recognize, and accept, that the chaplain was offering the prayer based on his or her own faith. "In a mandatory formation, there is some question as to whether *any* prayer is possible," the commission observed. (It may have been thinking of the implications of the recent Supreme Court decision, *Lee v. Weisman,* barring prayers at public school graduation ceremonies because they indirectly coerced religious observance.) Still, the commission continued, "based upon current practice, it would appear . . . that some form of prayer is possible with minimal or no offense."[61] But the commission did not venture any suggestions regarding the content of such prayer.

The NCMAF "Code of Ethics" and the NAE Commission on Chaplains statements were an important first step toward resolving issues raised in the debates over evangelism/proselytizing and public prayer. Neither organization offered a comprehensive solution. The NCMAF presented an official, written definition of the rule regarding proselytizing, and the NAE pointed out the tension between soldiers' and chaplains' right of free exercise. Much more reflection, analysis, and clarification were needed.

In 2002, the NAE Commission on Chaplains began work on another statement regarding public prayer. It went through several drafts, including one titled "Praying in Jesus' Name," and was finally issued on February 7, 2006, as

"The National Association of Evangelicals Statement on Religious Freedom for Soldiers and Military Chaplains."[62] By that time, chaplains' rights regarding evangelism and public prayer had become the focus of a civilian-dominated culture war that, in effect, hijacked chaplains' in-house discussion of those matters, in the navy and air force as well as the army. As we shall see in the next chapter, the culture war was fought primarily in the "public square" and mainly by advocates of the liberal-pluralist and conservative-sectarian cultural strategies, each group defining chaplains' rights and responsibilities in accord with its particular religious and cultural perspective. Not since Vietnam had chaplains undergone so bruising an experience.

10
The 2005–2006 Culture War

The 2005–2006 culture war developed in the wake of a much publicized "scandal" at the U.S. Air Force Academy (USAFA) that revealed numerous instances of religious intolerance and evangelical proselytizing by administrators, faculty, staff, chaplains, and cadets.[1] Liberal-pluralist and conservative-sectarian cultural strategists quickly mobilized their constituencies in the civilian community. Initially, they focused on the "religious climate" at the USAFA, but they soon turned their attention to a more heated, general debate regarding chaplains' evangelism, proselytizing, and public prayer in the armed forces.

The Culture War Begins

The first sign of impending culture war appeared in the midst of debate over a defense authorization bill in the U.S. House of Representatives two days before an air force task force released its report on the academy "scandal." On June 20, 2005, House members were considering an amendment to the bill asking the secretary of the air force to "develop a plan to ensure that the Air Force Academy maintains a climate free from coercive intimidation and inappropriate proselytizing." Speaking in its favor, the author of the amendment, David Obey (D-Wisc.), exhorted his colleagues to make a strong statement that all cadets at the academy "can practice their religion without fear." In the ensuing debate, Indiana congressman John Hostettler (R-Ind.) blasted the proposed legislation as part of a "long war on Christianity" being waged by "those who would eradicate any vestige of our Christian heritage." He specifically accused the Democrats of aiding and abetting that war by "denigrating and demonizing Christians." When he uttered those words, the House "erupted in shouts and finger-pointing," forcing a thirty-minute halt in deliberations. The Obey amendment failed; by a voice vote the House approved Republican language supporting "freedom for religious expression for all faiths" in the military.[2]

The culture war began in earnest in late August 2005, when the air force issued a document titled "Interim Guidelines concerning Free Exercise of Religion in the Air Force." The team that formulated the guidelines was headed by a born-again Christian, Lt. Gen. Roger A. Brady, air force deputy chief of staff, personnel.[3] The air force emphasized that these were "interim" recommendations, not official regulations, and that they were proposed not just for the academy but for all air force personnel. It encouraged discussion and commentary by air force personnel and invited civilian groups and organizations representing a wide spectrum of views to express their opinion of the guidelines.[4]

The "Interim Guidelines" contained a section regarding the officer corps and "religious expression," which concluded with the following declaration: "Abuse or disrespect of . . . our fellow Air Force people[,] including disrespect based on religious beliefs or the absence of religious beliefs, is unacceptable."

The most controversial sections of the guidelines were the ones addressed to chaplains. They did not mention the words "proselytizing" and "evangelizing," but they did remind chaplains that they were commissioned to minister to individuals of all faiths, "including those of no religious faith," that they "should respect the rights of others to their own religious beliefs, including the right to hold no beliefs," and that they "must be as sensitive to those who do not welcome offerings of faith, as they are generous in sharing their faith with those who do."

The guidelines set fairly explicit limits on chaplains' public prayer. First, "public prayer should not usually be included in official settings such as staff meetings, office meetings, classes or officially sanctioned activities such as sports events or practice sessions." Second, "consistent with longstanding military tradition, a brief nonsectarian prayer may be included in nonroutine military ceremonies or events of special importance, such as a change-of-command, promotion ceremonies, or significant celebrations, where the purpose of the prayer is to add a heightened sense of seriousness or solemnity, not to advance specific religious beliefs." Chaplains were advised that "they should respect professional settings where mandatory participation may make expressions of religious faith inappropriate."[5]

Liberal advocacy groups in the civilian community reacted quickly to the "Interim Guidelines." Americans United for the Separation of Church and State (AU) welcomed them as "an important step toward increasing religious tolerance in the military" but criticized the lack of an explicit prohibition against proselytizing by chaplains and officers. AU also deprecated the "vague" wording of the guidelines on public prayer. Abraham Foxman of the Anti-Defamation League (ADL) expressed concern as to how the guidelines would be implemented and whether individuals would be held responsible for violating them.[6]

Civilian evangelicals criticized the guidelines for being too restrictive. Throughout the culture war, their main concern was to prevent limitations on *their* religious expression, not to check religious intolerance at the academy or in the air force or the armed forces as a whole. They had the support of a number of powerful conservative-sectarian advocacy groups, including two Christian Right organizations, Focus on the Family (FOF) and the Christian Coalition of America. Both were politically active, tax-exempt organizations known to millions of Americans for their ideologically conservative, "pro-family" pronouncements on various social and political issues. Another conservative pro-life, religious-freedom organization, the American Center for Law and Justice (ACLJ), headed by Chief Counsel Jay Sekulow, assisted the evangelicals with legal issues relating to religion in the military. The three groups exhorted their followers to protest the "Interim Guidelines" by signing petitions, writing letters, or sending e-mails addressed to Congress, the secretary of the air force, and the White House.[7]

Debating Proselytizing and Evangelism

The culture war involved two main battles. In one, participants fought over chaplains' right to evangelize and proselytize; in the other, they battled over the propriety and content of chaplains' public prayer. The leader of the battle against proselytizing was an Albuquerque attorney, Michael L. "Mikey" Weinstein, a 1977 graduate of the USAFA, a Jew, a strong believer in separation of church and state, and a member of Americans United. In an interview published in the *Albuquerque Journal* in May 2005 explaining why he had decided to speak out about the USAFA "scandal," he said, "This is about the most basic, fundamental aspect of why this country was founded. We don't want government telling us what to do in our religion." He noted that "some people will . . . think I'm trying to deliver a blow to Christian evangelicals, but I'm not. This is about the law, and the academy is breaking the law."[8]

Weinstein did not accept the distinction conservative evangelicals made between proselytizing and evangelizing. He dismissed "evangelizing" and "evangelization" as "overt Christian terminology for the proselytizing process." He also questioned evangelicals' use of the terms "unchurched" and "unaffiliated" in declaring the propriety of evangelizing. "If I attend a synagogue or mosque, am I unchurched? . . . What if I never set foot in a recognized house of worship but gain inner peace and understanding hiking in the forest, reading the great philosophers, or sitting quietly in contemplation? Who defines 'unchurched'?" he asked.[9]

Weinstein presented this critique in reaction to a statement Air Force Deputy Chief of Chaplains Cecil R. Richardson had made in an interview with *New York Times* reporter Laurie Goodstein. Speaking as an evangelical, Richardson declared, "We will not proselytize, but we reserve the right to evangelize the unchurched." He explained the difference between the two actions by saying that proselytizing was "trying to convert someone in an aggressive way," while evangelizing was "more gently sharing the gospel." In the same interview, he denied that the impulse to evangelize compromised evangelical chaplains' ability to facilitate religious accommodation for all. He told Goodstein, "I am an Assemblies of God, pound-the-pulpit preacher, but I'll go to the ropes for the Wiccan" if they requested permission to hold a religious ritual.[10]

Civilian evangelicals involved in the culture war defended chaplains' "evangelizing" on similar grounds. Tom Minnery, vice president of public policy for Focus on the Family, told Alan Cooperman of the *Washington Post,* "It's the job of an evangelical Christian chaplain to evangelize." And, he added, evangelization is "protected by the First Amendment's guarantee of free exercise of religion."[11] He and other evangelicals also accepted and claimed to observe the distinction between evangelizing and proselytizing. Indeed, this distinction was widely acknowledged among Christians.[12]

Civilian evangelicals had significant support from Air Force Deputy Chief of Chaplains Richardson and Air Force Chief of Chaplains Charles C. Baldwin. Baldwin, a Southern Baptist, sent a videotaped message to active-duty and reserve chaplains and their assistants suggesting that the "Interim Guidelines" needed to be changed and inviting them to submit "feedback" to help the air force "get this right." Interpreting the guidelines for his chaplains, he advised them that they could share their faith in a noncoercive way and that "prayers, hymns, 'life lessons' or scripture readings" were still permissible at routine staff meetings. "This is America," he declared in the videotape, "and for those of us who come from belief systems that require us to tell others of our faith and what we believe, [it] is so important that we feel free to do this. Just have to put it in the right context and never again coerce anyone to believe something that they don't want to believe."[13]

Even before the air force released the "Interim Guidelines," some evangelicals had worried that chaplains' evangelizing would be limited or prohibited. Reverend Ted Haggard, president of the National Association of Evangelicals, told Eric Gorski of the *Denver Post* that he and other evangelicals feared the air force would erect "unconstitutional barriers" against "the practice of evangelical Christianity, which teaches that the faith should be shared with others."[14] When the guidelines made no mention of proselytizing or evangelizing, evangelicals felt relieved. But then they learned that on October 6, 2005, Mikey Weinstein had filed suit against the air force and its acting secretary, seeking an injunction to prohibit all members of the air force, including chaplains, from evangelizing and proselytizing or in any related way attempting "to involuntarily convert, pressure, exhort or persuade a fellow member of the USAF to accept their own religious beliefs while on duty."[15] In February 2006, Haggard filed a motion to intervene in the suit on behalf of the NAE Chaplains Commission. "The injunctions requested by the Plaintiffs directly impair the goals and interests of the NAE, its Commission, and its military chaplains" by "restricting the ability of its chaplains to minister according to the stated beliefs and doctrines of the NAE, their endorsing body," he declared. He specifically cited the necessity of "sharing . . . one's faith," advancing "the name of Jesus Christ through united evangelical action," and proclaiming "a biblical faith" throughout the world.[16]

Haggard's concern extended beyond the NAE chaplains. "This is a critical case," he observed. "Its outcome has ramifications for the future of religious expression and activity in the military context." He viewed both the "Interim Guidelines" and the Weinstein lawsuit as part of a general "attack upon the very existence of religious expression within the military context."[17] This became a major theme of conservative evangelical discourse during the culture war.

Weinstein's response to Haggard's declaration was, "He's the one who's really trying to suppress religious freedom by ensuring that one particular biblical worldview becomes the official biblical worldview of the U.S. government, and particularly the Department of Defense."[18] What made the culture war so intense was the conviction among individuals and groups on both sides that their opponents sought a truly radical transformation of the military. In the later stages of the war, Weinstein and his supporters maligned the evangelicals as "theocrats" and "Christian Taliban" who had effected an "evangelical coup" in the armed forces in order to create "a spiritually transformed military" made up of "ambassadors of Christ in uniform."[19] By the same token, evangelicals vilified their opponents as "secularists" who sought to eradicate religious expression in the military and make it a "wasteland of relativism."[20]

Debating Public Prayer

Many civilian and military evangelicals deemed the issue of chaplains' public prayer more serious than proselytizing and evangelizing. The "Interim Guidelines" advisory, that on certain occasions chaplains should offer "a brief non-sectarian prayer," especially rankled evangelicals. They contended that it forced evangelical chaplains to compromise their religious beliefs, because it prohibited them from praying in the name of Jesus, which their beliefs and endorsing agencies demanded they do, and which, furthermore, was their right, guaranteed by the First Amendment free exercise clause.

On October 25, 2005, seventy members of the House of Representatives sent a letter written by Walter B. Jones (R-N.C.) to President George W. Bush in which they expressed concern that "the right of military chaplains to pray according to their faith is in jeopardy" and asked him to issue an executive order protecting that right. As evidence of restrictions being placed on chaplains' praying, they cited the "Interim Guidelines." "We believe that the Air Force's suppression of religious freedom is a pervasive problem throughout our nation's Armed Forces," they wrote. "In all branches of the military it is becoming increasingly difficult for Christian chaplains to use the name of Jesus when praying. . . . We are deeply concerned that chaplains are now being instructed on what to say when they pray." They defended chaplains' right to "pray according to their faith" on the ground that it was their "constitutional right." Although the congressmen supported the right of all chaplains to "adhere to the religious expressions of their faith," they were clearly most concerned about the "Christian chaplains" who were having problems praying in Jesus's name. "For Christian chaplains," the congressmen explained, "praying in the name of Jesus is a fundamental part of their belief and to suppress this form of expression would be a violation of religious freedom."[21]

It is important to note that the congressmen who sought the executive order were protesting restrictions on chaplains "in all branches of the military." A *Washington Times* reporter quoted Representative Todd Akin (R-Mo.) as saying that they were not limited to the air force, had become "more widespread," and were part of "a pattern of hostility to freedom of speech."[22] With the help of the ACLJ, Jones and his fellow congressmen mounted a petition campaign in support of the executive order, ultimately collecting almost 160,000 citizens' signatures.[23]

In addition, twenty-nine prominent evangelical leaders, including Richard Land, president of the Ethics and Religious Liberty Commission of the Southern Baptist Convention, and Robert E. Reccord, president of the North American Mission Board, wrote to President Bush asking him to issue the order. In their letter they invoked "the freedoms that have made this country great," especially "the freedom of religion that was seminal at the settling and founding of this beacon of freedom called the United States."[24]

New Air Force and Navy Guidelines

Resistance on the part of the air force's Office of the Chief of Chaplains; the e-mail, letter-writing, and petition campaigns mounted by Focus on the Family, the Christian Coalition, and the ACLJ; and Congressman Jones's letter to President Bush and the media attention it received—all had their intended effect. On February 9, 2006, yielding to the pressure, the air force issued the "Revised Interim Guidelines concerning Free Exercise of Religion in the Air Force." The document softened the earlier advisory to officers about expressing their religious beliefs and eliminated the admonition to chaplains to respect the rights of others to their religious beliefs or no belief and to be "sensitive to those who do not welcome offerings of faith." It added to the section on chaplains the following declaration: "We will respect the rights of chaplains to adhere to the tenets of their religious faiths, and they will not be required to participate in religious activities, including public prayer, inconsistent with their faiths." The recommendations regarding public prayer, though more briefly stated, remained essentially unchanged. They read as follows:

> Public prayer should not imply government endorsement of religion and should not usually be a part of routine official business. Mutual respect and common sense should always be applied, including consideration of unusual circumstances and the needs of the command. Further, non-denominational, inclusive prayer or a moment of silence may be appropriate for military ceremonies or events of special importance when its primary purpose is not the advancement of religious beliefs.[25]

By this time, the U.S. Navy was publicly involved in the debate over public prayer. On January 19, 2006, Navy Chief of Chaplains Louis V. Iasiello had released the "Chief of Navy Chaplains Official Statement on Public Prayer in the Navy." On

February 21, 2006, less than two weeks after the air force released the "Revised Interim Guidelines," Secretary of the Navy Donald C. Winter issued a policy statement on public prayer, "SECNAV INSTRUCTION 1730.7C" (SECNAVINST 1730.7C), titled "Religious Ministry within the Department of the Navy."

Both documents promulgated the following guidance. Navy policy recognized a distinction between "Divine Services" and "command functions," such as changes of command or retirement ceremonies that might feature "religious elements" such as "invocations or benedictions." Based on its commitment to the free exercise and establishment clauses of the First Amendment, the navy expected its chaplains, "as a condition of appointment," to be "willing to function in a pluralistic environment in the military, where diverse religious traditions exist side-by-side with tolerance and respect." Therefore, they should "respect the rights of others to their own religious beliefs, including the right to hold no beliefs." Chaplains "will not be compelled to participate in religious activities inconsistent with their beliefs." Thus they may choose not to participate in a command function, and without suffering "adverse consequences." However, if they do participate, they should, in discussion with their commander, "assess the setting and context of the function," taking into consideration "the diversity of faith that may be represented among the participants and whether the function is mandatory for all hands." And finally, SECNAVINST 1730.7C declared, "Other than Divine/Religious Services, religious elements for a command function, absent extraordinary circumstances, should be non-sectarian in nature."[26]

The Reaction to the Air Force and Navy Directives

The civilian reaction to the air force and navy directives varied and divisions of opinion developed within some faith groups. Most of the attention focused on the revised air force guidelines. The directors of the Anti-Defamation League and Americans United regarded the "Revised Interim Guidelines" as a "step backward" from the original ones and charged the Air Force with softening them to "appease . . . powerful Religious-Right lobbyists." Barry Lynn especially faulted the new guidelines for emphasizing the rights of chaplains rather than the rights of adherents of minority faiths and nonbelievers. It appeared that "the Air Force does not understand that all prayer, including so-called 'inclusive prayer,' is an inherently religious activity for which not all staff and cadets wish to be subject to."[27]

Mikey Weinstein also denigrated the revised guidelines as an attempt to "pacify the Religious Right" and "a colossal step backward." According to his interpretation, the guidelines "would seemingly allow for public prayer at mandatory military formations and, likewise, apparently establish the 'special' rights of chaplains to pray in the name of Jesus Christ at these same mandatory military formations." However, they provided "no protection for the rights of those who don't practice a majority faith and those who choose to not worship at all."[28] He faulted the Navy's SECNAVINST 1730.7C for, among other things, failing to "protect naval officers, enlisted personnel and midshipmen from coercive proselytizing and evangelizing by their superiors."[29]

Although the ADL director sharply criticized the "Revised Interim Guidelines," the leaders of the American Jewish Congress, the American Jewish Committee, and the Religious Action Center of Reform Judaism generally approved them. In a joint statement they did admit a preference for the earlier guidelines, because they were "more detailed." But they commended the new guidelines on two grounds: "They make clear the duty of chaplains to serve all members of the Air Force regardless of faith, and appropriately warn against abusing the position of trust chaplains enjoy to engage in unwelcome proselytizing."[30]

A rift also appeared among conservative evangelicals and their supporters. In 2005 they had presented a united front against the "Interim Guidelines." But Tom Minnery, FOF senior vice president for government and public policy, declared that the "Revised Interim Guidelines," if "applied properly, will safeguard the free exercise of religion guaranteed to all citizens, both military and civilian."[31] According to NAE president Ted Haggard's assessment, the new guidelines had restored the proper balance between the free exercise and establishment clauses.[32]

A more striking indication of division among evangelicals was a document the NAE had released on February 7, 2006, two days before the air force issued the revised guidelines. "The National Association of Evangelicals Statement on Religious Freedom for Soldiers and Military Chaplains" was written by the NAE's Commission on Chaplains and Military Personnel, one of the largest of the armed forces' chaplain-endorsing bodies. Some of the guidelines it presented were similar to those presented in the commission's 1992 document, "Freedom of Expression in Public Prayer: Guidelines for Evangelical Chaplains," which was discussed in chapter 9. The 2006 document was, however, much longer than the earlier advisory, and it presented extensive legal documentation to support its recommendations. Some of them, especially on public prayer, were similar to those in the "Revised Interim Guidelines" and SECNAVINST 1730.7C.

For example, the NAE statement emphasized the importance of considering context when deciding what kind of prayer would be appropriate. "A military chaplain may preside, preach, or pray in sectarian language with a like-minded congregation that has voluntarily assembled," the commission declared. However, "when called upon to offer a blessing or conduct religious services for an interfaith group, or to offer a prayer at a mandatory military event, common courtesy, pastoral judgment, and constitutional principle commend offering a religious message or prayer respectful of all present." The purpose of "civic prayer" offered at military ceremonies was

> neither to favor one religion over another nor to proselytize. It is to dignify and mark a public occasion by reflecting upon the deeper significance of that which has or is about to transpire. It is to honor the most basic human impulses of giving thanks and of invoking God's protection, guidance, and blessing, and it is to reflect upon those religious values that unite the American people.

Civic prayer "should be non-sectarian," the commission advised. "One who would, in the setting of one's own congregation, pray 'in the name of Jesus,' or 'in the name of Allah,' or 'in the name of the Father, and of the Son, and of the Holy Spirit,' should consider if those formulations exclude believers from other faith traditions." It cited two court decisions indicating that such prayer would violate the establishment clause. The "preferred method of ceremonial prayer" was to use "wide and embracive terms" for God, but the statement also suggested alternative approaches. "A chaplain could identify his or her religious tradition and acknowledge that others present adhere to other religious traditions and then offer a prayer in keeping with civic traditions." Another suggestion was to take Psalm 19:14 as a guide: "Let the words of my mouth, and the meditation of my heart, be acceptable in thy sight, O Lord, my strength, and my redeemer." If a chaplain prayed in the spirit of the psalm, the statement explained, his words "could be religiously inclusive while the meditation of his or her heart would be deeply rooted in the chaplain's own faith tradition."[33]

The Hardline Conservative View

Congressman Walter Jones remained adamantly opposed to any restrictions on chaplains' public prayer. Interviewed by Alan Cooperman, Jones did cite two statements in the "Revised Interim Guidelines" that he regarded as "a step in the right direction": one about respecting "the rights of chaplains to adhere to the tenets of their religious faiths" and the other about allowing them not to "participate in religious activities, including public prayer, inconsistent with their faiths." However, he criticized the advisory that called for "nondenominational, inclusive prayer or a moment of silence" at military ceremonies. "There is some progress," he observed, "but it does not go as far as it needs to go in making sure that Christian chaplains can pray in the name of Jesus and other chaplains can pray according to their faiths."[34]

Jones and his supporters still hoped that President Bush would issue the executive order they had requested. Once they realized no such order was forthcoming, they decided to seek federal legislation to achieve their objective—a guarantee that Christian chaplains could pray, at any gatherings, in the name of Jesus Christ.[35] In May 2006, Jones and two other congressmen, Todd Akin (R-Mo.) and House Armed Services Committee chairman Duncan Hunter (R-Calif.), succeeded in attaching an amendment to the 2007 House of Representatives national defense authorization bill (Bill 5122) that granted chaplains "the prerogative to pray according to the dictates of the chaplain's own conscience, except as must be limited by military necessity, with any such limitation being imposed in the least restrictive manner feasible." The amendment did not distinguish between mandatory and voluntary gatherings or between worship services and military ceremonies. Nor did it explicitly state that chaplains could pray in Jesus's name. But the sponsors admitted that it was intended to give them that prerogative.[36]

On May 11, 2006, the House of Representatives passed Bill 5122, including the chaplain prayer amendment, by a vote of 396 to 31.[37] Explaining the need for the amendment, Representative Jones told reporters, "We felt there needed to be a clarification" of the rules "because there is political correctness creeping into the chaplains corps." ("Political correctness" presumably referred to the prayer guidelines issued by the air force and navy.) "I don't understand any one being opposed to a chaplain having the freedom to pray to God in the way his conscience calls him to pray," he added.[38]

The Senate passed Bill 5122 on June 22, but its version did not include the prayer amendment. So it was necessary for the Senate and House Armed Services Committees to meet in conference to reconcile the differences between the two versions. The chairmen of the two committees, Senator John Warner (R-Va.) and Representative Duncan Hunter (R-Calif.), negotiated the reconciliation.

During the time the prayer amendment was being formulated by the House Armed Services Committee and then when the House-Senate conferees considered it, several eminent military and religious officials voiced their objections. In May 2006, the Reverend Herman Keizer Jr., chairman of the National Conference on Ministry to the Armed Forces, and Navy Chief of Chaplains Louis Iasiello addressed letters or statements to the House Armed Services Committee. Both men expressed concern about the implications of the phrase in the amendment, "according to the dictates of the chaplain's own conscience."

Speaking in behalf of the NCMAF executive committee, Keizer pointed out that "chaplains represent their faith communities and we endorse them to represent that faith community with integrity and loyalty to that tradition, not to the dictates of their individual conscience." However, the prayer amendment phrasing "would violate the oversight of endorsers to maintain the accountability of chaplains to their faith community." Chief of Chaplains Iasiello expressed concern not only about this effect on the chaplain-endorsing agency relationship but also on the chaplain's responsibility to provide for or facilitate "the religious needs of all members of the command." In his opinion, the amendment's stress on "each chaplain doing that which is right in his or her own eyes" had the potential of driving a "wedge" between the chaplain on the one hand and the members of the command and his commander on the other. Since the chaplain's assignment was to help commanding officers administer the navy's religious ministries program, it was "essential that the chaplain possess the trust and respect of all the crew, not simply the members of his or her own faith group," he observed. But the amendment, "by allowing chaplains to lead prayers in nearly all situations, potentially independent of the endorsing faith group and legitimate concerns of the command and crew," would make them "independent agents operating outside the military command structure." Iasiello declared that "commanders, who must ensure good order and discipline in their commands, will have no choice but to limit chaplain access to the crew to preserve such good order, discipline and morale." That was why, in the conclusion of his letter, he observed that "this proposed legislation will, in the end marginalize chaplains and degrade their use and effectiveness to the crew and the commanding officer."[39]

During the House-Senate conference, Senator Warner invited the three chiefs of chaplains to speak to the conferees, only four of whom attended the meeting. Conference proceedings were confidential, but Warner did offer his "impression" of the chiefs' thinking—"that now is not the time to try to quickly put this one sentence [the amendment] into law by virtue of incorporating it into the final draft of the conference report." The conferees also heard from the

Department of Defense, which customarily transmitted its opinions of legislation affecting the military to members of the House and Senate Armed Services Committees. Warner quoted the DoD statement as follows: "This provision could marginalize chaplains who, in exercising their conscience, generate discomfort at mandatory formations. Such erosion of unit cohesion is avoided by the Military's present insistence on inclusive prayer at interfaith gatherings—something the House legislation would operate against. The Department urges exclusion of this provision."[40]

Ted Haggard sent a letter to the House-Senate conferees in which he asserted that the amendment was "unnecessary and likely counterproductive," insisting that "at this time our country needs no legislation in this area."[41] The Baptist Joint Committee for Religious Liberty joined with fourteen other religious organizations, including some Jewish groups and mainline Protestant denominations, in declaring their opposition to the amendment. They described it as "a clear attempt to undercut" air force guidelines drafted to curb religious intolerance. It was "unnecessary" because "under current law and regulations military chaplains are already permitted to pray in a manner fitting their individual religious tradition in the divine worship services they lead for armed forces members." When a prayer was called for in a "large-group setting or 'command ceremony' where attendance may not be voluntary," the interfaith group insisted that "chaplains should pray in a more inclusive manner." This was "common courtesy." If a chaplain did not "feel comfortable offering a non-sectarian, inclusive prayer in such a setting, he or she should have the right to refuse to participate without negative consequences." The group censured the language of the prayer amendment for being "deeply divisive" and for showing "a lack of respect for the diversity of religious beliefs in our military."[42]

The four Jewish groups that signed the Baptist Joint Committee statement were the American Jewish Committee, the Anti-Defamation League, the Jewish Council for Public Affairs, and the Union for Reform Judaism. Another Jewish organization, the Orthodox Union (OU), sent a letter to the armed services committees declaring its opposition to the prayer amendment. This action was especially significant because the OU was less inclined than other Jewish groups to take a public stand on religious debates involving the military and had in fact sided with Christian conservatives on some issues. The letter urged the armed services committees "to ensure that no language is included in the [national defense authorization] bill which upsets the delicate balance between the core religious liberties of either chaplains or the military personnel they serve." It advised them "to defer any legislation until your committees can have full and thorough hearings on the issues."[43]

The Republicans on the House Armed Services Committee seemed impervious to criticism of the prayer amendment and unimpressed by the knowledge and authority of the individuals and groups that offered it. They refused to delete the amendment or change its wording. Members of the Congressional Prayer Caucus, including Walter Jones and Todd Akin, helped them maintain their resolve. The caucus was a group of some forty members of the House of Representatives, founded by Representative J. Randy Forbes (R-Va.). One of its stated purposes was to "use the legislative process—both through sponsorship of affirmative legislation and through opposition to detrimental legislation—to assist the nation and its people in continuing to draw upon and benefit from [prayer,] this essential source of our strength and well-being."[44] On September 23, when the House Republicans were still stonewalling, Forbes was quoted in a Virginia newspaper saying, "There is clearly a well-funded, well-orchestrated movement that's anti-prayer and anti-religious in the military and society."[45]

Finally, on September 29, the House-Senate conference resolved the disagreement over the prayer amendment when its supporters agreed to eliminate it. The conferees then approved new language directing the secretaries of the air force and navy to rescind the "Revised Interim Guidelines" and Navy SECNAVINST 1730.7C and to reinstate the policies that were previously in effect. The House passed the revised Bill 5122 that evening, the Senate agreed to the conference report by unanimous consent, and President Bush signed the bill into law on October 17, 2006.[46]

The Significance of the Culture War

The 2005–2006 culture war provided the latest, and perhaps the most striking, illustration of chaplains' involvement in the momentous cultural/religious transition of the late twentieth century. The controversy over proselytizing and prayer was a much-publicized episode in the larger, ongoing culture war in the civilian community that engaged believers in the religion of democracy, inherited from the 1950s, as well as conservative-sectarian and liberal-pluralist cultural strategists and their supporters. The underlying issues were secularization and religious pluralism. The secularist/liberal religious side sought to accommodate increasing diversity of religious belief among armed forces personnel by prohibiting proselytizing, prescribing nonsectarian, inclusive prayer in settings other than religious worship, and requiring chaplains to respect individuals of all faiths, as well as those of no faith. The religious conservatives, civilians as well as chaplains, regarded secularization and religious pluralism as a major threat to their concept of America as a "Christian nation" and Protestant Christianity as the "majority faith." The "sinister" trend they detected in the military was emblematic of the assault on true religion they believed was taking place throughout America. Their solution was to legitimize—through legislation—military chaplains' absolute, unfettered right to "evangelize" and pray as they wished, regardless of any military or government regulations that provided for religious accommodation.

The culture war of 2005–2006 was also significant because it showed how vulnerable chaplains were to civilian criticism and controversy, especially when they themselves were divided on an issue. Just as during the Vietnam War, so during the culture war, chaplains became a "whipping boy" for cultural strategists and their supporters bent on achieving their own agenda. The culture war ended with congressional passage of the rescinding acts. But soon other issues arose that threatened a new outbreak of war.

Chaplain Charles Murphy teaches a class on "Religious and Moral Aspects of Citizenship" at Fort Knox, Kentucky, 1950. U.S. Army Chaplain Corps Museum, Fort Jackson, S.C.

Chaplain Alvin S. Bullen gives a Character Guidance lecture at Fort Jackson, South Carolina, 1950. U.S. Army Chaplain Corps Museum, Fort Jackson, S.C.

Baptism in the field, Korea. Courtesy of Rev. Bill Search.

Jewish field service during the Korean War. U.S. Army Chaplain Corps Museum, Fort Jackson, S.C.

Field service during the Vietnam War. U.S. Army Chaplain Corps Museum, Fort Jackson, S.C.

Catholic mass during the Vietnam War. U.S. Army Chaplain Corps Museum, Fort Jackson, S.C.

Baptism in the field, Vietnam. U.S. Army Chaplain Corps Museum, Fort Jackson, S.C.

Fit to Win

SPIRITUAL FITNESS

The cover of the U.S. Army's 1987 Spiritual Fitness program pamphlet depicts the Tomb of the Unknown Soldier at Arlington National Cemetery and a quotation from General George C. Marshall. Headquarters, Department of the Army, *Fit to Win: Spiritual Fitness* (Pamphlet 600-63-12; Washington, D.C., Sept. 1, 1987).

A unit ministry team (UMT): the chaplain and his assistant. Date and location unknown. U.S. Army Chaplain Corps Museum, Fort Jackson, S.C.

Chaplain Hart and Chaplain Kahn meet with Muslim clerics in Iraq, c. 2003. U.S. Army Chaplain Corps Museum, Fort Jackson, S.C.

Communion service in Iraq, 2005. U.S. Army Chaplain Corps Museum, Fort Jackson, S.C.

Chaplain Abdul-Rasheed Muhammad officiates during a Muslim field service. Date and location unknown. U.S. Army Chaplain Corps Museum, Fort Jackson, S.C.

In a setting familiar to Army chaplains, a Navy Muslim chaplain meets with indigenous religious leaders to promote cultural and religious understanding between the International Security Assistance Force (ISAF) and residents of Naw-Abad, Afghanistan. Photo by Mary E. Carlin, U.S. Marine Corps. Courtesy of the National Defense University Press, Fort Lesley J. McNair, Washington, D.C.

Chief of Chaplains Major General Douglas Carver meets with U.S. Army chaplains and chaplain assistants during a visit to Kandahar Airfield, Afghanistan, March 28, 2011. U.S. Navy photo by Ensign Haraz Ghanbari/Released. Courtesy of the *Military Review*, U.S. Army, Ft. Leavenworth, TRADOC.

Chaplain **Eric Keller, AN ARMY OF ONE.** In the United States Army.

WHEN SOLDIERS CALL FOR SUPPORT, THEY DON'T ALWAYS MEAN MORE FIREPOWER.

AN ARMY OF ONE

goarmy.com

U.S.ARMY

Through leadership and support, the U.S. Army gives its Soldiers the tools to succeed in any situation. Like taking a hill, securing a bridge or getting through the obstacle course we all face every day — life. Whether you're already ordained or still a seminary student, there are both full- and part-time ministry opportunities for you in the U.S. Army Chaplaincy. For more info or to find out how you can talk with an Army Chaplain, call **1-800-USA-ARMY, ext. 456 or log on to CHAPLAIN.GOARMY.COM**

Advertisement for opportunities in the U.S. Army Chaplaincy. *Army Chaplaincy* (Summer–Fall 2003), 106.

A U.S. Army poster promoting the Comprehensive Soldier Fitness program. U.S. Army image offered by Wikipedia, http://en.wikipedia.org/wiki/File:CSF.Spiritual.Fitness.Poster.png/.

The Comprehensive Soldier Fitness logo. Download from the CSF poster series, http://csf.army.mil/downloads.html/.

The regimental crest of the U.S. Army Chaplain Corps. U.S. Army Chaplain Corps Museum, Fort Jackson, S.C.

11

Developing a Culture of Pluralism

During the 2005–2006 culture war over religious expression in the military, the policies and practices of the navy and the air force, and their chaplains, were the main target. Occasionally, however, a newspaper reporter would ask about the army's position in the debate. The summary response, usually from spokesperson Martha Rudd, was, "We do not tell our chaplains how to pray." The *Washington Times* quoted her in October 2005 as saying that chaplains could "speak freely" during services for their own faith group but were expected to take a different approach in "general military assemblies or services." In those gatherings, she explained, "the Army wants chaplains to show respect for all faiths."[1] Several months later, on January 23, 2006, *Times* reporter Julia Duin reported Rudd saying that "sectarian prayer" had not been "an issue" among the army's 2,700 chaplains. "Chaplains are advised to consider their audience when they're speaking publicly and be considerate of a plurality of views in the audience," she observed. "But we don't tell them how they can and cannot pray."[2]

The Stertzbach Incident

On December 21, 2005, the *Times* had published a story about an army chaplain in Iraq who had called his endorsing agent to tell him that "he'd be hammered if he used Jesus' name" in his prayers. The endorser, a former army chaplain, was the Reverend Billy Baugham, executive director of the International Conference of Evangelical Christian Endorsers. He was also the president of Associated Gospel Churches, an endorsing agency for "independent evangelical fundamental churches."[3] During the culture war, his name was often linked with Representative Walter B. Jones and his associates in the campaign to persuade President Bush to issue an executive order allowing military chaplains to pray "in Jesus's name."[4]

The December story did not disclose the chaplain's name, but in a later story, published on January 23, Duin revealed that he was Chaplain Jonathan

Stertzbach, currently assigned to Camp Liberty, near Baghdad. In a telephone interview with Duin, he had complained that the brigade chaplain had asked him to "modify" his prayers. "You need to allow people to pray according to their faith group," Stertzbach told Duin. "Many faith groups do not pray in general and generic terms. . . . For Christian groups, the name of Jesus is from where all the power comes." He then volunteered other thoughts on the matter. He remembered an occasion at the Army Chaplain School when "a Muslim chaplain candidate . . . prayed in the name of 'Allah, the most blessed and beautiful,'" and "was talked to the next day." He said that in Iraq, "it seems like there is an unwritten expectation. We don't know what's going on. We are asking for answers and not getting them. . . . When we come to a monthly division training meeting, we are told we cannot do this, and if you do this, you will be punished." Stertzbach also cited a speech he had heard in May 2005 at Fort Drum, New York, by the division chaplain, Lt. Col. Philip Wright. He said Wright had expressed sympathy for chaplains who wanted to pray in Jesus's name but had counseled them not to insist on doing so "or we are going to lose our ability to pray at all." He then added, "If anyone does this without my knowledge, I will crunch you." When Duin asked Wright about the charge, he denied ever having said the words Stertzbach quoted. "I've never told chaplains they cannot pray in Jesus' name," he said. "Sometimes I've told them it's not a correct time to espouse their theology to everyone, [but] there is no regulation that says you cannot pray in Jesus' name."[5]

According to various accounts of the Stertzbach incident, his commanding officer had asked him to perform a memorial ceremony for a soldier who had been killed in action. Chaplain Stertzbach wrote two prayers and a meditation and submitted them for review by the division and brigade chaplains. The brigade chaplain criticized him for concluding one or both of his prayers "in Jesus' name," which he deemed inappropriate in "a public forum."[6] Invoking "his conscience and faith tradition,"[7] Stertzbach refused to "strike the words Jesus Christ."[8] He told the brigade chaplain that he could not pray a prayer unless he used Jesus's name. He explained that he belonged to "an independent Baptist church, and that's the way we pray."[9] The commanding officer intervened and "instructed that Chaplain Stertzbach would pray according to his faith tradition and the prayers he initially submitted." The brigade chaplain then told Stertzbach to qualify his prayer by saying, "Please pray according to your faith tradition, as I pray according to mine," and to close the prayer with the words, "In thy name we pray, and in Jesus' name I pray." Chaplain Stertzbach "followed orders," revised the prayer as suggested, and performed the memorial ceremony.[10] Afterward, he registered his complaint with his endorser, Billy Baugham, and then went public by telling reporter Duin about the incident.

Army officials responded quickly to Stertzbach's comments in the *Washington Times*. On January 30, they removed him from his preaching duties at chapel services and ordered him not to speak to the news media.[11] Baugham and Representative Jones came to his aid, using the incident to promote their cause. Speaking in Stertzbach's defense, Baugham asserted that "military regulations forbid restrictions on chaplains' prayers" and that "Department of Defense instructions" stated that a chaplain was to adhere to the tenets of his faith group. Baugham insisted that a chaplain "has no authority to pray any other prayer other than what his sending agency or his sending church allows him to pray." His duty was to represent his faith group. "It's not to represent another faith group, it's not to represent the prayers of the United States government or the United States Army," Baugham declared. "He can only speak for his endorsing agency, be that what it may be."[12] Jones wrote letters to the army inspector general and Secretary of Defense Donald Rumsfeld requesting an investigation to determine whether Stertzbach had been "illegally removed" from his preaching duties. In a speech delivered in the House of Representatives, during which he read his letter to the army inspector general, Jones cited the Stertzbach incident as one of many "documented" cases that showed that "suppression of religious freedom throughout our Armed Forces is a pervasive problem, affecting military chaplains from all denominations and religions." He concluded his speech with a pitch for his current crusade: "Since the beginning of our nation's military, chaplains have . . . prayed according to their faith tradition. It is in the best interest of our Armed Services and this nation to guarantee the constitutional right of military chaplains to pray according to their faith."[13]

The Stertzbach incident had little lasting effect on the debate over public prayer or, for that matter, on the army chaplaincy. It appears to have been an isolated event, in contrast to other, more numerous incidents and complaints about chaplain prayer in the Navy Chaplain Corps and the Air Force Chaplain Service. None of the army's official guidelines on chaplains' religious expression was rescinded.

The Culture of Pluralism and Its Sources

How did the Army Chaplain Corps escape being embroiled in the culture war? Over a period of decades, it had developed a culture of pluralism that enabled chaplains to meet the challenge of religious diversity with much less contentiousness than developed in the other two chaplaincies.[14] That culture of pluralism drew on three main sources: unwritten guidelines that had existed long before religious accommodation became a major issue, chaplain field manuals and army regulations that defined the roles, rights, and responsibilities of army chaplains, and speeches and written pronouncements in which army chiefs of chaplains elaborated a pluralist vision for both the chaplaincy and the army. The first two sources provided the framework for the culture of pluralism, the other, its guiding spirit.

The Unwritten Guidelines

One of the unwritten guidelines was based on the principle of "cooperative pluralism." On the one hand, it discouraged narrowly denominational or sectarian views and encouraged interfaith accommodation and mutual responsibility. Chaplains were expected not just to serve members of their own faith or denomination but also to make provision for those of other faiths and groups by arranging for the services of other chaplains, auxiliary chaplains, civilian clergy, or lay leaders. On the other hand, "cooperative pluralism" protected chaplains from being commanded to do anything that was in conflict with their conscience or the beliefs or practices of the religious body to which they belonged. This protection, embodied in the Navy Chaplain Corps motto, "Cooperation without Compromise," was accepted in all of the chaplaincies.[15] It allowed chaplains to claim exemption, which was generally granted, from various practices such as marrying divorced people or people of another denomination; serving communion or performing certain forms of baptism; wearing vestments or robes; dancing, drinking alcoholic beverages, or participating in activities where alcoholic beverages were served or consumed; and participating in joint, ecumenical, or interfaith services. It also allowed chaplains who objected on theological grounds to the ordination of women to refuse to acknowledge or accept female chaplains as "worship leaders, preachers, celebrants of sacraments, and chaplain colleagues."[16]

The other guideline was the no-proselytizing rule, which defined proselytizing as an effort to persuade an individual to leave a religion to which he or she was actively committed and to join another. It did not prohibit a chaplain from proselytizing the "unchurched," meaning a person who belonged to no church or faith group or an individual who, of his or her own volition, expressed interest in the chaplain's denomination or faith or voluntarily attended a denominational service, prayer meeting, religious retreat, or Bible study group led by the chaplain.[17] Until 1990, this rule was largely unwritten; it did not appear in any army chaplain field manuals. In 1990, however, the no-proselytizing rule was included in a "Code of Ethics for Chaplains of the Armed Forces" promulgated by the National Conference on Ministry to the Armed Forces. It read as follows: "I will not proselytize from other religious bodies, but I retain the right to evangelize those who are not affiliated."[18] Since the NCMAF was a private, interdenominational, interfaith association of chaplain-endorsing agencies, its rule, or "pledge," did not have the force of a DoD or service branch regulation. However, there is some evidence that it, or something similar, was taught or at least distributed in the chaplain schools of the armed forces.[19]

Chaplain Field Manuals and Army Regulations

In defining the roles, rights, and responsibilities of army chaplains, the field manuals and army regulations issued in the 1950s, 1960s, and 1970s described chaplains as the "representative" of a "recognized religious denomination"[20] or as "clergy of their particular religious faith."[21] Their "spiritual authority" to engage in various religious activities was said to be "imparted to them by their denomination or endorsing body,"[22] and the "ritual and rules" of those bodies were recognized as their "guide" in such matters.[23] By the 1980s, the manuals and regulations put considerable emphasis on chaplains' loyalty and obedience to the rules of their denominations, faith groups, and endorsers, as well as to "the dictates of their own conscience."[24] They also described the chaplains' relationship with those bodies in more specific terms. The November 1989 chaplain field manual asserted that the "chaplain's call, ministry, message, ecclesiastical authority, and responsibility come from the religious organization which the chaplain represents. Chaplains teach, preach, and conduct religious services in accordance with the tenets and rules of their tradition, the principles of their faith, and the dictates of conscience."[25]

In effect, if not in so many words, the emphasis placed on the chaplain's obedience to his or her denomination, faith group, endorser, and the dictates of conscience, validated the chaplain's right to the free exercise of religion. This emphasis conflicted with another new emphasis that also received increasing attention in the manuals and regulations—the free exercise right of the soldiers to whom the chaplain ministered. "The chaplain supports all soldiers in the free exercise of their religious beliefs," declared the 1989 field manual.[26] "Soldiers in an Army unit come from a multitude of faith groups, some of which are different from the chaplain's own faith group," observed the 1995 manual. Pointing to the constitutional guarantee of free exercise for all Americans, the manual advised chaplains that they were to provide "religious support for all soldiers in the unit."[27]

To accommodate the free exercise rights of both chaplains and soldiers, the manuals and regulations continued to assert chaplains' right to request exemption from performing activities they deemed to be in conflict with the requirements of their faith or conscience.[28] At the same time, however, they began to focus attention on the chaplain's dual responsibility to "perform" and "provide" what was now being called religious support.[29] When chaplains "performed" their numerous and various religious activities, they did so—were permitted to do so—in accord with "the tenets of their faith group."[30] But if they were unable because of "faith restrictions" to engage in any of those activities, they had the responsibility to "seek the required support from other chaplain sources."[31] The 2003 field manual offered a succinct statement of that responsibility on its very first page: "Chaplains provide and perform Religious Support (RS) in the army to ensure the free exercise of religion. Chaplains are obligated to provide for those religious services or practices that they cannot personally perform."[32] A few pages later, it elaborated on that guidance as follows: "When a conflict arises between the standards of a chaplain's faith tradition and the requirements of a religious support mission, chaplains are required to provide for the religious needs of the soldier by obtaining other chaplains or qualified people to perform the needed religious support."[33] By 2005, the "perform or provide" phrasing had become ubiquitous in descriptions of the chaplain's responsibility to ensure all soldiers' right to free exercise.[34] In effect, the stress on "providing" religious support revived the tradition of cooperative pluralism. The 2005 *Religious Support Handbook* declared, "Chaplains cooperate in ministry with each other without compromising their own faith, tradition, or ecclesiastical endorsement requirements. Together, chaplains perform and provide RS to ensure the most comprehensive RS possible for soldiers."[35]

Regulations increasing the number and variety of worship services offered on army installations facilitated the development of the pluralist culture. Traditionally, the General Protestant Service (also known as the Collective Protestant Service) had been regarded as the primary worship service.[36] But as we saw in chapter 9, in the late 1980s, OCCH declared that "distinctive faith groups" had "equal claim" on chaplaincy services. That guidance was later codified in an army regulation of 2004 regarding "religious services on military installations." It began with the statement that "the Army recognizes that religion is constitutionally protected and does not favor one form of religious expression over another." Therefore, all religious groups were to be viewed as "distinctive faith groups" and all soldiers were "entitled to chaplain services and support." According to the regulation, "When facilities are shared, scheduling priority will be given to worship services conducted by chaplains and services that minister to the largest number of soldiers and family members."[37]

Army chaplain field manuals and regulations addressed the matter of public prayer only occasionally. In the 1950s, the chaplaincy based its policy on a distinction between so-called patriotic and military ceremonies on the one hand and chaplain-conducted religious services on the other. The 1958 chaplain field manual included a fairly lengthy discussion of that distinction and its implications. It defined "patriotic ceremonies" as events "customarily held on days of national significance," such as Memorial Day, the Fourth of July, Thanksgiving, and the anniversaries of military victories. "These ceremonies *may* include the religious elements of invocation and benediction," the manual observed. "However, *they are not worship services* and *care should be exercised to exclude any emphasis which is strictly denominational* in order to permit *tri-faith participation* when appropriate and possible."[38] This guidance, with its mention of "tri-faith participation," may be seen as a reflection of the religious consensus of the 1950s. The manual advised against a "strictly denominational" emphasis so as to be inclusive of Protestants, Catholics, and Jews. The guidance also seemed to be based on the idea that since patriotic ceremonies were defined as something other than worship services, chaplains need not be considered (and should not expect) to have the right to follow the tenets of their denomination, faith group, or endorsing bodies in offering "religious elements" on such occasions.

In 1979, Army Regulation 165-20 rephrased the guidance presented in the 1958 manual (but omitted the advice to exclude any "strictly denominational" emphasis and to "permit tri-faith participation") to read as follows: "Chaplains may participate in military and patriotic ceremonies. Although these ceremonies may include an invocation, readings, prayer, or benediction, they will not be conducted as religious services but as military exercises."[39] The 1989 chaplain field manual made the distinction between religious services and patriotic and military ceremonies quite clear. Regarding the former, it said, "The chaplain conducts worship services, administers sacraments, and performs rites consistent with the tenets of the respective denomination and faith group." However, it pointed out, "patriotic and military ceremonies, at which attendance may be required by the command, are not considered or conducted as religious worship services."[40] In 2000, in the midst of increasing discussion of public prayer and praying in Jesus's name, the army issued a new regulation regarding "military and patriotic ceremonies [that] may require a chaplain to provide an invocation, reading, prayer, or benediction." It advised chaplains that they "will not be required to offer a prayer, if doing so would be in variance with the tenets and practices of their faith groups." This regulation applied to patriotic and military ceremonies the same "no compromise" exemption chaplains had long been granted with regard to religious services, which it also stated.[41]

In advising chaplains on public prayer, the chaplaincy generally avoided offering specific guidance regarding its content. The advisory in the 1958 chaplain manual against a "strictly denominational" emphasis was an exception. So was the statement in the 1995 field manual on memorial services, which read as follows: "The chaplain ensures that the content is sensitive to the deceased soldier's faith group and to the needs of the soldiers attending."[42] However, the manuals and regulations did present definitions and other commentary designed to prompt chaplains to reflect on the occasion and purpose of their prayers, the kind of audience they would be addressing, and the Chaplain Corps' strong commitment to soldiers' right to free exercise of religion. These advisories functioned, and were likely seen by chaplains, less as rules or regulations and more as "guidelines" that the chaplaincy trusted its chaplains to interpret correctly.

"Interpreting correctly" meant understanding the guidelines in terms of the free exercise and establishment clauses of the First Amendment, both of which received increasing attention and explication in the manuals and regulations of the late twentieth and early twenty-first century. The dual emphasis is clearly seen in "Constitutional Foundation for Religious Support," a section in the 1989 chaplain field manual:

> The First Amendment of the United States Constitution guarantees the right to the free exercise of religion. Being in the military does not deprive a soldier of this right. Limited only by compelling military necessity, commanders ensure the free exercise of religion for all soldiers in their units.
>
> Chaplains assist commanders in facilitating the soldiers' free exercise rights. The Chaplain Corps provides religious support to many faith groups. The Chaplain Corps does not constitute an establishment of any one religious viewpoint or doctrine.[43]

Sometimes the manuals and regulations added brief discussions of the implications of the chaplaincy's commitment to religious pluralism. For example, the 1995 manual pointed out that "soldiers in an Army unit come from a multitude of faith groups, some of which are different from the chaplain's own faith group," then added that "under the Constitution's provision for the 'free exercise of religion,' the MT [Ministry Team] provides religious support for all soldiers in the unit."[44] The emphasis on ministering to "all soldiers" was even stronger in two army regulations issued in May 2000 and March 2004, which declared, "Each chaplain will minister to the personnel of the unit and facilitate the 'free exercise' rights of all personnel, regardless of religious affiliation of either the chaplain or the unit member."[45]

The two regulations also stated the dual emphasis on the two clauses of the First Amendment. The following declaration appeared on page 1: "The First Amendment to the U.S. Constitution prohibits enactment of any law 'respecting the establishment of religion or prohibiting the free exercise thereof.' In striking a balance between the 'establishment' and 'free exercise' clauses, the Army chaplaincy, in providing religious services and ministries to the command, is an instrument of the U.S. Government to ensure that soldier's 'free exercise' rights are protected. At the same time, chaplains are trained to avoid even the appearance of any establishment of religion." Another statement, on page 5, also alluded to the establishment clause, declaring, "The Army recognizes that religion is constitutionally protected and does not favor one form of religious expression over another."[46]

Altogether, the guidelines presented in the chaplain manuals and army regulations formulated a pluralist approach to military ministry. But in their original form, they were isolated statements, sometimes rather vague, and enunciated in the flat, uninspired language characteristic of military manuals and regulations. They helped make the culture of pluralism work, but they did not play the main role in persuading chaplains to adopt a new and very challenging concept of their ministry.

The Pluralist Vision of the Chiefs of Chaplains

The army chiefs of chaplains performed that crucial role. During the late twentieth and early twenty-first century they presented a vision of pluralist ministry in language calculated to persuade and inspire even the most sectarian, tradition-bound chaplains. In written and oral pronouncements they sought to sensitize chaplains to the importance of being attuned to and respectful of religious diversity among army personnel as well as within the Chaplain Corps. This effort involved inculcating values and assumptions that would enable chaplains to interpret and follow the guidelines presented in the field manuals and army regulations and, even more important, to exercise a kind of self-regulation in dealing with the tension between their own and soldiers' rights to the free exercise of religion.

In a 1994 "Statement on Equal Opportunity," Chief of Chaplains Matthew A. Zimmerman described the army chaplaincy as "multi-faith, ethnically and religiously diverse, and supportive of the soldier's right to free exercise of religion." Affirming that diversity, he declared, "We recognize the intrinsic dignity and worth of each person and the right of each to receive just treatment and compassionate care which the exigencies of life require. We celebrate the immense diversity of religious practices, gender and racial differences, ethnic and cultural traditions, and various gifts and talents among all. We pursue a vision of caring individuals and communities committed to a partnership with the divine, creating a world of justice and peace."[47]

During Zimmerman's administration, the army chaplaincy recognized and showed respect for religious pluralism by adopting a new, more inclusive Chaplain Corps regimental crest. The old crest had portrayed a cross and a tablet, representing the Christian and Jewish faiths, whereas the new one depicted an open book "symbolizing God's word from all faiths for soldiers of many faiths."[48] Both Zimmerman and his successor, Chief of Chaplains Donald W. Shea, promoted the idea of the army chaplaincy as "the model for the Army," leading the way in accommodating and respecting not just religious pluralism but all forms of human diversity.[49] The "Army Chaplaincy Vision" formulated during Shea's administration (1994–99) described the Chaplain Corps as "a chaplaincy that is diverse, inclusive and responsive to Army leadership; a chaplaincy serving soldiers, families, and the nation."[50]

In 1999, when the Army Chaplain Corps promulgated the official listing of its six "key values," "Diversity" was one of them. These were values that Chief of Chaplains Gaylord T. Gunhus declared "nonnegotiable" and "the bedrock and foundation upon which we build our vision and mission." Addressing a senior leadership training conference in January 2000, he defined chaplains' commitment to diversity as follows: "We must be committed to a Chaplaincy that honors and enables the practices of religious expressions of an extremely diverse constituency. We must be a Chaplaincy that is not merely "politically correct" but is genuinely sensitive to the diverseness of race, gender, and culture."[51] During his administration, Gunhus challenged chaplains to exercise "prophetic leadership" and to serve as "the prophetic voice of the Army." Chaplains should "not hesitate to make a stand for righteousness." They must be willing "to confront the institutions, on matters of morals, ethics, morale and religion."[52] Modeling the values of diversity, toleration, and respect in the army, was, in his view, an important way of engaging in prophetic leadership.[53]

Chief of Chaplains David H. Hicks, who succeeded Gunhus, also viewed chaplains' pluralist ministry as a prophetic responsibility and a model for the army. This orientation is clearly seen in the white paper he issued in June 2004, "Taking Spiritual Leadership to the Next Level." Spiritual leaders "embrace diversity within the community," the paper declared. They "hold to a fundamental belief that all human beings are created in the image of God" and that their diversity is a consequence of "God's creativity." They "recognize that each person is gifted, has a contribution to make in the community and brings strengths precisely because of their diversity, not in spite of it." The chaplaincy "champions" freedom of religion "in a way few organizations can," the paper asserted. "We celebrate diversity of religious expression, we protect the views of the minority, this while representing our faith tradition with authenticity and integrity." Indeed, the chaplain as spiritual leader "not only practices respect and inclusion, but models it and advocates for it within the community." Spiritual leadership was described as "an intentional process by which personal example extends to positively influence a community for real change based on shared values."[54]

Addressing the 2005–2006 Culture War

At the beginning of the 2005–2006 culture war, the army was already in a strong position to defend itself against the conservative culture warriors if it needed to do so. In May 2005, it issued a *Religious Support Handbook for the Unit Ministry Team,* which on the very first page declared, "The establishment clause of the first amendment forbids any governmental authority from mandating a particular religion as the 'official' religion or way of prayer."[55]

As the culture war intensified, officials in the Office of the Chief of Chaplains, the Department of the Army, and the Defense Department engaged in "intense discussions" regarding chaplains and public prayer. According to Chaplain Glenn Bloomstrom, they decided there was no need to issue new "legal or policy statements" regarding the prayers used in "public command ceremonies," since the army already had an official position "outlined in Army regulation[s]." Bloomstrom described it as follows:

> The Army does not censor prayers in the Army chaplaincy. The Army had not experienced an institutional difficulty with it's [*sic*] method of dealing with the different prayer venues and circumstances. It had always stressed the responsibility of all Chaplains serving in a diverse and pluralistic setting during the accessioning process and in the formal training of Chaplains in both pluralism and religious accommodations. The resulting expectation is that Chaplains remain committed to their faith and also sensitive to the pluralistic culture in which they serve. This ensures that [a] soldier's "free exercise" rights are protected and Chaplains avoid even the appearance of any establishment of religion.[56]

During the culture war, the Army Chaplain Corps had a great advantage in being able to appeal to its longstanding culture of pluralism. It had evolved and strengthened over time, it offered a well articulated rationale, and it was firmly rooted in official regulations as well as traditional chaplain values and American democratic ideals. Challenged by dissident colleagues, civilian critics, or inquisitive newspaper reporters, army chaplains could point to an official working "strategy." Air force and navy officials scrambled to formulate guidelines and directives on public prayer in hopes of mollifying warring factions—chaplains as well as civilians—in the controversy, but the army felt no such pressure.

Although the army chiefs of chaplains avoided taking a public position in the culture war, they were not totally silent. In December 2005, Hicks offered an advisory on "Prayer and Pluralism" in his newsletter to chaplains. "The very definition of pluralism suggests that we fetter our own needs to enhance the needs of others," he wrote. "Therefore it is incumbent upon professional Chaplains to understand the needs of the audience before which they pray. In public ceremonies the needs of the audience may need to be considered over the needs of the Chaplain who stands as a representative of the command."[57]

In March 2006, Hicks sent another advisory assuring his chaplains that the army did not "censor prayers." Offering his own approach as a model for praying, he told them that he always prayed in the name of Jesus, "but I don't always use that name. It is a matter of the heart." He advised chaplains to be "sensitive" when praying in "public settings where there may be those whose religious viewpoints are disrespected by a narrowly focused, sectarian prayer."[58] The tone and approach of Hicks's advisories remind us of the guidance presented in army chaplain manuals and regulations. Rather than specifying the content of public prayers, Hicks urged chaplains to consider the occasion and purpose of those prayers and the kind of audience they would be addressing. Significantly, he did not feel impelled to issue an official directive on public prayer in the army. Both of Hicks's advisories may be seen as an effort to remind chaplains of the culture of pluralism as well as to bolster their morale in the midst of the culture war.

In December 2005, Chaplain Gaylord T. Gunhus, recently retired from OCCH, invoked elements of the culture of pluralism in a personal letter to Jay Sekulow, chief counsel of the American Center for Law and Justice, one of the conservative culture warriors' leading organizations. Gunhus began by saying that he had been "following the debate that is going on regarding prayer at ceremonies and in command functions" and that he was "concerned about the way you and the media are telling the story." He had heard the ACLJ chief counsel say, on a recent radio broadcast, that "chaplains can no longer pray in Jesus' name." Gunhus said, "You make it sound as if the chaplains are being told they cannot pray in Jesus' name at all. That is not true." He went on to point out the distinction between chaplains being able, even expected, "to pray as they believe in worship services, Bible studies, pastoral acts and activities, [and] during ministry of the sacraments and counseling," as opposed to praying "at command functions, where attendance is required or soldiers are standing mandatory formations." Gunhus wrote, "Jay, it is important that you make this distinction to the 170,000 people from whom you received signatures on your petition [for an executive order] and to the people at large who hear you talk about this on radio or wherever you speak." Then he added, "I am certain that you have consulted with the Chiefs of Chaplains and the Armed Forces Chaplains Board in regard to this issue on prayer. Their position on this issue is the official policy. . . . Before you take this thing to the President, seeking the executive order, be sure you have your facts straight and are accurately representing the position of the Chiefs of Chaplains."

Gunhus then recounted his own experience praying at command functions. He said that he had always regarded an invitation to pray at such gatherings "as a privilege" and as "an opportunity to honor the Lord." He "knew the soldiers were standing in the formation involuntarily, [and that] many were not Christians or religious at all." In his public prayer, "usually ending with 'To the most high God we pray' or 'In the name of our Lord we pray' or 'To the most Holy God we pray' or 'These things we pray together, Amen,' I was able to give honor to God and give a witness to literally thousands of soldiers, who would never attend worship in the chapel, or attend a field service in combat or in training exercises." Having said that, Gunhus told Sekulow, "Forcing the issue and directing that chaplains be allowed to pray 'in Jesus' name' at command functions, ceremonies and formations will only serve one purpose. Chaplains will no longer be invited to pray at these command ceremonies. They will be excluded from the program at these types of events and will forfeit the opportunity to be a witness for our Lord in these command functions."[59]

The Significance of the Culture of Pluralism

The Chaplain Corps culture of pluralism did shield army chaplains and their organization from civilian criticism and meddling during the 2005–2006 culture war. But that was not its main significance. More important was the fact that, over a period of many decades, it guided army chaplains in ministering to an increasingly diverse military population and facilitated cooperation without compromise within the Chaplain Corps. More than any of the other cultural transitions described in this book, army chaplains' experience developing a culture of pluralism would seem to have the most relevance for civilian society. At a time when civilian Americans are engaged in increasingly bitter religious and cultural conflict, they might discover something of value in the army chaplains' pluralist vision.

Epilogue

During some six and a half decades marked by great cultural, religious, political, and strategic-foreign policy shifts, Army chaplains wrestled with two fundamental issues: religious accommodation and morale building. Both tested chaplains' ability and determination to address increasing religious and cultural diversity within the army and to comply with the free exercise and establishment clauses of the First Amendment of the U.S. Constitution.

Religious Accommodation

In the mid-1960s, prodded by soldiers' complaints, the American Civil Liberties Union, and various army officials, the Office of the Chief of Chaplains worked, albeit somewhat grudgingly, to bring chaplains' character education programs into conformity with the establishment clause. Then, in 1979, chaplains faced another, much more serious challenge, the lawsuit filed by Joel Katcoff and Allen Wieder, which contended that the chaplaincy as an institution constituted a violation of the establishment clause of the First Amendment. In January 1985, the U.S. Court of Appeals for the Second Circuit cited both the establishment and free exercise clauses in deciding against the plaintiffs. The decision impelled the Chaplain Corps "to focus as never before" on its constitutional obligation to defend and provide for the free exercise right of soldiers.[1]

In retrospect, the *Katcoff* decision and chaplains' commitment to religious accommodation may have induced a false sense of security. Army chaplains may have believed that the decision affirmed the constitutionality of the chaplaincy and that they no longer needed to worry about meeting the requirements of the establishment clause but could focus their attention on free exercise. In fact, as Chaplains Israel Drazin and Cecil B. Currey pointed out, the *Katcoff* litigation produced "no clear constitutional mandate" in support of the chaplaincy. "Judge McLaughlin's decision that the chaplaincy was constitutional was only that of a district court," they wrote. "Neither it nor the appellate decision that followed was decisive, for only the Supreme Court can ultimately determine constitutional legality, and the case never reached that level." They added that "new suits, based either on the establishment or the free exercise clause of the Constitution, may one day be filed." They called specific attention to a statement in *Marsh v. Chambers* (1983) regarding the practice of having a chaplain open a legislative session with a prayer. In it the judges indicated their disapproval of "exploiting" the "prayer opportunity . . . to proselytize or advance any one, or to disparage any other, faith or belief." If this dictum were applied to military chaplains, Drazin and Currey noted, one could "readily see the danger to the very existence of the chaplaincy in the modern Army in the years ahead presented by chaplains who insist that proselytizing and evangelism is their mission within the chaplaincy."[2]

Drazin and Currey's book, *For God and Country,* was published in 1995. But army chaplains had been warned even earlier against viewing *Katcoff* as decisive. In the May 1986 *Yale Law Journal,* Julie B. Kaplan declared, "To date, no court, including the *Katcoff* court, has subjected the military chaplaincy program to the rigorous scrutiny required by the establishment clause." The *Katcoff* court had focused specifically on government funding of the chaplaincy, not on "elements" of the chaplaincy program that she believed "offended" two "central establishment clause principles, neutrality and nonentanglement."[3] In 1998, Maj. Michael J. Benjamin pointed out in the *Army Lawyer* that while *Katcoff* justified the chaplaincy as an institution, "individual religious activities in the military" were a different matter and needed to be examined in the light of the First Amendment. Among other things, he mentioned the "voluntariness requirement" established in *Anderson v. Laird* (1972), as well as chaplains' prayer at military ceremonies and proselytizing.[4] In 2000, Chaplain Terry A. Dempsey warned of the likelihood of future legal challenges to the chaplaincy. Besides Drazin and Currey's reference to *Marsh v. Chambers,* he also quoted a statement in the *Katcoff* decision, "No chaplain is authorized to proselytize soldiers or their families."[5]

Army chaplains began debating the issues of proselytizing and public prayer in the 1980s. The Chaplain Corps generally refrained from issuing any regulations or chief of chaplains' guidance regarding proselytizing and evangelism. Chaplains had only the traditional, unwritten no-proselytizing rule to rely on until 1990, when the National Conference on Ministry to the Armed Forces included a version of it in its "Code of Ethics for Chaplains of the Armed Forces." The Chaplain Corps did offer guidance and promulgate regulations on public prayer, but it generally avoided specific discussion of proper and improper content. It employed the language of "moral suasion," trusting in chaplains' devotion to the principle of free exercise, rightly understood. That approach worked well with many chaplains, but it had little or no effect on the chaplains and their civilian supporters who insisted that the chaplains' constitutional right to free exercise trumped that of the soldiers to whom they ministered. The culture war of 2005–2006 and the rescinding of air force and navy directives regarding proselytizing and public prayer showed how little regard those chaplains and their allies had for either moral suasion or official military regulations.

Avoiding any pronouncements regarding proselytizing and evangelism and maintaining its "moral suasion" approach to public prayer did save the Army Chaplain Corps from the rescinding action inflicted on the air force and navy chaplaincies. But conservative evangelical culture warriors had not given up. In 2008, 2009, and 2010, they proposed new legislation protecting chaplains' right to pray "according to the dictates of the chaplain's own conscience."[6] On the other side of the cultural/religious divide, an increasing number of army personnel protested chaplains' proselytizing and sectarian prayers. American troops in Iraq claimed they had been subjected to unwelcome proselytizing on the part of chaplains, felt pressured to attend religious services, or were present at official military gatherings that began with a prayer.[7] Following the 2005–2006 culture war, watchdog groups in the United States reported receiving thousands of letters and phone calls from armed forces personnel asserting that they had been forced to participate in religious services and prayers that were antithetical to their beliefs and had faced harassment or "career retaliation" if they expressed opposing views.[8]

At the U.S. Military Academy, some cadets, two officers, and a former chaplain observed that religion, especially evangelical Christianity, was "a constant at the academy." One of the cadets resigned from the academy, citing "the overt religiosity on campus," which he claimed fostered "open disrespect of non-religious new cadets." The other critics pointed specifically to mandatory banquets that began with a prayer or a reading from the Bible. Some of their complaints focused on Secretary of the Army Pete Geren and Maj. Gen. Robert L. Caslen, the commandant of cadets from 2006 to 2008. Geren had opened and closed a commencement address with a quote from Isaiah, and Caslen frequently mentioned God in speeches at mandatory events and urged cadets to "draw your strength in the days ahead from your faith in God." Interviewed by *New York Times* reporter Neela Banerjee, the West Point senior chaplain said of these prayers at mandatory gatherings, "No one is pushing them to believe. This is something we have done in the military for centuries. It is not designed to make people religious. The majority of people here are people of faith, and a prayer asks God's blessing on a gathering and on the food." Caslen cited the "leadership development model for West Point," which he said "recognizes that there is a supreme being" and that "the values of one's faith play an important role in moral development."[9]

During the administration of Chief of Chaplains David H. Hicks (2003–2007), the Chaplain Corps Directorate of Ministry Initiatives made arrangements to send thousands of free copies of *Experiencing God Day by Day* and *The Purpose Driven Life* to soldiers in Iraq and Afghanistan and around the world. The OCCH claimed that the purpose of the distribution was to build "spiritual fitness" among the troops and that it had been approved by the judge advocate's office.[10] The sectarian character of the two books was unmistakable. A *Newsweek* article characterized *Experiencing God* as a "deeply evangelical text" that "holds Christianity above other religions." The June 10 entry read as follows: "If you did not grow up in a Christian home, you can decide, as Joshua did, to reject your heritage of unbelief and be a generation that serves the Lord."[11] *The Purpose Driven Life,* by Rick Warren, the pastor of Saddleback Church in Lake Forest, California, was also clearly evangelical, invoking God, the Bible, and Jesus Christ throughout. The back cover featured testimonies from noted evangelical Christian leaders, including Billy Graham and Franklin Graham, who were quoted as saying, "*The Purpose-Driven Life* will guide you to greatness—through living the Great Commandment and the Great Commission." In the preface, Warren himself described it as "a guide to a *40-day spiritual journey* that will enable you to discover the answer to life's most important question: What on earth am I here for?" In the first chapter he explained, "You discover your identity and purpose through a relationship with Jesus Christ. If you don't have such a relationship, I will later explain how to begin one."[12]

In May and June 2009, the media reported at least two instances of army chaplains engaged in distributing Bibles translated into Arabic to civilians in Afghanistan and Iraq. Evidence of one such effort had appeared in film footage shown on Al Jazeera television in 2008, which showed soldiers with a stack of Bibles at a church service at the Bagram Air Base and the army command chaplain in Afghanistan, Lt. Col. Gary Hensley, exhorting them to fulfill their responsibility to be "witnesses for Christ."[13] The other proselytizing effort, reported by *Newsweek,* involved forty active-duty chaplains in Iraq distributing translated Bibles to the civilian population. It was coordinated by Jim Ammerman, founder and director of the Chaplaincy of Full Gospel Churches, working with the International Missions Network Center (IMNC), an evangelical group based in Arkansas. The Pentagon declared that it had confiscated and destroyed the Bibles to be distributed in Afghanistan. However, the effort in Iraq apparently succeeded; the IMNC claimed that somewhere between one hundred thousand and five hundred thousand Bibles were distributed. In an interview on the subject of Ammerman's proselytizing, Art Schulcz, an attorney who was representing the Chaplaincy of Full Gospel Churches chaplains in a lawsuit against the navy, declared that "the Constitution protects that kind of activity." Asked about "General Order Number 1 (GO-1)," issued by the Commanding General of the Multi-National Corps–Iraq, which prohibited "proselytizing of any religion, faith or practice," Schulcz contended that it was "overly vague" and a "violation of religious freedom," and that chaplains should be exempt since "they are not military representatives but representatives of their faith groups." He said, "The Constitution prohibits absolutely the government from proselytizing, but it protects the proselytizer to do so, unless they're harming the public good."[14]

Mikey Weinstein and his Military Religious Freedom Foundation (MRFF) and the American Civil Liberties Union continued to expose, denounce, and file or threaten to file suits challenging actions by evangelical chaplains, commanders, and army officials that violated the First Amendment establishment and free exercise clauses. Following the 2005–2006 culture war, several other organizations, including some recently established ones, joined that effort. They were non-Christian, nontheistic, atheist, agnostic, humanist or secularist groups whose consciousness may have been raised by the culture war, as well as by the increasing number of soldiers who identified themselves as belonging to either the "Nones" or "Other religions" category. During the first decade of the twenty-first century, many of those soldiers felt impelled to assert their right to practice and express their nontraditional views, as well as to protest any interference with that right. Besides the MRFF and the ACLU, the advocacy groups that took up their cause included JewsOnFirst.org, the Secular Coalition for America, the Military Association of Atheists and Freethinkers (MAAF), the Freedom From Religion Foundation, and the American Humanist Association.[15]

As of 2008, the MRFF claimed to have received complaints about proselytizing from "nearly 11,000 soldiers," of whom "about 96 percent" identified themselves as Christians.[16] It also aided "numerous" atheist and Jewish soldiers.

The organization filed at least two lawsuits against military officials, in both of which the plaintiffs identified themselves as atheists. In 2008, the MRFF and Spc. Jeremy Hall filed suit against Maj. Freddy J. Welborn and Secretary of Defense Robert M. Gates, claiming that Hall's constitutional rights had been violated, "including being forced to submit to a religious test to qualify as a soldier." Two incidents that took place at Camp Speicher in Iraq had prompted the suit. On the one occasion, a staff sergeant had rebuked him for refusing to join in a prayer another soldier had offered at a Thanksgiving meal. When Hall explained that he did not believe in God, the sergeant became angry and told him to sit at another table. On the other occasion, having recently started an MAAF chapter at Camp Speicher, Hall had received permission from a chaplain to post notices of a meeting for atheists and other nonbelievers. Once the meeting was underway, Major Welborn "began to berate Specialist Hall and another soldier about atheism," according to Hall's sworn statement. He quoted Welborn as saying, "People like you are not holding up the Constitution and are going against what the founding fathers, who were Christians, wanted for America!" Hall also said that Welborn threatened to block the soldiers' reenlistment and bring charges against them.[17]

The other MRFF lawsuit, filed in 2008 against Defense Secretary Gates and others, claimed that soldiers were being required to attend military events that included "fundamentalist Christian prayer sessions." The plaintiff, Spc. Dustin Chalker, alleged that "on at least three occasions, beginning in December 2007," he was directed to attend such events, where "an Army battalion chaplain led a Christian prayer ceremony for military personnel." On each occasion, Chalker requested permission to leave the prayer sessions, and each request was denied. Despite his objections, his superiors continued to "force" him to attend other military events that included such sessions.[18]

Morale Building

During the Vietnam War, chaplains and their morale-building role were subjected to harsh criticism. Antiwar critics charged chaplains with "sanctifying" war making, "using religion as a rationale for the righteousness of our cause" and thereby "sacrificing faith to the fighting spirit." Their very presence on the battlefield was said to constitute an endorsement of an immoral, unjust war.[19] When chaplains abandoned combat motivation and developed the pastoral-care approach to "religious support," such criticism dissipated. During the first decade of the twenty-first century chaplains' renewed emphasis on combat motivation—

"spiritual readiness"—prompted some disapproval, but nothing like that of the Vietnam era. No doubt one reason was that the so-called Global War on Terror did not inspire intense debate among the American people and because antiwar, antimilitarist thinking was practically nonexistent.

One expression of concern about chaplains' morale building came from an army chaplain, Scott R. Borderud, in a paper read at a 2007 symposium sponsored by the International Society for Military Ethics (ISME). Borderud observed that "in the last twenty years, the Army Chaplaincy has trafficked in terms such as 'combat multiplier' and 'spiritual battle-proofing' as a way of telling commanders how chaplains actually make a significant and positive difference in the spiritual and emotional preparation of soldiers for combat. This approach insists that soldiers whose spiritual lives are in order (as the result of effective chaplain ministry) will face the fear and uncertainty of battle with a type of confidence or resolve not shared by the spiritually unprepared." Borderud pointed out that no studies had been made or statistics produced showing a correlation between chaplains' ministry and soldiers' battlefield performance. "Do solid believers (of any faith group) march further, carry greater loads, kill more enemy, or sustain fewer injuries than those soldiers who never darken the chapel door?" he asked. The chaplaincy, he observed, "has been unable to establish such metrics or correlations." More importantly, in Borderud's opinion, "it is not really the chaplain's duty to multiply the combat power of the organization nor to 'battle-proof' (whatever *that* means!) soldiers preparing for violence in the workplace of battle."[20]

Mission- and combat-focused morale building on the part of chaplains also prompted the charge that they are, in effect, working as a "religious instrument of the state," in violation of the First Amendment and contrary to their religious "calling."[21] A British army chaplain has pointed out the "dilemma" chaplains face in reconciling their pastoral role with that of "a 'force multiplier,'" which contributes to combat effectiveness.[22]

Another type of criticism targets the increasing use, in discussions of chaplains' morale role, in both the Chaplain Corps and the army, of words such as "spiritual" and "spirituality." In recent years, some chaplains and army leaders have defined or interpreted those words in nonreligious terms. However, most people, including many chaplains, continue to associate them with and/or define them in terms of religion.[23]

Two individuals who presented papers at the 2007 ISME symposium commented on this problem. Erik Wingrove-Haugland, a professor of humanities at the U.S. Coast Guard Academy, addressed the question of whether it was appropriate for military institutions "to develop spiritual or cosmological awareness in [their] members." In his view, the answer depended on whether they followed certain guidelines. He presented the following analysis:

> If spirituality is equated with a particular religious perspective, then the answer to this question is "no," since promoting a particular religious perspective, or even religious perspectives in general over non-religious perspectives, violates the primary principle of accommodation of religious diversity. Furthermore, promoting a particular religious perspective, or promoting religion in general, is likely to be counterproductive and detrimental to unit cohesion, since it is likely to be offensive to those who do not share the religious perspective being promoted, or at least to atheists and agnostics. . . . [Not only would it] decrease unit cohesion and morale, [it would also be] immoral, in that it would violate the fundamental right to freedom of religious thought and expression.
>
> If, however, the term "spirituality" is taken in the broad sense, as referring to all varieties of what Tillich called "ultimate concern," then we must answer "yes" to the question . . . , with the added proviso that the specific path this development takes must be left up to the individual members of the military.

Regarding this last guideline, Wingrove-Haugland insisted that military institutions "must scrupulously avoid any sense of dictating a particular path of spiritual development, or of placing restrictions on the paths of spiritual development available to its members," since "doing so would involve failing to accommodate religious diversity." He concluded his paper by observing that only by allowing members to follow their own paths of spiritual development "can we assist in their development as the unique human beings they are, while respecting the diversity of religious views that they hold."[24]

At the same 2007 ISME symposium, military ethicist Martin L. Cook took issue with the growing tendency in the Chaplain Corps and the army to label morale-building programs "spiritual." In his opinion, calling a program "spiritual" gained little and might actually be cause for apprehension, "especially in a climate where we know many of our officers and leaders are looking for a loophole through which to insert their own personal religious convictions into their official communications."[25] A case in point was the much-publicized "incident" of 2008, in which Gen. David Petraeus, then commander of the Multi-National Force–Iraq, endorsed *Under Orders: A Spiritual Handbook for Military Personnel* (2005), written by Army Chaplain William McCoy. McCoy claimed to speak for and to the "heroic post 9-11 generation who are prosecuting the war against terror." He believed they needed a "handbook on spirituality and spiritual things" to help them "understand both themselves and the larger reality of God around them." He wrote of his hope that "the servicemen and servicewomen who read this book will pick up on my Christian hope" and help bring "the good news of the Gospel" to "our world."[26] In a published endorsement of the book, Petraeus declared that "it should be in every rucksack for those times when soldiers need spiritual energy."[27] Both the McCoy handbook and Petraeus's endorsement enraged the Military Religious Freedom Foundation, which accused the general of "betraying 21% of our troops," a reference to "the men and women fighting [in Iraq] who define themselves as atheists or having no religious preference." Chris Rodda, MRFF senior research director, wrote that "for Petraeus to endorse a book disparaging this segment of our military population is a reprehensible betrayal of all of the non-theists who are putting their lives on the line for our country with every bit as much bravery and dedication as their religious comrades."[28]

A civilian evangelical publicist, Stephen Mansfield, took advantage of the spirituality "loophole." In *The Faith of the American Soldier* (2005), Mansfield vehemently criticized the morale role chaplains were playing in Iraq and Afghanistan. During the latter part of 2004, he had spent several weeks talking to chaplains and troops in Iraq. He claimed to have discovered that many soldiers and Marines wanted and needed a kind of counseling from chaplains they were not receiving. He quoted a soldier beseeching his chaplain, "Tell me that our enemies are the enemies of God. Explain to me how this is a war between good and evil. . . . I need to be sure that I am a soldier of Christ." Mansfield also contended that many chaplains in Iraq were frustrated because, serving as "an arm of the secular state," they were prohibited from "framing their nation's wars in spiritual terms or suggesting that the religions of their enemies are inferior to their own." Mansfield concluded that what the troops needed and wanted, and what chaplains should provide, was a "faith-based warrior code" that would include statements declaring that "America is a Christian nation, . . . with a calling under the hand of God," and that "her battles . . . are spiritual and should be fought by men and women who comprise a 'Christian army.'"[29]

In 2009, the establishment of a new army program called Comprehensive Soldier Fitness (CSF) prompted new criticism of chaplains' morale building.

Launched in October of that year, CSF was a much publicized, very expensive program developed in consultation with what Army Chief of Staff Gen. George W. Casey Jr. called "some of the nation's top behavioral health experts."[30] Its stated mission was to serve as a "holistic fitness program" to "enhance the resilience, readiness and potential of Soldiers, family members and Army civilians." The anticipated result was "an Army of balanced, healthy, self-confident people whose resilience and total fitness enables them to excel in an era of high operational tempo and persistent conflict." All soldiers were required to take an online self-assessment test immediately, and then once a year. (Family members and army civilians were "encouraged" but not required to take the test.) This so-called Global Assessment Tool (GAT) was a 240-question survey designed to measure soldiers' emotional, social, family, and spiritual resilience. If soldiers did not do well on the survey, they were referred to online, self-help "Comprehensive Resilience Modules" as a means of improving their "resilience skills." In addition, "resilience training" was to be provided "at every Army school and at every level of military training from initial entry training through the Army War College." The instructors, "master resilience trainers," were noncommissioned officers and army civilians who had completed a ten-day MRT-C course developed by the army.[31]

The army established CSF to achieve a long-range and an immediate objective. As mentioned above, the long-range objective was the enhancement of soldiers' "resilience, readiness and potential" to enable them to "excel in an era of high operational tempo and persistent conflict." In an article published in a 2011 special issue of the *American Psychologist,* General Casey defined "persistent conflict" as "protracted confrontation among state, nonstate, and individual actors who are increasingly willing to use violence to accomplish their political and ideological objectives." The army could not predict when this era of persistent conflict would end. But, he continued, "we in the Army" do "know that—for the foreseeable future—American servicemen and women will continue to be in harm's way defending our way of life." The army's senior leadership therefore had the responsibility of ensuring that "our soldiers, family members, and Army civilians are prepared—both physically and psychologically—to continue to serve and/or to support those in combat for years to come."[32]

The immediate objective of CSF was to address the new crisis the army faced. In his *American Psychologist* article, Casey described it as follows:

> Many of our soldiers have already served multiple combat tours in Iraq and Afghanistan, enduring long separations from loved ones while simul-

> taneously operating in complex, high-stress situations for months on end. . . . When faced with the intense stress of repeated combat deployments and multiple missions, some soldiers struggle and sometimes—without help from others—fail to cope effectively. The results—in these cases—can be devastating, fueling some disturbing trends we are seeing in the army.

The "disturbing trends" Casey referred to included rising rates of soldier suicide, not only in Iraq and Afghanistan but also in the United States, the growing number of soldiers suffering from post-traumatic stress, and increasing spousal abuse and divorce among combat veterans. He believed that the CSF program, with its emphasis on psychological resilience, would help alleviate these problems by, among other things, changing the "prevailing view among many within our ranks . . . that having problems with stress or seeking help is not only inconsistent with being a warrior but also a sign of weakness." (He cited data collected from soldiers serving in Afghanistan revealing the existence of this "stigma.")[33]

Another CSF objective was the prevention or mitigation of those "disturbing trends." The director of the program, Brig. Gen. Rhonda Cornum, emphasized this goal in her article in the 2009 *Army Chaplaincy* special issue on spiritual resilience. She observed that instead of waiting until a soldier exhibited a "negative event or symptoms" that indicated the need for therapy, the army was now in a position, with the establishment of CSF, to train soldiers in "positive life coping skills and resilient thinking skills" before they experienced "a traumatic event, not after." Assessing soldiers' psychological fitness "continuously over the life cycle of the Soldier" was a means to that end.[34]

The "Spiritual Dimension" of CSF

The CSF program focused on five "dimensions" of soldier fitness: physical, emotional, social, family, and spiritual. The spiritual dimension was the one that engaged most chaplains, although they were also involved in strengthening "the Army family" by helping soldiers to improve their "relationship resiliency."[35] In her *Army Chaplaincy* article, Cornum briefly defined spiritual fitness as "strength of the human spirit" and noted that it "must be inclusive of everybody's faith group (or lack of faith group)."[36] The official definition, presented on the CSF website, was "Strengthening a set of beliefs, principles or values that sustain a person beyond family, institutional, and societal sources of strength."[37] It was intentionally nonreligious. As a psychologist and a U.S. Military Academy professor explained in the 2011 *American Psychologist,* CSF used terms such as "spirit, spirituality, and spiritual fitness in the human rather than theological sense." Thus the spiritual dimension of CSF was "not based on a particular stance or position on the ontological truth or validity of philosophical, nonreligious, or religious frameworks of belief and practice." Indeed, they observed, "Department of Defense leaders are not in a privileged position to answer ontological questions about God's existence or the truth of religious claims." What those leaders could do, however, was "facilitate the search for truth, self-knowledge, purpose, and direction in life as group members define it."[38]

Since much of the response to CSF from army personnel and in the civilian community focused on terms such as spirit, spirituality, and spiritual fitness, it will be helpful to quote the amplified definitions presented in an article published in 2012 in the *Joint Force Quarterly*:

> The *human spirit* is the essence and animating force of the individual. It is the deepest part of the self, which includes one's core values and beliefs, identity, purpose in life, vision for creating a meaningful life, knowledge and truth about the world (perspective), autonomy to lead one's life, connection with others, and the quest to realize potential. . . .
>
> *Spirituality* refers to the continuous journey people take to discover and develop their human spirit. It is the process of searching for the sacred in one's life; discovering who one is; finding meaning and purpose; establishing interconnectedness with others and, if one so believes, with the divine; and charting a path to create a life worth living.

The authors of the *Joint Force Quarterly* article emphasized that "spirituality" and "religion" were "two distinct concepts." Thus "spirituality is both a process and path people use to discover their inner selves and develop their human spirit," whereas religion "refers to institutions that propose and promote specified belief systems." Religion was, however, "one approach people can use in the process of developing their spirit."

"Spiritual fitness," according to the authors, referred to "an individual's overall spiritual condition." A person who is "spiritually fit" exhibits all of the following abilities:

> To continuously gain understanding of who one is in terms of core values and identity; live in accordance with core values; find purpose and mean-

> ing in life; be open to and continuously seek education and experiences that broaden one's view of the world; manage thoughts, emotions, and behavior; be uplifted by strong connections with others; demonstrate the strength of will and resilience to persevere when faced with challenges and adversity; make meaning out of their experiences; and exercise the autonomy to create a meaningful life that will realize one's full potential.[39]

Like the earlier Spiritual Fitness program, CSF was a command program. Chaplains assisted the commander in promoting the "spiritual dimension" by conducting prayer meetings, Bible studies, and prayer breakfasts and luncheons. Their purpose was to enhance soldiers' "spiritual fitness" or "spiritual resiliency."[40] A sample of articles published on the websites of various army installations offers a glimpse of the orientation and content of such activities. While the chaplains clearly interpreted spiritual fitness and resiliency in religious terms, some also pointed out the nonreligious definition. Speaking at a Fort McCoy Prayer Luncheon in December 2009, Chaplain Ray Woolridge observed that "people who pray" or "pay attention to their religious values" tend to be more resilient. He urged the people attending the luncheon to "make sure your connection with God is what it should be, and you'll be prepared for any trauma that comes your way." Two years later, addressing Fort McCoy's senior leadership, he noted the distinction between the army's and the Chaplain Corps' definition of spiritual resiliency. "The Army describes spiritual strengthening as a set of beliefs, principles or values that sustain a person," he said. "This applies to everybody. But the Army's Chaplain Corps defines it as an art of providing spiritual care for the Army Family in their search for purpose, identity and meaning as they discover and live their sacred story, connect to sacred communities and engage in sacred formation."[41] Another chaplain, talking about "spirituality," defined it in both religious and nonreligious terms. Speaking for himself, he said that "we as chaplains see spiritual fitness to be a religious issue." Paraphrasing the CSF definition, he said, "We consider spirituality as that which you value most deeply or hold sacred, as well as spiritual struggles, the tension of conflict within yourself, with others, or with a higher power about your deepest values or what you hold sacred."[42]

A New Culture War in the Making

The army's CSF program and chaplains' spiritual fitness and resiliency activities aroused vehement opposition on the part of some soldiers as well as secularist civilian advocacy groups. Army leaders and the Chaplain Corps discovered that the "Nones," agnostics, and atheists had now become a force to be reckoned with, within as well as outside the army.

Two well-publicized "incidents" dramatized the issues involved. In August 2010, the media reported that the army was investigating a claim that dozens of soldiers who refused to attend a Christian band concert at Fort Eustis had been sent to their barracks and ordered to clean them up. The concert was one of the "Commanding General's Spiritual Fitness Concerts." According to Pvt. Anthony Smith, a staff sergeant told him and some two hundred other men in their barracks that they could either attend the concert or remain in their barracks. Eighty to one hundred soldiers decided not to attend, and they were not only ordered to clean the barracks but also told not to use their cell phones or personal computers. Smith estimated that about twenty of the men, including several Muslims, refused to attend the concert because of their religious beliefs.[43]

In September 2012, SSgt. Victoria Gettman, an instructor at Fort Sam Houston who identified herself as an atheist, told the *Huffington Post* that some eight hundred soldiers undergoing "resilience training" had been subjected to improper religious indoctrination by a "Christian chaplain." The soldiers were "fresh out of boot camp and in training for their first job in the Army," she observed. The chaplain was conducting the "spiritual fitness" part of the training. As a seventeen-year Army veteran, Gettman knew that that part of the training was optional, so she left the room, but stood just outside, where she was able to hear what the chaplain was saying. She pointed out that the soldiers were not informed that they had the option of "stepping out." According to Gettman, "The chaplain said we have to have something bigger than ourselves. We need, and he stresses *need,* to have something divine in our life." He had the lights turned off and battery-operated candles were passed around "as the soldiers were told to bow their heads." Then, Gettman said, "The entire theater was forced into a mass Christian prayer. I heard him refer to his 'Heavenly Father' and 'Lord.'"

Sergeant Gettman complained to her supervisors about the chaplain's activity and filed written complaints with the base equal opportunity office and with the inspector general of the army, "charging that the chaplain violated Army regulations and the U.S. Constitution by forcing trainees to sit through a sectarian prayer." She also contacted Army sergeant Justin Griffith, the military director for a civilian group, American Atheists. He referred to the chaplain's prayer as "mandatory Christian privileged prayer" and declared it to be "one of the most jaw-dropping and blatant violations I've seen in a very long time." Mikey Weinstein of the Military Religious Freedom Foundation reported that thirty-eight service members at the base, including eleven of the trainees, had informed him that they were willing to sign a federal complaint against "this unconstitutional disaster." According to Weinstein, the group comprised "24 Christians (Protestants and Catholics), two Jews and 12 agnostics or atheists." Weinstein himself told the *Huffington Post* that he had "contacted the Army to demand that the chaplain be disciplined and the Army make a written apology to the troops."[44]

In 2010, less than a year after the launching of the CFS program, the Secular Coalition for America and the Military Association of Atheists and Freethinkers had joined forces to protest it. Both organizations disputed the army's contention that the spiritual dimension questions on the GAT were not religious in nature but pertained to the "human spirit." They insisted that the "term 'Spirituality' calls immediately to mind supernatural 'spirits' and religious piety" and they expressed "concern that the military is officially endorsing and even requiring a supernatural or traditional religious viewpoint." The fact that the GAT was mandatory for soldiers made that endorsement even more questionable. The Secular Coalition suggested that it constituted the kind of "religious test" prohibited by the U.S. Constitution in Article VI, paragraph 3. In a notice posted on its website on January 7, 2010, MAAF expressed the hope that "the military will reach out to us to ensure the nontheist perspective is represented prior to the publication of new doctrine and training" and contended that the "unconstitutional promotion of religion or Christianity specifically must be avoided." It also mentioned the possibility of litigation: "If the military does not respond to MAAF and other representatives of the nontheist community, these issues will continue to end up in EO [Equal Opportunity] and legal battles."[45]

The CSF suicide-prevention training program also aroused the concern of the MAAF, the Secular Coalition for America, and the Military Religious Freedom Foundation. In 2009, the MRFF filed a federal lawsuit against Secretary of Defense Robert Gates and the Defense Department challenging statements in the 2008 *Army Suicide Prevention Manual* and a "Suicide Awareness" PowerPoint presentation for soldiers. One of the statements in the manual advised chaplains to promote "religiosity" as a way of deterring soldiers from taking their own lives. Another instructed behavioral health providers "to openly advocate spirituality and religiosity as resiliency factors." One of the PowerPoint slides read, "Spirituality looks outside of oneself for meaning and provides resiliency for failures in life experiences. Religiosity adds the dimension of a supportive community to help one deal with crises. Both embed themselves in a relationship with God, or a higher power, that provides an everlasting relationship." Another PowerPoint slide featured "an image of a group of silhouetted soldiers with one soldier up in the clouds looking at a large cross." Other slides stated that "connectivity to the Divine is fundamental to developing resiliency" and stressed the importance of soldiers having "a relationship with God." The MRFF lawsuit alleged that these statements and the PowerPoint presentation showed "a pattern and practice of constitutionally impermissible promotion of religion by the military." It specifically declared the PowerPoint presentation "not only an unconstitutional promotion of Christianity for the soldiers who are mandated to attend it, but for the behavioral health providers and non-Christian chaplains who must present it."[46]

In October 2012, Jason Torpy, president of MAAF, expressed similar objections to the "personal religious expressions" chaplains used in providing counseling or training as part of the army's suicide prevention program. He was reacting to reports about chaplains' activities during a recent army-wide "stand down" that focused soldiers' attention on suicide prevention. "Chaplain-led sessions on 'resilience' showed a priority on religion, prayer, and 'god's plan,'" Torpy explained. In Torpy's opinion, such "personal religious expressions may be appropriate in private, optional settings, but they have no place in mandatory settings and even violate the army's own guidance that 'spiritual fitness' training be optional." Indeed, he added, "in a mandatory briefing, they only serve to ostracize the nonreligious and give the appearance that the chain of command prefers theistic viewpoints and does not accommodate the nontheistic." He went on to say that "organizations like MAAF and American Atheists are advising soldiers to report issues to their command to see if the military can resolve these issues internally."[47]

Whether an in-house solution could be effected remained to be seen. In the meantime, a recently organized conservative advocacy group pounced on the issue. The Chaplain Alliance for Religious Liberty described itself as an organization of chaplain endorsers, speaking for more than two thousand chaplains serving in the armed forces, as well as an organization "comprised of veteran service members, primarily chaplains" whose purpose was "to defend and maintain religious liberty, and freedom of expression and conscience that the Constitution guarantees our chaplains and those whom they serve." It apparently was formed during the time the military was proceeding with the abolition of "Don't ask, don't tell," which the members had opposed. Since then the alliance had been speaking out against the "punitive" effect of the "Don't ask, don't tell" repeal on chaplains, "censorship" of chaplains' prayer, Mikey Weinstein, and the MRFF—and now Jason Torpy's criticism of chaplains' "personal religious expressions." Ron Crews, a retired U.S. Army Reserve chaplain, the organization's executive director, censured Torpy's statement as "one more example of intolerance toward faith that runs counter to everything we hold dear in this nation" and expressed the hope that "military leaders will not bow toward this intolerance." The Christian Alliance was in a good position to foment a new culture war. Its website proclaimed its partnership with the Alliance Defense Fund and the Speak Up Movement.[48]

The Army Chaplain Corps in a Quandary

As the twenty-first century entered its second decade, a new culture war over religion in the military seemed to be brewing. At the same time, army officials and the Office of the Chief of Chaplains seemed reluctant to take decisive action to prevent future "incidents" and forestall adverse litigation. This is understandable, given what happened to the air force and navy when they issued specific, official guidelines and directives. The Army Chaplain Corps apparently believed it faced a true dilemma, which is to say a choice between two equally distasteful options—either issuing official regulations prohibiting such things as proselytizing, sectarian public prayers at mandatory events, and "personal religious expressions" in resiliency and suicide-prevention training, or inviting, by the lack of such regulation, congressional or court action that would adversely affect the Chaplain Corps or chaplains' ministry.

Forestalling congressional action might be difficult, given the continuing culture war in the civilian community. But several legal scholars who have analyzed the First Amendment issues facing the Chaplain Corps have expressed confidence that the army chaplaincy, as an institution, would likely survive any attempt to have it declared unconstitutional. The 1985 *Katcoff* decision had placed it on a strong, constitutional footing and was not likely to be overturned "anytime soon." At the same time, these scholars have expressed concerns about certain "specific practices" of the chaplaincy or on the part of chaplains, which they think would not pass constitutional muster. Ira C. Lupu and Robert W. Tuttle of the George Washington University Law School point to practices that "seem more to reflect government promotion, rather than accommodation, of religion," such as "official religious professions" or statements that "warrant" chaplains' religious authority or even "the reality of divine presence." On the matter of public prayer, they contend that "a variety of powerful constitutional themes . . . coalesce to support a military policy precluding sectarian prayer at [military] ceremonies." And they especially lament the "vacuum of regulation, and even of official guidance," regarding proselytizing.

Lupu and Tuttle recognize the possibility that evangelical chaplains might bring suit against the army chaplaincy, as some of their counterparts have done in the navy. But they characterize as "weak" the free exercise and free speech claims such chaplains would likely make in support of their right to proselytize or to offer sectarian prayers at other-than faith-group gatherings. In a similar vein, Richard D. Rosen of the Texas Tech University School of Law has declared that such claims, as well as "proposed legislation attempting to secure that 'right,'" have been "ill-advised and constitutionally problematic." They paid little, if any, attention to the free exercise rights of service members. "Ultimately," Rosen asserts, "the chaplaincy exists to accommodate the free exercise of religion by members of the military, not the free exercise or free speech rights of military chaplains." In discounting the likelihood that evangelical chaplains' suits would succeed, he also points out the military's "compelling interest in maintaining good order and discipline in their ranks," something the courts have consistently supported. "Although a chaplain may ground his religious discourse in denomination-specific religious beliefs and deliver it in the context of a worship service, the government has the indisputable obligation to control the speech if it undermines good order and military discipline," Rosen explains. These observations by Lupu, Tuttle, and Rosen suggest that the Army Chaplain Corps need not refrain from formulating official regulations as a means of strengthening the system of religious accommodation and chaplains' compliance with the First Amendment's establishment and free exercise clauses. Such action might prompt lawsuits, but their chance of succeeding appears to be slight. Of course, another congressional sanction, this time including the army chaplaincy, is also a possibility. But taking action against existing problems in the Chaplain Corps should have priority over something that has not yet happened and might never happen. And as Lupu and Tuttle observe, the "strategy of avoidance in this context is not constitutionally defensible."[49]

Addressing the Current Army Crisis

The recent protests against army chaplains' "personal religious expressions" during mandatory spiritual fitness, spiritual resiliency, and CSF activities call attention to another problem the Chaplain Corps faces in the early twenty-first century: chaplains' faith-based morale building. This kind of morale building—called religious support in the combat environment—was a product of chaplains' adaptation to the perceived "religious needs" of soldiers and the requirements of the AirLand Battle doctrine. Now, some four decades later, chaplains need to consider another adaptation. The current army crisis is preeminently a crisis of morale. Chaplains have a choice. They can continue to play their "unique" faith-based morale role, engaging in voluntary spiritual fitness and CSF activities such as prayer luncheons, prayer meetings, Bible studies, and the like.[50] Or they can embrace an inclusive morale-building role, one attuned not just to soldiers' religious diversity but also to the growing number of soldiers who profess no religion and/or identify themselves as atheists or agnostics.

A 2012 survey by the Pew Forum on Religion & Public Life confirms the wisdom of the second option. The survey report, "'Nones' on the Rise," not only shows a significant increase in the percentage of "Nones" in the general population since 2007–2008 but also offers a more nuanced analysis of that group than was presented in the Pew Forum and ARIS surveys for those years. According to the 2012 Pew report, the percentage of "Nones" increased from 15.3 percent in 2007 to 19.6 percent in 2012. In analyzing this change, the report clarifies our understanding of the "Nones" by referring to them as "the unaffiliated" and breaking them down into three subgroups: atheist, agnostic, and "nothing in particular." In the general population, the percentage of each of these subgroups was as follows: atheist, 2.4 percent; agnostic, 3.3 percent; and "nothing in particular," 13.9 percent. Significantly, 35 percent of the unaffiliated were eighteen to twenty-nine years old, 42 percent of that age group were atheist/agnostic, and 32 percent were "nothing in particular." Perhaps one of the most revealing findings of the Pew survey was that the unaffiliated are, in some ways, more religious and spiritual than has generally been assumed. However, the report also points out that most of them "say they are *not* looking for a religion that would be right for them."[51] Army chaplains would do well to consider the implications of this survey and its report for their current and future ministry to the military, as army personnel and their families continue to reflect religious and cultural change in the general population.

In effect, chaplains face the problem of deciding whether they want to be "integral" to the military mission (in this case, remedying the current crisis) or "unique." In some contexts, as we have seen, they have been able to be—or claimed to be—both. But in the current context, adhering to a faith-based approach to morale building will limit their ability to help remedy the army crisis. A nonreligious, inclusive approach is imperative. Consider two lessons gleaned from Chaplain Corps history. One is how Chief of Chaplains Gerhardt W. Hyatt helped the army resolve its Vietnam-era crisis by developing a "humanizing ministry" for chaplains that used secular "people skills," taught by the behavioral sciences, to combat drug and alcohol abuse and improve human relations in the army. The other is the warning Navy Chief of Chaplains Louis V. Iasiello and Gaylord T. Gunhus, the retired army chief of chaplains, issued during the 2005–2006 culture war regarding the likely result of passage of the chaplain prayer amendment. In his letter to members of the House Armed Services Committee, Iasiello predicted that the amendment would "marginalize chaplains." In a similar vein, Gunhus told Jay Sekulow, a powerful supporter of the amendment, that legislation allowing chaplains to pray "in Jesus' name" at command functions, ceremonies, and formations would result in their no longer being invited by commanders to participate in such events. In the midst of the current army crisis, chaplains who refuse to adopt a nonreligious approach to morale building, including spiritual resilience and spiritual fitness training and counseling, also risk marginalization.

Notes

1. Teaching the Religion of Democracy

1. *New York Times* (hereafter *NYT*), Nov. 18, 1946, p. 5; July 29, 1947, p. 14; Oct. 5, 1947, p. 2; Nov. 16, 1947, p. 21. On the VD crisis in the army, see John Costello, *Virtue Under Fire: How World War II Changed Our Social and Sexual Attitudes* (Boston: Little Brown, 1985), 251–54; and John Willoughby, "The Sexual Behavior of American GIs During the Early Years of the Occupation of Germany," *Journal of Military History* 62 (Jan. 1998): 160–61.

2. *NYT,* June 21, 1946, p. 11; "Criticize Occupation Force," *Army and Navy Journal* 84 (Jan. 4, 1947): 423; Rodger R. Venzke, *Confidence in Battle, Inspiration in Peace: The United States Army Chaplaincy, 1945–1975* (Washington, D.C.: Office of the Chief of Chaplains, Department of the Army, 1977), 25–27; *NYT,* Apr. 3, 1946, p. 14.

3. On the military's VD and prostitution policies, see Costello, *Virtue Under Fire,* 85, 86, 246.

4. Edward A. Fitzpatrick, *Universal Military Training* (New York: McGraw-Hill, 1945), 229, 258.

5. Portions of the discussion of Character Guidance (CG) in this chapter and in chapter 2 first appeared in Anne C. Loveland, "Character Education in the U.S. Army, 1947–1977," *Journal of Military History* 64 (July 2000): 795–818; and Anne C. Loveland, "From Morale Builders to Moral Advocates: U.S. Army Chaplains in the Second Half of the Twentieth Century," in *The Sword of the Lord: Military Chaplains from the First to the Twenty-First Century,* ed. Doris L. Bergen (Notre Dame, Ind.: Univ. of Notre Dame Press, 2004), 233–49. The 2004 essay also briefly discussed topics covered in chapters 4, 5, 6, and 7 of this book. For an interesting discussion of CG as a component of universal military training, see James Gilbert, *Redeeming Culture: American Religion in an Age of Science* (Chicago: Univ. of Chicago Press, 1997), 100–106.

6. Venzke, *Confidence in Battle,* 40–41.

7. Patterson quoted in ibid., 41; Roy J. Honeywell, *Chaplains of the United States Army* (Washington, D.C.: Office of the Chief of Chaplains, Department of the Army, 1956), 324.

8. The Chaplain School, Carlisle Barracks, Pennsylvania, *The Army Character Guidance Program* (ST 16-151, Mar. 1, 1950), 3.

9. Chaplain (Brig. Gen.) Luther G. [*sic*] Miller, "Moral Effect of Military Service," *Army and Navy Chaplain* 16 (July–Aug. 1945): 13; "Strengthening Spiritual

Foundations," *Chaplain* 5 (Aug. 1948): 7; Chaplain (Maj. Gen.) Luther D. Miller, "The Chaplains in the Army," *Army and Navy Journal* 85 (Sept. 20, 1947): 74.

10. Major General John M. Devine, *Address to First Army Chaplains, New York, N.Y., 21 June 1950* (Washington, D.C.: U.S. Government Printing Office, 1950), 3.

11. General Jacob L. Devers, "Training the Army of Today," *Army Information Digest* 4 (Apr. 1949): 7, 8.

12. On the appointment and activities of the two committees, see Robert David Ward, "The Movement for Universal Military Training in the United States, 1942–1952" (Ph.D. diss., Univ. of North Carolina, 1957), 228–42; Gilbert, *Redeeming Culture,* 101–3, 111–16; *NYT,* Oct. 28, 1948, p. 14.

13. The President's Advisory Commission on Universal Training, "Report to the President on Moral Safeguards for Trainees to Be Inducted Under the Selective Service Act," mimeographed typescript, Sept. 13, 1948, 1–2, 6, in Government Documents Department, Louisiana State University Library, Baton Rouge.

14. President's Advisory Commission on Universal Training, *A Program for National Security* (Washington, D.C.: U.S. Government Printing Office, 1947), 2, 62, 63, 71–75.

15. Gilbert, *Redeeming Culture,* 113.

16. President's Committee on Religion and Welfare in the Armed Forces, *Community Responsibility to our Peacetime Servicemen and Women: First Report of the President's Committee on Religion and Welfare in the Armed Forces, March 24, 1949* (Washington, D.C.: U.S. Government Printing Office, 1949), 3, 7.

17. *Universal Military Training: Hearings before the Select Committee on Postwar Military Policy, House of Representatives, Seventy-ninth Congress, First Session, Pursuant to H. Res. 465* (Washington, D.C.: U.S. Government Printing Office, 1945), 159, 281, 359, 283, 146. On pacifist criticism of universal military training and Selective Service, see Lawrence S. Wittner, *Rebels Against War: The American Peace Movement, 1933–1983* (Philadelphia: Temple Univ. Press, 1984), 162–64; on liberal mainline Protestants and antimilitarism, see Joe Pender Dunn, "The Church and the Cold War: Protestants and Conscription, 1940–1955" (Ph.D. diss., Univ. of Missouri, 1973), and Jerrold Lee Brooks, "In Behalf of a Just and Durable Peace: The Attitudes of American Protestantism Toward War and Military-Related Affairs Involving the United States, 1945–1953" (Ph.D. diss., Tulane Univ., 1977).

18. Headquarters, Department of the Army, "Personnel-General: Character Guidance Program" (Army Regulation 600-30; Washington, D.C., Oct. 15, 1958), 1.

19. Venzke, *Confidence in Battle,* 41; Martin H. Scharlemann, "The Theology of the Chaplaincy," *Chaplain* 32 (Third Quarter 1975): 11. Later the U.S. Army Chaplain Board wrote the CG lectures.

20. "Personnel-General: Character Guidance Program" (Army Regulation 600-30; Oct. 15, 1958), 2.

21. Headquarters, Department of the Army, "Personnel-General: Character Guidance Program" (Army Regulation 600-30; Washington, D.C., Aug. 9, 1961), 2;

Headquarters, Department of the Army, "Personnel-General: Character Guidance Program" (Army Regulation 600-30; Washington, D.C., Mar. 1, 1965), 2.

22. Lawrence K. Brady, Chaplain (Major), USA, Assistant for Plans, Programs and Policies, "Narrative Description of the Character Guidance Program, 4 January 1962," 6, typescript, File 721-01, Stennis Subcommittee Study, Acc. No. 68-A-3353, RG 247, Washington National Records Center, Suitland, Md.

23. These terms are a variation on ones Martin E. Marty used in *The New Shape of American Religion* (New York: Harper & Row, 1958), 67, 78.

24. For examples of cultural historians writing about "cultural strategies" (without necessarily using that term), see Warren I. Susman, *Culture as History: The Transformation of American Society in the Twentieth Century* (New York: Pantheon Books, 1984), 185, 194, 208, 288; David A. Hollinger, "Science as a Weapon in *Kulturkampfe* in the United States during and after World War II," *Isis* 86 (Sept. 1995): 444, 447; James Davison Hunter, *Culture Wars: The Struggle to Define* America (n.p.: Basic Books, 1991), 346; and Ralph H. Gabriel, "Traditional Values in American Life," in *American Values: Continuity and Change,* by Ralph H. Gabriel (Westport, Conn.: Greenwood Press, 1974), 149.

25. Will Herberg, *Protestant-Catholic-Jew: An Essay in American Religious Sociology,* 2nd ed. (1955; reprint, Garden City, N.Y.: Anchor Books, 1960), 87.

26. Robert N. Bellah described civil religion as existing "alongside of and rather clearly differentiated from the churches" and "neither sectarian nor in any specific sense Christian." Robert N. Bellah, "Civil Religion in America," in *Religion in America,* ed. William G. McLoughlin and Robert N. Bellah (Boston: Houghton Mifflin, 1968), 3, 10, and see also 9, 17, 20.

27. Eisenhower quoted in Marty, *New Shape of American Religion,* 83.

28. "Covenant Nation," *United Evangelical Action,* July 1, 1957, 7; Edward L. R. Elson, "A Nation Under God," in *Representative American Speeches: 1957–1958,* ed. A. Craig Baird (New York: H. W. Wilson, 1958), 164.

29. "Text of Eisenhower Speech [before the Freedoms Foundation]," *NYT,* Dec. 23, 1952, p. 16.

30. Eisenhower, speaking for the American Legion's "Back to God" campaign, Feb. 1955, quoted in *The Quotable Dwight D. Eisenhower,* ed. Elsie Gollagher (New York: Grosset and Dunlap, 1967), 85.

31. Charles E. Wilson, president of General Electric and co-chairman of the National Conference of Christians and Jews, quoted in *NYT,* June 24, 1948, p. 10.

32. Harry S. Truman, "Address Before the Midcentury White House Conference on Children and Youth (December 5, 1950)," in *Public Papers of the Presidents of the United States: Harry S. Truman: 1950: Containing the Public Messages, Speeches, and Statements of the President, January 1 to December 31, 1950* (Washington, D.C.: U.S. Government Printing Office, 1965), 734–35.

33. Dulles quoted in *NYT,* Oct. 12, 1953, p. 16; Lilienthal in *NYT,* Jan. 17, 1949, p. 7; Eisenhower in *NYT,* Dec. 23, 1952, p. 16.

34. Eisenhower quoted in *NYT,* Apr. 10, 1953, p. 12.

35. Dr. Daniel Poling and Rev. Carl McIntire, both quoted in *Universal Military Training: Hearings* (1945), 509 and 475, and 472, respectively.

36. George C. Marshall, "The Obligation to Serve," *Army Information Digest* 6 (Apr. 1951): 3–4.

37. Adm. Arthur W. Radford, Chairman of the Joint Chiefs of Staff, quoted in *NYT,* Nov. 30, 1956, p. 5.

38. "One Nation Under God," in Department of the Army, *Character Guidance Discussion Topics: Duty, Honor, Country* (Department of the Army Pamphlet 16-5; Washington, D.C., Jan. 14, 1966), 1, 8, 3, 11. The content of the 1966 pamphlet is much the same as that of the original pamphlet, published in February 1957.

39. "Worship in Life," in Departments of the Army and the Air Force, *Character Guidance Discussion Topics: Duty, Honor, Country. Series I* (Department of the Army Pamphlet 16-5; Washington, D.C., 1951), 68.

40. "Religion in Our Way of Life," in Departments of the Army and Air Force, *Character Guidance Discussion Topics: Duty, Honor, Country. Series V* (Department of the Army Pamphlet 16-9; [Washington, D.C.], Apr. 1952), 67.

41. "The Nation We Serve," in *Character Guidance Discussion Topics: Duty, Honor, Country. Series I* (Department of the Army Pamphlet 16-5; Washington, D.C., 1951), 29.

42. Chaplain School, *Army Character Guidance Program,* 19.

43. "The Concept of Authority," in Departments of the Army and Air Force, *Character Guidance Discussion Topics: Duty, Honor, Country. Series III* (Department of the Army Pamphlet 16-7; [Washington, D.C.], 1951), 3.

44. Headquarters, Department of the Army, *Character Guidance Manual* (Field Manual 16-100; [Washington, D.C.], Mar. 1961), 9.

45. Chaplain School, *Army Character Guidance Program,* 6.

46. Headquarters, Department of the Army, *Character Guidance Manual* (Field Manual 16-100; [Washington, D.C.], June 26, 1968), 7.

47. Chaplain School, *Army Character Guidance Program,* 4; "Chaplains," *Chaplain* 5 (Dec. 1948): 16. See also "CG for GI's," *Newsweek,* Jan. 21, 1952, 85.

48. Willoughby, "Sexual Behavior of American GIs," 160–63.

49. *Investigation of War Department Publicity and Propaganda in Relation to Universal Military Training. Hearings before the Committee on Expenditures in the Executive Departments, House of Representatives, Eightieth Congress, Second Session, January 14, 1948* (Washington, D.C.: U.S. Government Printing Office, 1948), Appendix, 58.

50. "New VD Film," *Army and Navy Journal* 85 (Feb. 28, 1948): 683.

51. *Character Guidance Manual* (FM 16-100, Mar. 1961), 9.

52. "Our Moral Defenses," in Departments of the Army and Air Force, *Character Guidance Discussion Topics: Duty, Honor, Country, Series IV* (Department of the Army Pamphlet 16-8; [Washington, D.C.], 1951), 36, 45, 46, 49.

53. See, for example, editorial, *NYT,* Oct. 23, 1953, p. 22.

54. On the public furor, see H. H. Wubben, "American Prisoners of War in Korea: A Second Look at the 'Something New in History' Theme," *American Quarterly* 22 (Spring 1970): 5–6; Albert D. Biderman, *March to Calumny: The Story of the American POW's in the Korean War* (New York: Macmillan, 1963); Louis Menand, "Brainwashed," *New Yorker,* Sept. 15, 2003, pp. 90–91.

55. "The President's News Conference of October 21, 1953," in *Public Papers of the Presidents of the United States: Dwight D. Eisenhower: 1953: Containing the Public Messages, Speeches, and Statements of the President. January 28 to December 31, 1953* (Washington, D.C.: U.S. Government Printing Office, 1960), 706; *NYT,* Oct. 22, 1953, pp. 1, 3.

56. Dwight D. Eisenhower, "Remarks at the 22d Annual Convention of the Military Chaplains Association. May 6, 1954," *Public Papers of the Presidents of the United States: Dwight D. Eisenhower: 1954: Containing the Public Messages, Speeches, and Statements of the President. January 1 to December 31, 1954* (Washington, D.C.: U.S. Government Printing Office, 1960), 461–62.

57. "Report of the Defense Advisory Committee on Prisoners of War," in Department of Defense, Office of Armed Forces Information and Education, *The U.S. Fighting Man's Code* ([Washington, D.C.], Nov. 1955), 7, 12, 13.

58. Department of Defense, Office of Armed Forces Information and Education, *The U.S. Fighting Man's Code* ([Washington, D.C., 1959]), iii.

59. Admiral Arthur Radford, "The Mind and the Spirit in National Security," address before the Second National Conference on Spiritual Foundations, Washington, D.C., Oct. 25, 1955, in Department of Defense, Office of Armed Forces Information and Education, *The U.S. Fighting Man's Code* ([Washington, D.C., Nov. 1955]), 53.

60. Department of Defense, Office of Armed Forces Information and Education, *The U.S. Fighting Man's Code* ([Washington, D.C., Nov. 1955]), 12–13, 15.

61. *Character Guidance Manual* (FM 16-100, Mar. 1961), 6, 14–16.

62. Ibid., 9. See also *Character Guidance Manual* (FM 16-100, June 26, 1968), 7.

63. Headquarters, Department of the Army, "Education and Training, Code of Conduct" (Army Regulation 350-30; Washington, D.C., Dec. 30, 1957), 2–3.

64. See, for example, *Character Guidance Manual* (FM 16-100, Mar. 1961), 3, 28–29; U.S. Army Chaplain School, *Character Guidance Common Subjects: Instructional Guide for U.S. Army Service Schools* (Fort Hamilton, N.Y.: U.S. Army Chaplain School, Dec. 1968), 14.

65. *Character Guidance Manual* (FM 16-100, Mar. 1961), Appendix.

66. Ibid., 2. See also *Character Guidance Manual* (FM 16-100, June 26, 1968), 12.

67. *Character Guidance Manual* (FM 16-100, Mar. 1961), 14. See also *Character Guidance Manual* (FM 16-100, June 26, 1968), 10.

68. Department of the Army, Office of the Chief of Chaplains (OCCH), *Summary of Major Events and Problems 1 July 1960 to 30 June 1961* (Washington,

D.C.: n.p., n.d.), 26; Brady, "Narrative Description of the Character Guidance Program," 6.

69. Chaplain (Maj. Gen.) Luther D. Miller, "Post-War Program," *Army and Navy Journal* 85 (Sept. 20, 1947): 74.

70. Edward M. Coffman, *The Old Army: A Portrait of the American Army in Peacetime, 1784–1898* (New York: Oxford Univ. Press, 1986), 391.

71. Richard M. Budd, *Serving Two Masters: The Development of American Military Chaplaincy, 1860–1920* (Lincoln: Univ. of Nebraska Press, 2002), 74.

72. Nancy K. Bristow, *Making Men Moral: Social Engineering During the Great War* (New York: New York Univ. Press, 1996), 7–17.

73. Earl F. Stover, *Up from Handymen: The United States Army Chaplaincy, 1865–1920* (Washington, D.C.: Office of the Chief of Chaplains, Department of the Army, 1977), 196, 198–99. See also Gynther Storaasli, "The Evolution of the Chaplaincy," *Chaplain* 8 (May–June 1951): 36–37.

74. Robert L. Gushwa, *The Best and Worst of Times: The United States Army Chaplaincy, 1920–1945* (Washington, D.C.: Office of the Chief of Chaplains, Department of the Army, 1977), 131, 186, 187.

75. Daniel B. Jorgensen, *The Service of Chaplains to Army Air Units, 1917–1946* (Washington, D.C.: U.S. Government Printing Office, n.d.), 146, 152.

76. Chaplain James L. McBride, "The Chaplain's Handicaps," *Chaplain* 2 (Oct. 1945): 31, 34; "The Chaplain and His C.O.," *Chaplain* 2 (Nov. 1945): 40; Presbyterian Church in the USA, *General Assembly Minutes,* May 28, 1946, pp. 182–84, photocopy received from Diana Ruby Sanderson, Administrator, Presbyterian Study Center, Montreat, N.C., Aug. 22, 1988; and see also *NYT,* May 26 and May 29, 1946, pp. 30 and 2, respectively.

77. Samuel McCrae Cavert and C. Oscar Johnson, "Stand Behind the Chaplains!" *Chaplain* 4 (Jan. 1947): 27. The two church leaders made their overseas tour under appointment by the chiefs of chaplains of the army and navy and with the approval of the secretary of war.

78. Frank Pace Jr., "Protection of Moral Standards," June 18, 1951, in George C. Marshall Papers, SD 000.3 (1951) (Uncl) Box 607, George C. Marshall Research Library, Lexington, Va. See also *NYT,* Oct. 19, 1952, p. 49.

79. *NYT,* May 13, 1948, p. 14, July 25, 1952, p. 5, June 6, 1955, p. 25, July 24, 1955, p. 28, July 25, 1955, p. 15, July 2, 1956, p. 23; Devers, "Training the Army of Today," 6–7; Venzke, *Confidence in Battle,* 58; Eisenhower, "Remarks at the 22d Annual Convention of the Military Chaplains Association," 462; Bruce C. Clarke, "The Layman Looks at the Chaplain," *Chaplain* 20 (Aug. 1963): 10–13.

80. On the public relations campaign, see Gilbert, *Redeeming Culture,* chap. 5; on chaplains as "publicity agents," Venzke, *Confidence in Battle,* 42.

81. See, for example, *NYT,* Sept. 16, 1950, p. 12, July 25, 1952, p. 5, Jan. 11, 1953, p. 3, Mar. 4, 1956, p. 82, Jan. 1, 1958, p. 22, Apr. 15, 1958, p. 35; Rabbi Aryeh Lev, "The Rabbinate Reports for Duty," *Chaplain* 9 (Nov.–Dec. 1952): 20; "GIs Go to Church,"

National Council Outlook 4 (Jan. 1954);14; Douglas G. Scott, "Our Sons and Daughters in This Military Era," and James DeForest Murch, "Responsibility in Leadership and Service," *United Evangelical Action,* Mar. 1, 1955, 9–10, and May 15, 1955, 21, respectively.

82. "Lutheran Service Commission, National Lutheran Council and the Lutheran Church–Missouri Synod, Annual Report—1956," 82, 88, 89, typescript, Archives of the Evangelical Lutheran Church in America, Chicago.

83. See, for example, John M. Swomley Jr., "End Conscription in 1959!" *Christian Century* (hereafter *CC*) 76 (Jan. 7, 1959): 14–17; "Is Conscription Becoming a National Habit?" *CC* 76 (Feb. 18, 1959): 189; "UMT and Our Spiritual Heritage," *National Council Outlook* 9 (Feb. 1959): 18.

84. William V. Kennedy, "Can a Catholic Be Anti-Military?" *Catholic World* 187 (Apr. 1958): 14–15.

85. John C. Bennett, "The Draft and Christian Vocation," *CC* 73 (Aug. 22, 1956): 970–72; Bruce Morgan, "Christian Vocation and Military Service," *Theology Today* 14 (Jan. 1958): 519.

86. OCCH, *Summary of Major Events and Problems,* 26; Brady, "Narrative Description of the Character Guidance Program," 5.

87. "Interview with Major General Gerhardt W. Hyatt (U.S. Army Chief of Chaplains, Ret.) by Chaplain Colonel Jay H. Ellens, 22 September 1976, Detroit Metro Airport," 1–2, typescript, Gerhardt W. Hyatt Papers, Manuscript Department, U.S. Army Military History Institute, Carlisle Barracks, Pa.; Army Chief of Chaplains Patrick J. Ryan quoted in *NYT,* Oct. 7, 1956, p. 75; Navy Chief of Chaplains (Rear Admiral) George A. Rosso quoted in "Three Chiefs of Chaplains Discuss Moral Training Throughout Armed Forces," *Military Chaplain* (hereafter *MC*) 34 (May–June 1961): 8.

2. The Sixties Watershed

1. On the 1960s as a "watershed," see Wade Clark Roof and William McKinney, *American Mainline Religion: Its Changing Shape and Future* (New Brunswick, N.J.: Rutgers Univ. Press, 1987), 8, 9, 11; Sydney E. Ahlstrom, *A Religious History of the American People,* 2 vols. (Garden City, N.Y.: Doubleday, 1975), 2:600; Sydney E. Ahlstrom, "The Radical Turn in Theology and Ethics: Why It Occurred in the 1960's," *Annals of the American Academy of Political and Social Science* 387 (Jan. 1970): 1. Portions of this chapter first appeared in Anne C. Loveland, "Prophetic Ministry and the Military Chaplaincy during the Vietnam Era," in *Moral Problems in American Life: New Perspectives on Cultural History,* ed. Karen Halttunen and Lewis Perry (Ithaca, N.Y.: Cornell Univ. Press, 1998), 245–58.

2. Martin E. Marty, "Foreword," in *Understanding Church Growth and Decline, 1950–1978,* ed. Dean R. Hoge and David A. Roozen (New York: Pilgrim Press, 1979), 10.

3. Jackson W. Carroll, Douglas W. Johnson, and Martin E. Marty, *Religion in America: 1950 to the Present* (San Francisco: Harper & Row, 1979), 13; Dean R. Hoge and David A. Roozen, eds., *Understanding Church Growth and Decline, 1950–1978* (New York: Pilgrim Press, 1979), 22; Wade Clark Roof, "America's Voluntary Establishment: Mainline Religion in Transition," *Daedalus* 111 (Winter 1982): 170.

4. George Gallup Jr. and Jim Castelli, *The People's Religion: American Faith in the 90's* (New York: Macmillan, 1989), Table 2.1, 24, 13; Carroll, Johnson, and Marty, *Religion in America,* Table 1, 9.

5. In 1954 Congress passed legislation that inserted the words "under God" in the Pledge of Allegiance, put "In God We Trust" on all U.S. coins and paper money, and made "In God We Trust" the nation's official motto. Ahlstrom, *Religious History of the American People* 2:450–51; Mark Silk, *Spiritual Politics: Religion and America Since World War II* (New York: Simon and Schuster, 1988), 99.

6. Peter L. Berger, "From the Crisis of Religion to the Crisis of Secularity," in *Religion and America: Spiritual Life in a Secular Age,* ed. Mary Douglas and Steven Tipton (Boston: Beacon Press, 1982), 15.

7. *NYT,* June 1, 1969, p. 39. On the decline in church attendance, see also Gallup and Castelli, *People's Religion,* 30–32.

8. Gallup and Castelli, *People's Religion,* Tables 2.11 and 2.14, pp. 36, 41.

9. Rodney Stark and Charles Y. Glock, *American Piety: The Nature of Religious Commitment* (Berkeley and Los Angeles: Univ. of California Press, 1968), 204–5.

10. Martin E. Marty, "Religion in America Since Mid-century," *Daedalus* 111 (Winter 1982): 149, 151, 152, 157; Jeffrey K. Hadden, "Religious Broadcasting and the Mobilization of the New Christian Right," *Journal for the Scientific Study of Religion* 26 (Mar. 1987): 2–4.

11. Roof and McKinney, *American Mainline Religion,* 15–16, 95, 96.

12. Ibid., 16, 96–97, 237.

13. Richard G. Hutcheson Jr., *Mainline Churches and the Evangelicals: A Challenging Crisis?* (Atlanta: John Knox Press, 1981); Roof, "America's Voluntary Establishment," 170.

14. The term "moral vision" is from Hunter, *Culture Wars,* 107. In writing this section on the two strategies, I found three books especially helpful: Hunter's *Culture Wars*; Robert Wuthnow, *The Restructuring of American Religion: Society and Faith Since World War II* (Princeton, N.J.: Princeton Univ. Press, 1988); and Andrew R. Murphy, *Prodigal Nation: Moral Decline and Divine Punishment from New England to 9/11* (New York: Oxford Univ. Press, 2009).

15. On the "secular constituency," see Roof and McKinney, *American Mainline Religion,* 115.

16. I use the term "conservative evangelicals" to distinguish them from the "radical evangelicals," since the two groups differed significantly with regard to social and cultural issues. On that matter and for the NAE as the "voice of

evangelical Christianity," see Anne C. Loveland, *American Evangelicals and the U.S. Military, 1942–1993* (Baton Rouge: Louisiana State Univ. Press, 1996), 133–36, and 245.

17. On "exclusivism" and "inclusivism," see Ronald L. Massanari, "The Pluralisms of American 'Religious Pluralism,'" *Journal of Church and State* 40 (Summer 1998): 589–601. On the sectarian, exclusivist character of evangelicalism, see Christian Smith, *American Evangelicalism: Embattled and Thriving* (Chicago: Univ. of Chicago Press, 1998), chap. 5.

18. Hutcheson, *Mainline Churches and the Evangelicals,* 33.

19. Lausanne Covenant quoted in ibid., 96.

20. Homer Duncan quoted in Grant Wacker, "Uneasy in Zion: Evangelicals in Postmodern Society," in *Evangelicalism and Modern America,* ed. George Marsden (Grand Rapids, Mich.: William B. Eerdmans, 1984), 297; and see also Jerry Falwell quoted in Roof and McKinney, *American Mainline Religion,* 187. On "the legacy of the 1960s," according to Pat Robertson, see Murphy, *Prodigal Nation,* 99.

21. See, for example, Murphy, *Prodigal Nation,* 92, 93.

22. Ed Dobson, Richard John Neuhaus, and Charles Colson quoted in ibid., 88. See also Hunter, *Culture Wars,* 145–46.

23. Falwell in *USA Today,* June 7, 1985, quoted in Roof and McKinney, *American Mainline Religion,* 187, 268n2.

24. Pat Robertson quoted in Hunter, *Culture Wars,* 113. On the restoration theme, see Roof and McKinney, *American Mainline Religion,* 30–31.

25. Murphy, *Prodigal Nation,* 96, 101, 102, 104; Hunter, *Culture Wars,* 109–10.

26. Describing the late-nineteenth-century "golden age of liberal theology," Sydney Ahlstrom wrote that "liberalism was, first of all, a point of view which, like the adjective 'liberal' as we commonly use it, denotes both a certain generosity or charitableness toward divergent opinions and a desire for intellectual 'liberty.'" Ahlstrom, *Religious History of the American People* 2:243.

27. Wuthnow, *Restructuring of American Religion,* 305.

28. Hutcheson, *Mainline Churches and the Evangelicals,* 21, 96, 104.

29. Martin E. Marty, "Protestantism Enters Third Phase," *CC* 78 (Jan. 18, 1961): 72, 73.

30. Hutcheson, *Mainline Churches and the Evangelicals,* 21. As Hutcheson's use of the pronoun "we" suggests, he was himself a mainline, liberal Protestant. While serving as a U.S. Navy chaplain, he played a very active role in the leadership of the Presbyterian Church in the United States. See Richard G. Hutcheson Jr., *The Churches and the Chaplaincy* (Atlanta: John Knox Press, 1975), 8–9. For an interesting discussion of theological diversity within the mainline churches, see James H. Smylie, "Church Growth and Decline in Historical Perspective: Protestant Quest for Identity, Leadership, and Meaning," in Hoge and Roozen, *Understanding Church Growth and Decline,* 89–90.

31. Thomas C. Berg, "'Proclaiming Together'? Convergence and Divergence in Mainline and Evangelical Evangelism, 1945–1967," *Religion and American Culture* 5 (Winter 1995): 51, 61, 63.

32. Hutcheson, *Mainline Churches and the Evangelicals,* 117, 177, and see also 138.

33. Sidney E. Mead, "The Post-Protestant Concept and America's Two Religions," *Religion in Life* 33 (Spring 1964): 197; J. E. W. Jr. [James E. Wood Jr.], "Editorial: The United States as a Pluralistic Society," *Journal of Church and State* 8 (Autumn 1966): 332, 335; Marty, "Protestantism Enters Third Phase," 72, 73.

34. Franklin H. Littell, "Religious Liberty in a Pluralistic Society," *Journal of Church and State* 7 (Autumn 1966): 431–32, 438. Littell was the author of *From State Church to Pluralism: A Protestant Interpretation of Religion in American History* (1962).

35. Ibid., 434. See also Mead, "Post-Protestant Concept," 199. For a Roman Catholic's celebration of religious freedom and secular government, see John Courtney Murray, S.J., *We Hold These Truths: Catholic Reflections on the American Proposition* (New York: Sheed and Ward, 1960).

36. Robert McAfee Brown, "Theology and the Gospel: Reflections on Theological Method," in *Theology and Church in Times of Change,* ed. Edward LeRoy Long Jr. and Robert T. Handy (Philadelphia: Westminster Press, 1970), 19; NCC Commission on Evangelism quoted in Berg, "'Proclaiming Together'?" 66.

37. For a good summary of this view and its spokesmen, see Charles Y. Glock, Benjamin B. Ringer, and Earl R. Babbie, *To Comfort and to Challenge: A Dilemma of the Contemporary Church* (Berkeley: Univ. of California Press, 1967), 3–7.

38. This summary view of the origins and content of the secular theology is based on the following: Ahlstrom, *Religious History of the American People* 2:599–604, 608–9; Ahlstrom, "Radical Turn in Theology and Ethics," and Richard Henry Luecke, "Protestant Clergy: New Forms of Ministry, New Forms of Training," in *Annals of the American Academy of Political and Social Science* 387 (Jan. 1970): 1–13 and 87–95, respectively; Langdon Gilkey, "Social and Intellectual Sources of Contemporary Protestant Theology in America," and Harvey G. Cox, "The 'New Breed' in American Churches: Sources of Social Activism in American Religion," *Daedalus* 96 (Winter 1967): 69–98 and 135–50, respectively; and Daniel Callahan, "The Quest for Social Relevance," in *Religion in America,* ed. William G. McLoughlin and Robert N. Bellah (Boston: Houghton Mifflin, 1968), 339, 343–441.

39. The two decisions were *Engel v. Vitale* (1962) and *Abington Township School District v. Schempp* (1963).

40. Loveland, *American Evangelicals,* chap. 2.

41. "Why We Fight," *Eternity* 3 (Aug. 1952): 15.

42. Clyde W. Taylor, "Evangelicals Examine Ecumenicity," *United Evangelical Action* (Apr. 1963): 20.

43. Headquarters, Department of the Army, *The Chaplain* (Field Manual 16–5; Washington, D.C., Apr. 15, 1958) 13; Headquarters, Department of the Army, *The Chaplain* (Field Manual 16-5; Washington, D.C., Aug. 27, 1964), 7.

44. Roy J. Honeywell, "The New Religion," *MC* 31 (Apr. 1958): 7.

45. Robert M. Rutan, "We CAN Preach Doctrinal Sermons," *Chaplain* 15 (Apr. 1958): 5–7; Glenn J. Witherspoon, "Chaplains' Queries," *Chaplain* 16 (Feb. 1959): 17–18.

46. Venzke, *Confidence in Battle*, 109–10.

47. Loveland, *American Evangelicals*, 89–91.

48. Venzke, *Confidence in Battle*, 112.

49. DoD Judge Advocate General opinion, Aug. 3, 1964, addressed to the Armed Forces Chaplain Board, quoted in Martin H. Scharlemann, *Air Force Chaplains, 1961–1970* (Washington, D.C.: U.S. Government Printing Office, 1975), 113.

50. Loveland, "Character Education in the U.S. Army," 806–7; A. Ray Appelquist, ed., *Church, State and Chaplaincy: Essays and Statements on the American Chaplaincy System* (n.p.: General Commission on Chaplains and Armed Forces Personnel, 1969), 21–24, 44.

51. Loveland, "Character Education in the U.S. Army," 808.

52. Ibid., 807–8; *Character Guidance Manual* (FM 16-100, June 26, 1968), 10.

53. Loveland, "Character Education in the U.S. Army," 809.

54. U.S. Army Office of the Chief of Chaplains (hereafter OCCH), *Historical Review, 1 January 1967 to 30 June 1968* (Washington, D.C.: U.S. Government Printing Office, 1969), 89, 90; OCCH, *Historical Review, 1 July 1968 to 30 June 1969* (Washington, D.C.: U.S. Government Printing Office, 1970), 72–73.

55. OCCH, *Historical Review*, 1967–68, 90.

56. OCCH, *Historical Review*, 1968–69, 73–75.

57. Ibid., 73–75.

58. Ibid., 75.

59. Ibid., 75, 77, 78–79.

60. Ibid., 79.

61. "Pentagon Piety," *Nation* 208 (Mar. 17, 1969): 325, 326; Bray quoted in *Congressional Record*, 91st Cong., 1st sess., 1969, 115, pt. 6:7343–44; "Chaplains Affected as Army Prohibits Religious References in Character Guidance," *Religious News Service*, Mar. 28, 1969, 28–30; Lawrence L. Knutson, "Lectures Revised: Army Bans 'God' in Talks," *Evening Star*, Mar. 28, 1969, p. A-18; *NYT*, Mar. 29, 1969, pp. 1, 4; *Washington Post* (hereafter *WP*), Mar. 29, 1969, p. C-9.

62. "Chaplains Affected as Army Prohibits Religious References in Character Guidance," 29, 30; William Willoughby, "Chaplains' Role Under New Scrutiny," *Christianity Today* (hereafter *CT*), Apr. 25, 1969, pp. 22–23, 32; Loveland, "Character Education in the U.S. Army," 811.

63. OCCH, *Historical Review*, 1968–69, 82; *NYT*, Apr. 4, 1969, p. 4.

64. See, for example, "National Award Honoree" and "MCA Convention Resolutions, 1969," *MC* 42 (Nov.–Dec. 1969): 3, 21.

65. OCCH, *Historical Review*, 1968–69, 83–87.

66. Loveland, "Character Education in the U.S. Army," 812–17; John W. Brinsfield Jr., *Encouraging Faith, Supporting Soldiers: The United States Army*

Chaplaincy, 1975–1995 (Washington, D.C.: Office of the Chief of Chaplains, Department of the Army, 1997), pt. 1, 136–37.

3. Chaplains under Fire

1. R. G. Hutcheson Jr., "Should the Military Chaplaincy Be Civilianized?" *CC* 90 (Oct. 31, 1973): 1072. Portions of this chapter first appeared in Loveland, "Prophetic Ministry and the Military Chaplaincy," 245–58.

2. James H. Smylie, "American Religious Bodies, Just War, and Vietnam," *Journal of Church and State* 11 (Autumn 1969): 383–408; David E. Settje, *Lutherans and the Longest War: Adrift on a Sea of Doubt About the Cold and Vietnam Wars, 1964–1975* (Lanham, Md.: Lexington Books, 2007); David E. Settje, *Faith and War: How Christians Debated the Cold and Vietnam Wars* (New York: New York Univ. Press, 2011).

3. "Text of the President's Address on U.S. Policies in Vietnam," *NYT,* Apr. 8, 1965, p. 16.

4. See, for example, John Cooney, *The American Pope: The Life and Times of Francis Cardinal Spellman* (New York: Times Books, 1984), 238, 245, 293–94, 306–7; Edward L. R. Elson, "War and Peace in Vietnam and the Church's Message," *Congressional Record,* 90th Cong., 1st sess., 1967, 113, pt. 8:10143–44; Reverend George R. Davis, "The Vietnam War: A Christian Perspective," in *The Vietnam War: Christian Perspectives,* ed. Michael P. Hamilton (Grand Rapids, Mich.: William B. Eerdmans, 1967), 45–60; James T. Johnson, "Just War in the Thought of Paul Ramsey," *Journal of Religious Ethics* 19 (Fall 1991): 183–207; Loveland, *American Evangelicals,* chap. 9; "Text of Bishops' Pastoral Statement on Peace and Vietnam," *NYT,* Nov. 22, 1966, p. 18. Scharlemann, *Air Force Chaplains,* 124–25, contended that most of the church bodies that issued declarations on the war "either ignored or rejected the concept of a just war as it had been formulated since the days of Saint Augustine." See also, on this point, Settje, *Lutherans and the Longest War,* 93, 103.

5. Venzke, *Confidence in Battle,* 127, 146, 148.

6. *NYT,* June 22, 1971, p. 37.

7. As we shall see in the next chapter, U.S. Army chaplains' memoirs of service in Vietnam, published several decades after the war, provide fuller and more candid explanations of their views on the war.

8. See, for example, Eugene F. Klug, "Christianity, The Chaplaincy, And Militarism," *Springfielder* 31 (Winter 1968): 26; and Chaplain William G. Devanny, "The Ecumenical Movement and the Military," *Military Review* 47 (Mar. 1967): 28–34.

9. OCCH, *Historical Review, 1 July 1969 to 30 June 1970* (Washington, D.C.: U.S. Government Printing Office, 1971), 52–53.

10. *NYT,* June 22, 1971, p. 70. On Luther, see Lewis W. Spitz Jr., "Impact of the Reformation on Church-State Issues," in *Church and State Under God,* ed. Albert G.

Huegli (St. Louis, Mo.: Concordia Publishing House, 1964), 73. Hyatt's view was similar to that of two leading prowar Lutherans; see Settje, *Lutherans and the Longest War,* 101–5, 107.

11. *WP,* Sept. 2, 1969, p. A6.

12. *NYT,* June 22, 1971, p. 70.

13. "Wounded Chaplain Tells of Ministry in Vietnam War," *Mighty Fortress* 16 (Jan. 1967): n.p.; Chaplain R. J. Wood, "Citizenship in a Democracy: It's Your Heritage to Defend," *Army Digest* 21 (Dec. 1966): 12.

14. For one such example, see Rabbi Bruce M. Freyer, "Reflective Analysis of the Chaplain's Role," in *Military Chaplains: From Religious Military to a Military Religion,* ed. Harvey G. Cox Jr. (New York: American Report Press, 1971), 121–22.

15. OCCH, *Historical Review,* 1969–70, 47.

16. Philip Caputo, *A Rumor of War* (New York: Ballantine Books, 1977), 168–70.

17. Mangham quoted in Rosemary Sawyer, "The BDU-pulpit: Ministering to the Military," *Stars and Stripes,* July 26, 1989, 16.

18. William J. Hughes, "A Chaplain's Dilemma About Viet Nam," *Christian Advocate,* Sept. 30, 1971, quoted in Cox, *Military Chaplains,* iv.

19. Billy Libby, "The Chaplain's Allegiance to His Church," *Military Chaplains' Review* (hereafter *MCR*) (Fall 1983): 34–36.

20. *NYT,* Jan. 30, 1972, sec. 4, p. 6. Hughes's statement appeared in Cox, *Military Chaplains,* iv.

21. Charles DeBenedetti and Charles Chatfield, *An American Ordeal: The Antiwar Movement of the Vietnam Era* (Syracuse, N.Y.: Syracuse Univ. Press, 1990), 87, 95, 96, 100; Mitchell K. Hall, *Because of Their Faith: CALCAV and Religious Opposition to the Vietnam War* (New York: Columbia Univ. Press, 1990), 9.

22. Smylie, "American Religious Bodies," 389, 389n22; Hall, *Because of Their Faith,* 5, 7, 9.

23. Hall, *Because of Their Faith,* 7, 9, 10.

24. Ibid., 11, 13–25.

25. On the founding and interfaith orientation of CALCAV, as well as biographical information on its prominent members, see ibid., chapters 1 and 2. Miller was a member of the United Church of Christ, who had served as assistant editor, then managing editor of *Fellowship,* the journal of the Fellowship of Reconciliation, 1956 to 1962. (Ethan Vesely-Flad, Communications Director, FOR, to author, e-mail, July 16, 2009.) Gordon Zahn was a World War II conscientious objector, professor of sociology at Loyola University, Chicago, in the mid-1960s, and at the University of Massachusetts, Boston, in the late 1960s (Paul Boyer, *By the Bomb's Early Light: American Thought and Culture at the Dawn of the Atomic Age* [New York: Pantheon Books, 1985], 346). For one of Zahn's statements on pacifism, see "Pacifism and America," *Sign* (Aug. 1966): 15–16. On the cooperation between pacifists and liberal, mainline Protestants in the antiwar movement, see, for example, John C. Bennett,

"Christian Realism in Vietnam," *America* 114 (Apr. 30, 1966): 616; Richard R. Fernandez, "Guest Editorial: The Air War in Indochina: Some Responses," *CC* 88 (Dec. 1, 1971): 1404; Hall, *Because of Their Faith,* 98.

26. Robert McAfee Brown, Abraham J. Heschel, and Michael Novak, "Introduction," in *Vietnam: Crisis of Conscience,* by Robert McAfee Brown, Abraham J. Heschel, and Michael Novak (New York: Association Press, Behrman House, Herder and Herder, 1967), 7.

27. Robert McAfee Brown, "An Appeal to the Churches and Synagogues," in Brown, Heschel, and Novak, *Vietnam,* 67–68.

28. For the quoted words and phrases, in the order they appear in the text, see Brown, Heschel, and Novak, *Vietnam,* 48, 65, 8, 9, 7, 62, 8, 105, 65, 105.

29. Richard R. Fernandez, "Foreword," in Cox, *Military Chaplains,* iii–iv.

30. Harvey G. Cox Jr., "Introduction: The Man of God and the Man of War," in Cox, *Military Chaplains,* vi, vii, x, xi.

31. Waldo W. Burchard, "Role Conflicts of Military Chaplains," *American Sociological Review* 19 (Oct. 1954): 528–35. Burchard received his Ph.D. from the University of California at Berkeley in 1953. The title of the dissertation was "The Role of the Military Chaplain."

32. Burchard quoted in Alex B. Aronis, "A Summary of Research Literature on the Military Chaplain," *Chaplain* 29 (Summer Quarter 1972): 7; Burchard, "Role Conflicts of Military Chaplains," 529–30. Burchard defined the "Christian philosophy" in terms of "the doctrines of love, of universal brotherhood, of peace, and of non-resistance to evil, and the commandment, 'You shall not kill.'" In effect, he was defining the Christian pacifist philosophy.

33. Two members of Burchard's dissertation committee were prominent sociologists and they referred to his dissertation in their own publications (Aronis, "Summary of Research," 8). Burchard's *American Sociological Review* article was reprinted in *Religion, Society, and the Individual,* ed. J. Milton Yinger (New York: Macmillan, 1957), and in 1959 and 1960 brief references to the article appeared in *Sociology and the Military Establishment,* published by the Russell Sage Foundation, and *The Professional Soldier,* by Morris Janowitz.

34. Pierre Berton, *The Comfortable Pew: A Critical Look at Christianity and the Religious Establishment in the New Age,* U.S. ed., 4th printing (Philadelphia: J. B. Lippincott, 1965), vii, 67. For a critical view of the Burchard thesis, which notes its pacifist bias, see Clarence L. Abercrombie III, *The Military Chaplain* (Beverly Hills, Calif.: Sage, 1976), 107–8. In the 1980s, when the Army Office of the Chief of Chaplains was developing new Chaplain Corps doctrine, two field manuals published during that decade pointedly observed that the two roles were "basically complementary." Headquarters, Department of the Army, *The Chaplain and Chaplain Assistant in Combat Operations* (Field Manual 16-5; Washington, D.C., Dec. 10, 1984) , 7; Headquarters, Department of the Army, *Religious Support*

Doctrine: The Chaplain and Chaplain Assistant (Field Manual 16-1, Final Coordinating Draft; Washington, D.C., Apr. 1989), p. 1.14.

35. For a general application of the role-conflict thesis, see Richard John Neuhaus, "The Anguish of the Military Chaplain," *Lutheran Forum* (Nov. 1967): 16.

36. Robert E. Klitgaard, "Onward Christian Soldiers," *CC* 87 (Nov. 18, 1970): 1378–79. According to an editor's note (1337), Klitgaard had recently completed a period of active duty in the army and was currently working on a doctorate at Harvard University's Kennedy School of Government.

37. OCCH, *Historical Review,* 1967–68, 37. In the late 1960s and early 1970s, a number of religious groups supported selective conscientious objection, including the National Council of Churches, the United Presbyterian Church in the USA, the Disciples of Christ, the Council for Christian Social Action of the United Church of Christ, the American Baptist Convention, the North American Area of the World Alliance of Reformed Churches, the Roman Catholic Bishops, and the United Methodist Church. *NYT,* Feb. 5, 1967, sec. 4, p. 5, Feb. 24, 1967, p. 5, Oct. 23, 1971, p. 9, Mar. 28, 1972, p. 16, Apr. 27, 1972, p. 16.

38. "United States v. Seeger," in *Draftees or Volunteers: A Documentary History of the Debate Over Military Conscription in the United States, 1787–1973,* ed. John Whiteclay Chambers II (New York: Garland, 1975), 525.

39. *NYT,* June 26, 1970, p. 1, Sept. 1, 1971, p. 1.

40. Peter Barnes, *Pawns: The Plight of the Citizen-Soldier* (New York: Alfred A. Knopf, 1972), 141–42. See also Headquarters, Department of the Army, "Personnel Separations: Conscientious Objection" (Army Regulation 635–20; Washington, D.C., Aug. 15, 1970).

41. OCCH, *Historical Review,* 1967–68, 37. See also Scharlemann, *Air Force Chaplains,* 124.

42. Gordon Zahn, "What Did You Do during the War, Father?" *Commonweal* 90 (May 2, 1969): 199. In 1969, Zahn published *The Military Chaplaincy: A Study in Role Tension in the Royal Air Force* (Toronto: Univ. of Toronto Press, 1969), in which he advanced conclusions similar to those Burchard had presented in his dissertation.

43. Gordon Zahn, "Sociological Impressions of the Chaplaincy," in Cox, *Military Chaplains,* 80, and see also 157n35.

44. United Presbyterian Church in the USA, *Minutes of the General Assembly* (New York: Office of the General Assembly, 1975), 477; United Church of Christ, "Ministries to Military Personnel," in *Advance Reports for General Synod Nine, United Church of Christ, 1973, St. Louis, MO,* by United Church of Christ (n.p.: n.p., [1973]), 76.

45. United Church of Christ, "Ministries to Military Personnel," 76. For a similar view, by an official of the UCC Board for Homeland Ministries, see Rev. Ralph Weltge, "An Honest Letter to the Not-Yet-Drafted," *Focus* (Winter 1970): 4.

46. Peter Berger and Daniel Pinard, "Military Religion: An Analysis of Education Materials Disseminated by Chaplains," in Cox, *Military Chaplains,* 88, 91, 93, 94, 99. Berger and Pinard explained that they chose to write about voluntary religious instruction, rather than Character Guidance, because they believed an analysis of its materials would prove even "more revealing" of the chaplains' bias in favor of the military (88). They did contend that the chaplain's role in the CG program was that of "an indoctrination agent," acting "in behalf of the military," but they said that was to be expected, that the chaplain was "supposed to represent an official military viewpoint." In their article, they cited both Burchard's *American Sociological Review* article and Zahn's book on Royal Air Force chaplains.

47. United Church of Christ, "Ministries to Military Personnel," 81, 82.

48. Ibid., 82.

49. Cox, "Introduction," in Cox, *Military Chaplains,* x. See also Zahn, "Sociological Impressions of the Chaplaincy"; and Berger and Pinard, "Military Religion," 63 and 90.

50. Cox, "Introduction," x.

51. Gordon C. Zahn, "The Scandal of the Military Chaplaincy," *Judaism* 18 (Summer 1969): 313, 314, 315; Zahn, "What Did You Do During the War, Father?" 196, 197, 198. *Judaism* was published by the American Jewish Congress, *Commonweal* by lay Catholics.

52. United Church of Christ, "Ministries to Military Personnel," 82, 86.

53. Clergy and Laymen Concerned About Vietnam (CALCAV), *In the Name of America: The Conduct of the War in Vietnam by the Armed Forces of the United States as Shown by Published Reports, Compared with the Laws of War Binding on the United States Government and on Its Citizens* (Annandale, Va.: Turnpike Press, 1968), 1, 10.

54. On March 16, 1968, in My Lai (4), a sub-hamlet of the village of Son My, some five hundred Vietnamese noncombatant men, women, and children were murdered by soldiers of Charlie Company, Task Force Barker, 11th Brigade, Americal Division. Commanders and their staffs within the division concealed the incident; not until November 13, 1969, were the first reports of the massacre published in the *New York Times* and other newspapers. Three chaplains were involved in the investigation of the incident, two of whom were criticized in a report resulting from an inquiry conducted by Lt. Gen. William R. Peers one and a half years after the massacre. See David L. Anderson, ed., *Facing My Lai: Moving Beyond the Massacre* (Lawrence: Univ. Press of Kansas, 1998); Ronald H. Spector, *After Tet: The Bloodiest Year in Vietnam* (New York: Free Press, 1993), 203–6; Seymour M. Hersh, *Cover-Up: The Army's Secret Investigation of the Massacre at My Lai 4* (New York: Random House, 1972); Joseph Goldstein, Burke Marshall, and Jack Schwartz, *The My Lai Massacre and Its Cover-up: Beyond the Reach of Law? The Peers Commission Report with a Supplement and Introductory Essay on the Limits of Law* (New York: Free Press, 1976); Richard Hammer, *One Morning in the War: The Tragedy at Son My* (New York: Coward-McCann, [1970]); James S. Olson and Randy Roberts, eds., *My*

Lai: A Brief History with Documents (Boston: Bedford Books, 1998); Deborah Nelson, *The War Behind Me: Vietnam Veterans Confront the Truth About U.S. War Crimes* (New York: Basic Books, 2008).

55. Robert McAfee Brown, "Military Chaplaincy as Ministry," in Cox, *Military Chaplains,* 144–45.

56. Gordon Zahn, "Communications: The Military Chaplaincy," *Judaism* 18 (Fall 1969): 490.

57. Zahn, "Sociological Impressions of the Chaplaincy," 84–85.

58. Ibid., 84.

59. Brown, "Military Chaplaincy as Ministry," 144 (emphasis added).

60. Zahn, "Sociological Impressions of the Chaplaincy," 63 (emphasis added).

61. Zahn, "What Did You Do During the War, Father?" 198.

62. Cox, "Introduction," vi.

63. Brown, "Appeal to the Churches and Synagogues," 105.

64. Cox, "Introduction," vii, xi.

65. Ibid., vi.

66. "Commentary by Religious Leaders on the Erosion of Moral Constraint in Vietnam," in CALCAV, *In the Name of America,* 2. The signers of the commentary included John C. Bennett, Robert McAfee Brown, Harvey G. Cox, Robert Drinan, Joseph Fletcher, Abraham Heschel, Martin Luther King Jr., Martin Marty, Robert V. Moss Jr., and John Sheerin.

67. Richard Falk, "The Circle of Responsibility," *Nation* 210 (Jan. 26, 1970): 81.

68. Abraham J. Heschel, "The Moral Outrage of Vietnam," in Brown, Heschel, and Novak, *Vietnam,* 57.

69. Cox, "Introduction," vii–viii.

70. Unidentified persons quoted in Venzke, *Confidence in Battle,* 129; Rabbi Martin Siegel, "Notes of a Jewish Chaplain," in *Military Chaplains: From Religious Military to a Military Religion,* ed. Harvey G. Cox Jr. (New York: American Report Press, 1971), 111, 113; Klitgaard, "Onward Christian Soldiers," 1379; United Church of Christ, "Ministries to Military Personnel," 82; William Robert Miller, "Chaplaincy vs. Mission in a Secular Age," *CC* 83 (Nov. 2, 1966): 1336; Weltge, "Honest Letter to the Not-Yet-Drafted," 4; Zahn, "Scandal of the Military Chaplaincy," 313–19; Richard Neuhaus quoted in "Honest to God—or Faithful to the Pentagon?" *Time,* May 30, 1969, 49.

71. Klug, "Christianity, The Chaplaincy, And Militarism," 26; Clarence L. Reaser, "The Military Chaplain: God's Man or the Government's?" *Princeton Seminary Bulletin* 62 (Autumn 1969): 72, 73; Chester A. Pennington, "The Ministry of Our Chaplains," *Chaplain* 28 (Mar.–Apr. 1971): 7; Hutcheson, "Should the Military Chaplaincy Be Civilianized?" 1072, 1074; [General Commission on Chaplains and Armed Forces Personnel], "Armed Forces Chaplains: *All* Civilians? (A Feasibility Study)," *Chaplain* 29 (Spring Quarter 1972): 73.

72. "Why the Chaplaincy?" *Lutheran Witness* (May 1971): 9.

73. John R. Himes, "More on Chaplains," *Lutheran Quarterly* 19 (May 1967): 189.

74. Klitgaard, "Onward Christian Soldiers," 1379; William E. Austill and William L. Wells, "More on the Military Chaplaincy," *CC* 88 (Apr. 14, 1971): 467.

75. The General Commission on Chaplains and Armed Forces Personnel was established in 1917. In the 1960s and early 1970s, it was an independent chaplain-endorsing agency for some thirty-five Protestant denominations, which included members of the National Council of Churches as well as several evangelical churches. Among its many and varied purposes, it served as a liaison between its members and the chiefs of chaplains and Department of Defense as well as an advocate of chaplains' interests and professional improvement. It also acted, on occasion, as an advisor to chaplains and their denominations, issuing statements and "guidelines" on various chaplaincy matters, most of which were published in its journal, the *Chaplain*. In 1976–77, it changed its name to the Conference of Ecclesiastical Endorsing Agents for the Armed Forces, and in 1982 the name changed to the National Conference on Ministry to the Armed Forces (NCMAF). See Appelquist, *Church, State and Chaplaincy*, 98–99n47, and, for a list of its member denominations and consultative and contributing bodies, 122–25; Reverend Dr. S. David Chambers, "History of the National Conference on Ministry to the Armed Forces," Dec. 6, 1988, http://www.ncmaf.org/policies/PDFs/ChambersSpeech88.pdf/ (accessed Mar. 20, 2012); NCMAF website, http://www.ncmaf.org/ (accessed Jan. 20, 2009).

76. "Chaplains' Guidelines for Free and Responsible Expression of Conscience in the Military," *Chaplain* 27 (May–June 1970): 1, 3. The guidelines were written by a committee appointed by the General Commission's executive committee and subsequently adopted by the commission.

77. Sam Lamback, Dan Garrett, and John Brinsfield, "The Ministry in the Military," *Reflection: Journal of Yale Divinity School* 67 (Jan. 1970): 7, 8. Brinsfield was the YDS graduate, Lamback and Garrett the students.

78. Hughes, "Chaplain's Dilemma About Viet Nam," iv; Austill and Wells, "More on the Military Chaplaincy," 467; Reaser, "Military Chaplain," 73; Clarence L. Reaser, "Why This Ministry?" *Chaplain* 29 (Winter Quarter 1972): 42.

79. [General Commission on Chaplains and Armed Forces Personnel], "Armed Forces Chaplains," 45; Lamback, Garrett, and Brinsfield, "Ministry in the Military," 7; Hutcheson, "Should the Military Chaplaincy Be Civilianized?" 1073, 1075.

80. Reaser, "Why This Ministry?" 39, 40. On the chaplain's "essentially pastoral" role, see also Reaser, "Military Chaplain," 73.

81. [General Commission on Chaplains and Armed Forces Personnel], "Armed Forces Chaplains," 83, 45.

82. Lamback, Garrett, and Brinsfield, "Ministry in the Military," 8.

83. Albert F. Ledebuhr, "Military Chaplaincy: An Apologia," *CC* 83 (Nov. 2, 1966): 1332. See also Pennington, "Ministry of Our Chaplains," 8.

84. Hutcheson, "Should the Military Chaplaincy Be Civilianized?" 1075, 1073.

4. Navigating the Quagmire

1. According to Henry F. Ackermann, *He Was Always There: The U.S. Army Chaplain Ministry in the Vietnam Conflict* (Washington, D.C.: Office of the Chief of Chaplains, Department of the Army, 1989), 149, "The majority of [army] chaplains who ministered to the soldiers in Vietnam did so in one of the seven U.S. Army divisions, or in one of the separate infantry brigades," which "contained most of the U.S. soldiers who were actively engaged in combat operations in the Republic of Vietnam." According to Venzke, *Confidence in Battle,* 139, U.S. military personnel had been in Vietnam since December 1961, and the first army chaplains arrived in late February 1962.

2. *Chaplain* (FM 16-5, Aug. 27, 1964), 40–42; Headquarters, Department of the Army, *The Chaplain* (Field Manual 16-5; Washington, D.C., Dec. 26, 1967), 39–41. Army chaplains' "life-sharing" ministry of presence is one of the main themes of Ackerman's *He Was Always There,* based on surveys of 1,350 chaplains and 1,171 soldiers (see 220, 241). See also Venzke, *Confidence in Battle,* 152, 153.

3. Jerry Autry, *Gun-Totin' Chaplain* (San Francisco: Airborne Press, 2006), 88. Thirteen army chaplains and eight chaplain assistants lost their lives during the Vietnam conflict. "The Army Chaplaincy: Vietnam Remembered," http://www.usachcs.army.mil/SPECFEAT/Vietnam1.htm/ (accessed May 11, 2012).

4. Joseph P. Dulany, *Once a Soldier: A Chaplain's Story* (n.p.: n.p., 1971), 58; Claude D. Newby, *It Took Heroes: One Chaplain's Story and Tribute to Combat Veterans and Those Who Waited for Them,* 2nd ed., rev. (Bountiful, Utah: Tribute Enterprises, 2000), 9, 56; James D. Johnson, *Combat Chaplain: A Thirty-year Vietnam Battle* (Denton: Univ. of North Texas Press, 2001), 3, 16, 18, 29. In the introduction to his memoir (3), Johnson points out that it was written "almost verbatim from my diaries and extensive detailed journals that I kept" and that he also drew extensively from letters and tapes his wife had saved. It seems likely that other chaplains also relied on diaries, journals, and letters. Johnson's memoir gives the greatest sense of immediacy, mainly because it, unlike the other memoirs, is written in the present tense.

5. Dulany, *Once a Soldier,* 41; Autry, *Gun-Totin' Chaplain,* 3, 29, 54, 59; David E. Knight, "Supreme Six," in *Vietnam: The Other Side of Glory,* ed. William R. Kimball (New York: Ballantine Books, 1988), 43, 44; Johnson, *Combat Chaplain,* 9; Curt Bowers, *Forward Edge of the Battle Area: A Chaplain's Story* (Kansas City: Beacon Hill Press of Kansas City, 1987), 41–43.

6. *Chaplain* (FM 16-5, Aug. 27, 1964), 43; *Chaplain* (FM 16-5, Dec. 26, 1967), 42.

7. Colonel Samuel W. Hopkins Jr., *A Chaplain Remembers Vietnam* (Kansas City, Mo.: Truman Publishing, 2002), 231; Earl C. Kettler, *Chaplain's Letters: Ministry by "Huey" 1964–1965, the Personal Correspondence of an Army Chaplain from Vietnam* (Cincinnati: Cornelius Books, 1994), 3.

8. J. Robert Falabella, *Vietnam Memoirs: A Passage to Sorrow* (New York: Pageant Press International, 1971), 21; news story by Irene Holt, *Joplin Globe,* in "Chaplain Jim Young Returns from Vietnam," *Congressional Record,* 92nd Cong., 1st sess., 1971, 117, pt. 17:22877. See also Dulany, *Once a Soldier*, 50, 93, 94.

9. Newby, *It Took Heroes,* 57–58.

10. Dulany, *Once a Soldier,* 44, 51, 52.

11. Bowers, *Forward Edge of the Battle Area,* 41.

12. Johnson, *Combat Chaplain,* 29.

13. Autry, *Gun-Totin' Chaplain,* 115, 116, 118.

14. Johnson, *Combat Chaplain,* 29, 31. During most of his time in Vietnam, June 1967–June 1968, Johnson was in the 9th Infantry Division, 3/60th Battalion, which was engaged with naval forces in riverine operations (20).

15. Ackerman, *He Was Always There,* 226, and see also 241, 220.

16. Campbell quoted in Ackerman, *He Was Always There,* 149; Autry, *Gun-Totin' Chaplain,* 122. See also Falabella, *Vietnam Memoirs,* 53, 84–85.

17. *Chaplain* (FM 16-5, Aug. 27, 1964), 1, 40, 41, 42; *Chaplain* (FM 16-5, Dec. 26, 1967), 3, 41, 42; *Character Guidance Manual* (FM 16-100, Mar. 1961), 9; "Education and Training, Code of Conduct" (AR 350-30, Dec. 30, 1957), 2–3. The phrase about enabling the soldier to be "a devoted defender of the nation" did not appear in the 1967 field manual.

18. "Convention Addresses, Chiefs of Chaplains," *MC* 41 (May–June 1968): 26–27.

19. Vietnam Veterans Against the War, *The Winter Soldier Investigation: An Inquiry into American War Crimes* (Boston: Beacon Press, 1972), 8. See also Citizens Commission of Inquiry, *The Dellums Committee Hearings on War Crimes in Vietnam: An Inquiry into Command Responsibility in Southeast Asia* (New York: Vintage Books, 1972), 212. There is considerable evidence that soldiers and Marines serving in Vietnam did agonize over the moral and religious questions raised by the war. See, for example, Bernard Edelman, ed., *Dear America: Letters Home from Vietnam* (New York: W. W. Norton, 1985), 118–19, 131–32, 198; Al Santoli, *Everything We Had: An Oral History of the Vietnam War* (New York: Random House, 1981), 71; A. D. Horne, *The Wounded Generation: America After Vietnam* (Englewood Cliffs, N.J.: Prentice-Hall, 1981), 44, 165, 179–80; Myra MacPherson, *Long Time Passing: Vietnam & the Haunted Generation* (New York: New American Library, 1984), 238.

20. Michael Herr, *Dispatches* (New York: Alfred A. Knopf, 1978), 45; *WP,* Dec. 4, 1970, p. 18.

21. Kettler, *Chaplain's Letters,* 81, 157–58; Dulany, *Once a Soldier,* 56; Autry, *Gun-Totin' Chaplain,* 188. For chaplains' comments on the effect of civilian protest against the war, see Dulany, *Once a Soldier,* 91; Knight, "Supreme Six," 42; Hopkins, *Chaplain Remembers Vietnam,* 168.

22. Kettler, *Chaplain's Letters,* 165; Autry, *Gun-Totin' Chaplain,* 187, 189; Johnson, *Combat Chaplain,* 3, 72, 76, 135.

23. Kovic quoted in Timothy J. Lomperis, *"Reading the Wind": The Literature of the Vietnam War* (Durham, N.C.: Duke Univ. Press, 1987), 34, and see also 52; and Lt. Brian Sullivan, quoted in Edelman, *Dear America,* 132. For other chaplains' commentary on soldiers' nihilism, see Larry Haworth, *Tales of Thunder Run* (Eugene, Ore.: ACW Press, 2004), 67, 117; William P. Mahedy, *Out of the Night: The Spiritual Journey of Vietnam Vets* (New York: Ballantine Books, 1986), 125, 148. Richard Holmes, *Acts of War: The Behavior of Men in Battle* (New York: Free Press, 1985), 240–43, 286–91, presents a useful perspective on soldiers' spirituality and "fatalism" in combat.

24. Haworth, *Tales of Thunder Run,* 133. See also Knight, "Supreme Six," 42.

25. Kettler, *Chaplain's Letters,* 205, 75, 200, 152, 100, 78, 79, 101, 158, 187.

26. Falabella, *Vietnam Memoirs,* 17; Johnson, *Combat Chaplain,* 99, 106, 214; Dulany, *Once a Soldier,* 81, 93; Newby, *It Took Heroes,* 63, 87.

27. Based on questionnaires sent to chaplains and commanding officers who served in Vietnam, Clarence L. Abercrombie found a significant disparity in their views of the chaplain's role as advisor and consultant on morale. Each group was asked to rate eleven chaplain functions or tasks in order of their importance. The commanders deemed the following the most important: Helping "men gain the spiritual strength that will enable them to perform their duties more effectively despite the suffering and hardships of military operations." The chaplains listed the most important tasks as follows: first, administering the sacraments and conducting worship services; second, visiting "the sick and wounded"; and third, counseling troops on "personal problems." The task that the commanders rated first, promoting combat effectiveness, was rated fourth in importance by the chaplains. Abercrombie points out that while the commanders gave priority to tasks or functions that "legitimated" military service, the chaplains did not believe that their main function was "to enhance the effectiveness of the fighting machine and thus contribute to the chances of obtaining victory." Abercrombie, *Military Chaplain,* 74, 91.

28. Falabella, *Vietnam Memoirs,* 34, 132–33.

29. Johnson, *Combat Chaplain,* 180.

30. Dulany, *Once a Soldier,* 78.

31. *Chaplain* (FM 16-5, Aug. 27, 1964), 40; *Chaplain* (FM 16-5, Dec. 26, 1967), 39–40 (emphasis in original). For additional guidance promulgated by Chief of Chaplains Charles E. Brown Jr. in 1966, see OCCH, *Historical Review, 1 July 1965 to 31 December 1966* (Washington, D.C.: U.S. Government Printing Office, 1969), 64–66; and Venzke, *Confidence in Battle,* 149. A more elaborate statement of the regulation appeared in the 1984 chaplain manual:

> The Geneva Conventions recognize the chaplain as a non-combatant. The chaplain has protected status and, when detained, is not considered a prisoner of war. Chaplains neither bear arms nor receive training in the use of weapons. Violation of this non-combatant

> principle by a chaplain would endanger the protected status of all chaplains captured by the enemy. Chaplain assistants, however, are combatants. They provide security for the chaplain and serve as a combat resource for the unit. (*Chaplain and Chaplain Assistant in Combat Operations* [FM 16-5, Dec. 10, 1984], 4)

A subsequent update of the regulation, in which the chief of chaplains personally barred chaplains from carrying weapons, provoked considerable protest on the part of chaplains who had served in Vietnam. See Brinsfield, *Encouraging Faith,* pt. 1, 183, 196, 218n84; Dulany, *Once a Soldier,* 65; Newby, *It Took Heroes,* 25.

32. Autry, *Gun-Totin' Chaplain,* 137; Dulany, *Once a Soldier,* 64; Johnson, *Combat Chaplain,* 32; Newby, *It Took Heroes,* 22, 24; Kettler, *Chaplain's Letters,* 172; Falabella, *Vietnam Memoirs,* 72; Freyer, "Reflective Analysis of the Chaplain's Role," 117.

33. Rev. William P. Mahedy, "Sermon: July 4, 2004, St. David's, San Diego," http://www.ecusa-chaplain.org/072004.html/ (accessed Nov. 19, 2008); Newby, *It Took Heroes,* 23, 24; Dulany, *Once a Soldier,* 64; Johnson, *Combat Chaplain,* 8; Hopkins, *Chaplain Remembers Vietnam,* 109.

34. See, for example, Dulany, *Once a Soldier,* 65; Newby, *It Took Heroes,* 23.

35. Dulany, *Once a Soldier,* 65; Curt Bowers quoted in Venzke, *Confidence in Battle,* 149, and in OCCH, *Historical Review,* 1965–66, 64–66.

36. Autry, *Gun Totin' Chaplain,* 73, 137.

37. Falabella, *Vietnam Memoirs,* 15, 16, 72, 73–74, 82–85.

38. Hutcheson, *Churches and the Chaplaincy,* 105. For criticism and recommendations regarding chaplains' lack of instruction on issues of war and peace, see Zahn, "Sociological Impressions of the Chaplaincy," 80–83; "Demilitarization of the Chaplain Corps?" *America* 126 (Feb. 12, 1972): 137; United Church of Christ, "Ministries to Military Personnel," 76.

39. Ike C. Barnett, "The United States Air Force Chaplain School," *Chaplain* 20 (Feb. 1963): 3.

40. James T. Wilson, "United States Army Chaplain School," *Chaplain* 15 (June 1958): 12. Gus Engelman, "Army Padre Aids Morale In Viet Nam," *New York World Journal Tribune,* Feb. 12, 1967, p. 10, reporting an interview with the commandant of the Army Chaplain School, Chaplain (Col.) Edward J. Saunders, described the school's nine-week basic course as including "administration, combined arms, map and aerial photo reading, organization and drill, effective writing, officers indoctrination, character guidance, psychology and counseling." See also Ackerman, *He Was Always There,* 24, 221–23; Dulany, *Once a Soldier,* 5–7.

41. Knight, "Supreme Six," 43; Dulany, *Once a Soldier,* 23–24; Autry, *Gun-Totin' Chaplain,* 107, and see also 108, 59.

42. Bowers, *Forward Edge of the Battle Area,* 34–38.

43. Johnson, *Combat Chaplain,* 135; Bowers, *Forward Edge of the Battle Area,* 77.

44. Dulany, *Once a Soldier,* 103.

45. Mahedy, *Out of the Night,* 133, 7, 9. On the chaplain as a "facilitator," see Chaplain Orris E. Kelly, quoted in Venzke, *Confidence in Battle,* 164, 165.

46. Falabella, *Vietnam Memoirs,* 72, 73. Falabella's commentary on the morality of killing in war generally followed the position taken by military people and chaplains of varying faiths, as well as theologians and ethicists. See, for example, Colonel Robert B. Rigg, USA Ret., "Killing or Murder?" *Military Review* (Mar. 1971): 3–9; Loveland, *American Evangelicals,* 148, 157.

47. [A. Ray Appelquist], "Editor's Notes: Dissent Over Vietnam," *Chaplain* 25 (May–June 1968): 1. For a chaplain's statement of this position, see William Greider, "Pastors in Uniform," *WP,* Sept. 2, 1969, p. A6.

48. Mahedy, *Out of the Night,* 133.

49. [Appelquist], "Editor's Notes," 1.

50. Hersh, *Cover-Up*; Spector, *After Tet,* 203–6; Goldstein, Marshall, and Schwartz, *My Lai Massacre and Its Cover-up*; Hammer, *One Morning in the War*; Anderson, *Facing My Lai*; Olson and Roberts, *My Lai.* In the introduction to *Facing My Lai,* ix, Anderson says that the number killed was 504. According to Olson and Roberts (44), the Americal Division (another name for the Twenty-third Infantry Division) had been deactivated in April 1956 then reassembled and reactivated on September 25, 1967.

51. Erik Blaine Riker-Coleman, "Reflection and Reform: Professionalism and Ethics in the U.S. Army Officer Corps, 1968–1974," 3n1, 20, Chapel Hill, N.C., 1997, http://www.unc.edu/~chaos1/reform.pdf/ (accessed Mar. 31, 2012); Hersh, *Cover-Up,* 138, 177; Goldstein, Marshall, and Schwartz, *My Lai Massacre and Its Cover-up,* 267–68, 312, 334, 337–38; Creswell quoted in Venzke, *Confidence in Battle,* 159, and in Earl F. Stover, "Army Sponsors Ethics Workshop," *CC* 93 (Oct. 20, 1976): 904; and see also Newby, *It Took Heroes,* 52. Venzke, *Confidence in Battle,* 159, points out that following the My Lai investigation, the Army Chaplain School began scheduling classes on "the means of properly reporting alleged atrocities." See also Dulany, *Once a Soldier,* 71–72.

52. Olson and Roberts, *My Lai,* 146.

53. Fr. Patrick Burke, S.C.C., "'Chaplains React to Massacre,'" *Boston Pilot,* Dec. 27, 1969, p. 13.

54. OCCH, *Historical Review,* 1965–66, 149–50; OCCH, *Historical Review,* 1969–70, 45–46. During basic training, soldiers received instruction in the laws of war as formulated in the provisions of the Geneva Conventions, the "Nine Rules of Conduct," specific "rules of engagement" in Vietnam, and the U.S. Army Field Manual 27-10, *The Law of Land Warfare.*

55. OCCH, *Historical Review,* 1969–70, 47. See also OCCH, *Historical Review,* 1968–69, 42.

56. See, for example, Wendell T. Wright, "The Problems and Challenges of a Ministry in Vietnam," *Chaplain* (Oct. 1970): 44.

57. Martin Gershen, *Destroy or Die: The True Story of Mylai* (New Rochelle, N.Y.: Arlington House, 1971), 272, 273, 276, 280. See also Robert Jay Lifton, "Advocacy and Corruption in the Healing Professions," in *The Social Psychology of Military Service,* ed. Nancy L. Goldman and David R. Segal (Beverly Hills: Sage, 1976), 54; and David L. Anderson, "Introduction: What Really Happened?" in Anderson, *Facing My Lai,* 6.

58. Goldstein, Marshall, and Schwartz, *My Lai Massacre and Its Cover-up,* 8. According to a 1971 article in *Newsweek,* a few days after news of the massacre broke, Army Secretary Stanley Resor and Chief of Staff William Westmoreland ordered a review of training in the laws of land warfare. At that time, it consisted of "an hour-long, perfunctory rundown on the Geneva Conventions." In October 1970, the army implemented a new, more rigorous three-hour training program which was compulsory for all draftees, recruits, and officer candidates and "specifically tailored to the combat GI in Vietnam." "Ounce of Prevention," *Newsweek,* Apr. 19, 1971, 30, 35. Riker-Coleman, "Reflection and Reform," 39, points out the shortcomings of the training officer candidates received in the laws of war. The "one-hour lecture focused largely on rules and failed to show how they might come into play in the field." A 1967 study had revealed that the training was inadequate, but the army did little to address the problem "until My Lai brought [it] into sharp, tragic focus."

59. Citizens Commission of Inquiry, *Dellums Committee Hearing on War Crimes,* 271–72. See also Olson and Roberts, *My Lai*; Kyle Longley, *Grunts: The American Combat Soldier in Vietnam* (Armonk, N.Y.: M. E. Sharpe, 2008), 62–64, 80–81.

60. Hammer, *One Morning in the War,* 65. See also Robert Debs Heinl Jr., "The U.S. Fighting Man's Dilemma: Brutality in Vietnam," *Navy Magazine* 9 (Aug. 1966): 15–19, 62–63; "William L. Calley, Testimony at Court-Martial, 1970," in Olson and Roberts, *My Lai,* 182–86.

61. William Broyles Jr., *Brothers in Arms: A Journey from War to Peace* (New York: Alfred A. Knopf, 1986), 254.

62. Hammer, *One Morning in the War,* 197. See also Spector, *After Tet,* 202–3; and, for recent exposes, Nick Turse, "The Tip of the Iceberg," http://www.zmag.org/ (accessed Dec. 21, 2009); Nick Turse, "My Lai a Month," *Nation,* Dec. 1, 2008, pp. 13–14, 16, 18–20; Joe Mahr, "Army Makes Adjustments in Effort to Prevent Abuses," *Toledo Blade,* Dec. 21, 2009, http://www.toledoblade.com/ (accessed Dec. 21, 2009); Nelson, *War Behind Me.*

63. Daniel Lang, *Casualties of War* (New York: McGraw-Hill, 1969), 61–65, 72–77; Venzke, *Confidence in Battle,* 159. Newby, *It Took Heroes,* provided an account of his experience in chapter 6 of the memoir.

64. Abercrombie, *Military Chaplain,* 50–53, 97–98. Abercrombie based his study on the response to questionnaires submitted in 1972, to 984 chaplains, 447 commanders, and 400 civilian clergy.

65. Ackerman, *He Was Always There,* 182. For a similar experience, recounted by a navy chaplain in Vietnam, see Rabbi Arnold E. Resnicoff, "The Three Pillars of

Leadership," *Reporter of the Judge Advocate General's Corps, United States Air Force* 35, no. 4 (2008): 163.

66. Johnson, *Combat Chaplain,* 134–35.

67. Dulany, *Once a Soldier,* 54, 73, 74.

68. Anthony E. Hartle, *Moral Issues in Military Decision Making,* 2nd ed., rev. (Lawrence: Univ. Press of Kansas, 2004), 17.

69. Johnson, *Combat Chaplain,* 65. According to Johnson, Kit Carson scouts were "former VC who have come over to the Vietnamese government side by way of the Chieu Hoi program" (62). See Hartle, *Moral Issues in Military Decision Making,* 116, on the specifications of Article 3 of the 1949 Geneva Convention regarding the treatment of prisoners. Among the actions it "absolutely prohibited" was "violence to life and person, in particular murder of all kinds, mutilation, cruel treatment and torture."

70. Autry, *Gun-Totin' Chaplain,* 188.

71. Johnson, *Combat Chaplain,* 157; Dulany, *Once a Soldier,* 53–54.

72. Johnson, *Combat Chaplain,* 198, 199, and see also 205–6.

73. Hopkins, *Chaplain Remembers Vietnam,* 261.

74. Dulany, *Once a Soldier,* 71.

75. Johnson, *Combat Chaplain,* 216.

76. For the adjective "life-sharing," see Ackerman, *He Was Always There,* 220, 241.

5. Ministering to the Military Institution

1. Haynes Johnson and George C. Wilson, *Army in Anguish* (New York: Pocket Books, 1972), 9. Johnson's and Wilson's book originated as a series of reports in the *WP,* Sept. and Oct. 1971. See also B. Drummond Ayres Jr., "Army Is Shaken by Crisis in Morale and Discipline," *NYT,* Sept. 5, 1971, pp. 1, 36. Portions of this chapter first appeared in Loveland, "Prophetic Ministry and the Military Chaplaincy," 245–58.

2. Col. Robert D. Heinl Jr., "The Collapse of the Armed Forces," *Armed Forces Journal* 108 (June 7, 1971): 30.

3. Johnson and Wilson, *Army in Anguish,* 20, 22, 101.

4. Heinl, "Collapse of the Armed Forces," 30–31, 34, 35; Sherie Mershon and Steven Schlossman, *Foxholes & Color Lines: Desegregating the U.S. Armed Forces* (Baltimore: Johns Hopkins Univ. Press, 1998), 174.

5. Heinl, "Collapse of the Armed Forces," 31, 33.

6. Adam Yarmolinsky, *The Military Establishment: Its Impacts on American Society* (New York: Harper and Row, 1971), 363, 364; James R. Hayes, "The Dialectics of Resistance: An Analysis of the GI Movement," *Journal of Social Issues* 31, no. 4 (1975): 125–39; David Cortright, *Soldiers in Revolt: The American Military Today* (Garden City, N.Y.: Anchor Press/Doubleday, 1975), 70–73, 80, 88, 89, 91, 110; Andy Stapp, *Up Against the Brass* (New York: Simon and Schuster, 1970), 88–91;

Larry G. Waterhouse and Mariann G. Wizard, *Turning the Guns Around: Notes on the GI Movement* (New York: Praeger, 1971), 132–39; *NYT,* Feb. 22, 1968, p. 10, Feb. 24, 1968, p. 6. For an early military assessment of the movement, see "The GI Antiwar Movement: Little Action and Money and Few GIs," *Armed Forces Journal* 108 (Sept. 7, 1970): 32–33, 39.

7. On the two organizations, see Cortright, *Soldiers in Revolt,* 108–10; *NYT,* June 3, 1970, p. 11, Aug. 31, 1970, p. 3, Sept. 27, 1970, p. 19; "GI Antiwar Movement," 39; Hayes, "Dialectics of Resistance," 127.

8. Heinl, "Collapse of the Armed Forces," 31; William M. Hauser, "The Impact of Societal Change on the US Army," *Parameters* (Winter 1972): 9; Gen. Bruce Palmer Jr., "Challenges Give Unique Chance to Better Army," *Army* 20 (Oct. 1970): 29, 30.

9. William M. Hauser, *America's Army in Crisis: A Study in Civil-Military Relations* (Baltimore: Johns Hopkins Univ. Press, 1973), 162; Johnson and Wilson, *Army in Anguish,* 18; General Lucius Clay, "Interview with General Lucius Clay by Colonel Rogers, New York, February 28, 1973," Lucius Clay Papers, Manuscript Department, U.S. Army Military History Institute, Carlisle Barracks, Pa., 38; Palmer, "Challenges Give Unique Chance to Better Army," 29.

10. Davison quoted in Maureen Mylander, *The Generals* (New York: Dial Press, 1974), 17.

11. Heinl, "Collapse of the Armed Forces," 37; Riker-Coleman, "Reflection and Reform," 16–19; Ronald H. Spector, "The Vietnam War and the Army's Self-Image," in *The Second Indochina War: Proceedings of a Symposium Held at Airlie, Virginia, 7–9 November 1984,* ed. John Schlight (Washington, D.C.: Center of Military History, United States Army, 1985), 172; Richard A. Gabriel and Paul L. Savage, *Crisis in Command: Mismanagement in the Army* (New York: Hill and Wang, 1978), 9.

12. Davison quoted in Johnson and Wilson, *Army in Anguish,* 26. See also Col. Samuel H. Hays, "The Growing Leadership Crisis," *Army* 20 (Feb. 1970): 39.

13. On military leaders' views of the "new breed," see Johnson and Wilson, *Army in Anguish,* 27, 29, 69.

14. "Edited Transcript: Keynote Address by General W. C. Westmoreland, Chief of Staff, U.S. Army, Army Commanders' Conference, . . . The Pentagon, . . . 30 November 1970," typescript, Franklin M. Davis, Speech File, 1970–74, Franklin M. Davis Papers, Manuscript Department, U.S. Army Military History Institute, Carlisle Barracks, Pa.; Westmoreland quoted in Hauser, *America's Army in Crisis,* 139.

15. Hughes quoted in "Humanizing the U.S. Military," *Time,* Dec. 21, 1970, 20; Lt. Gen. W. T. Kerwin Jr., Deputy Chief of Staff for Personnel, "Youth's 'Why?' Key Challenge in Today's Army," *Army* 20 (Oct. 1970): 69; Hays, "Growing Leadership Crisis," 39; Secretary of the Army Robert F. Froehlke, quoted in Johnson and Wilson, *Army in Anguish,* 156–57; Lt. James H. Toner, "Leaders Must Reply When Soldiers Ask," *Army* 20 (Aug. 1970): 56; Westmoreland quoted in Beth Bailey,

America's Army: Making the All-Volunteer Force (Cambridge: Belknap Press of Harvard Univ. Press, 2009), 51.

16. See, for example, Hays, "Growing Leadership Crisis," 40, 41; Hauser, *America's Army in Crisis,* 135; Edward Brubaker, "Issues from Europe," *Chaplain* 28 (Jan.–Feb. 1971): 11; Gen. Bruce Palmer Jr., "The American Soldier In An Equivocal Age," *Army* 19 (Oct. 1969): 31.

17. Davison quoted in Johnson and Wilson, *Army in Anguish,* 26.

18. Training manual quoted in ibid., 127.

19. Brinsfield, *Encouraging Faith,* pt. 1, 8; Ackerman, *He Was Always There,* 199.

20. "Interview with Major General Gerhardt Hyatt (U.S. Army Chief of Chaplains, Ret.) by Chaplain Colonel J. H. Ellens, 19 July 1976," 55, typescript, Gerhardt W. Hyatt Papers, Manuscript Department, U.S. Army Military History Institute, Carlisle Barracks, Pa; and see also "Interview with Major General Gerhardt Hyatt," Sept. 22, 1976, 60–61.

21. OCCH, *Historical Review,* 1968–69, 105–6, and see also Venzke, *Confidence in Battle,* 161.

22. Department of Defense directive quoted in Colonel George Walton, USA (Ret.), *The Tarnished Shield: A Report on Today's Army* (New York: Dodd, Mead, 1973), 80. Stuart H. Loory, *Defeated: Inside America's Military Machine* (New York: Random House, 1973), chap. 10, contended that as drug abuse increased in Vietnam, commanders tried to ignore or cover it up. He argued (213) that much of the military's drug problem "resulted from a lack of any genuine interest of commanders in morale, troop health and welfare. They were more interested in the institution—in keeping it well stocked with a basic resource, manpower—than in the well-being of the individuals themselves." Gabriel and Savage, *Crisis in Command,* 60–61, asserted that initially, the army ignored the drug crisis. For an early statement by army leaders denying the existence of widespread narcotics usage in the armed forces, see *NYT,* June 19, 1966, p. 25.

23. *NYT,* Nov. 3, 1971, p. 42; Chaplain (MAJ) H. Harrell Hicks, "The Chaplain and the Army's Drug and Alcohol Abuse Program," *MCR* (Winter 1977): 25.

24. Office of Chief of Chaplains, "Historical Review 1 July 1971 to 30 June 1972," 50, 37, Department of the Army, Washington, D.C., [1972], typescript, U.S. Army Center of Military History, Washington D.C.; Office of Chief of Chaplains, "Annual Report of Major Activities: Historical Review of the Office of Chief of Chaplains 1 July 1972 to 30 June 1973," 59, Department of the Army, Washington, D.C., [1973], typescript, U.S. Army Center of Military History, Washington, D.C.; Ackerman, *He Was Always There,* 199–203; Venzke, *Confidence in Battle,* 161–62; Brinsfield, *Encouraging Faith,* pt. 1, 31–32; "Interview with Major General Gerhardt Hyatt," Sept. 22, 1976, 16–17; "Interview with Major General Gerhardt Hyatt," July 19, 1976, 56.

25. Chaplain (Col.) Thomas A. Harris, "Identifying the Drug Problem," *MCR* (Apr. 1972): 1–3; OCCH, "Historical Review," 1971–72, 39, and see also 40; *NYT,*

June 27, 1971, p. 42; OCCH, "Annual Report," 1972–73, 60, 61; "Interview with Major General Gerhardt Hyatt," Sept. 22, 1976, 15; Ackerman, *He Was Always There,* 208.

26. Sampson quoted in Ackerman, *He Was Always There,* 202.

27. OCCH, "Annual Report," 1972–73, 59; "Why the Chaplaincy?" 9. For brief descriptions of specific substance abuse presentations or programs, see Ackerman, *He Was Always There,* 199–201; Venzke, *Confidence in Battle,* 161–62.

28. OCCH, "Historical Review," 1971–72, 39.

29. In his Military History Institute interviews, Hyatt asserted that "one of the things that first attracted the senior people in the Armed Forces, the commanders and the senior staff, was that during the beginnings of the drug program the chaplains were the only ones that had any answers." He went on to declare, "The only thing that the medics, the medical corps, contributed to the drug program . . . was the identification and detoxification. They had no program of rehabilitation. That was left entirely to the chaplains. There was no program for preventive medicine in it or anything. The whole drug education program was developed by the chaplains. . . . What successes the Army has had in [drug control] are directly attributable to the chaplaincy. And to no one else. Some . . . senior commanders will tell you that, too." "Interview with Major General Gerhardt Hyatt," Sept. 22, 1976, 16, 17; "Interview with Major General Gerhardt Hyatt," July 19, 1976, 56. See also Ackerman, *He Was Always There,* 199, 202–4.

30. OCCH, "Historical Review," 1971–72, 40, 50; "Interview with Chaplain (Major General) Orris Kelly by Chaplain (Colonel) Jay H. Ellens, 20 July 1976," 16; "Interview by Chaplain (Colonel) Jay H. Ellens with Chaplain (Colonel) Charles Kriete," 5–6, typescripts of both in Gerhardt W. Hyatt Papers, Manuscript Department, U.S. Army Military History Institute, Carlisle Barracks, Pa.

31. OCCH, "Annual Report," 1972–73, 60, 61, 62; "Interview with Major General Gerhardt Hyatt," Sept. 22, 1976, 10.

32. Gerhardt W. Hyatt, "The Army Chaplaincy, Today and Tomorrow," *Chaplain* 31 (Fall Quarter 1974): 3, 4; "Interview with Major General Gerhardt Hyatt," Sept. 22, 1976, 5; OCCH, "Historical Review," 1971–72, 40.

33. "Interview with Major General Gerhardt Hyatt," July 19, 1976, 12, 55; "Interview with Major General Gerhardt Hyatt," Sept. 22, 1976, 13–14; Chaplain (MG) Gerhardt W. Hyatt, "Men of Faith," *MCR* (Spring 1975): 2.

34. "Interview with Major General Gerhardt Hyatt," July 19, 1976, 12.

35. OCCH, *Historical Review,* 1968–69, 106.

36. Office of the Chief of Chaplains, "Historical Review 1 July 1970–30 June 1971," 109, Department of the Army, Washington, D.C., [1971], typescript, U.S. Army Center of Military History, Washington, D.C.; Office of the Chief of Chaplains, "Annual Report of Major Activities: Historical Review 1 July 1973 to 30 June 1974," 107, Department of the Army, Washington, D.C., [1974], typescript, U.S. Army Center of Military History, Washington, D.C.; Thomas A. Harris and Robert D. Crick, "CPE in the United States Army," *Journal of Pastoral Care* 29 (Mar. 1975): 23–31; Thomas A. Harris, "Clinical

Pastoral Education in the U.S. Army," *Chaplain* 32 (Second Quarter 1975): 20–27; Venzke, *Confidence in Battle,* 122; Brinsfield, *Encouraging Faith,* pt. 1, 91, 159, 166, 263; E. Brooks Holifield, *A History of Pastoral Care in America: From Salvation to Self-Realization* (Nashville: Abingdon Press, 1983).

37. "Interview with Major General Gerhardt Hyatt," Sept. 22, 1976, 10–11, 18–20.

38. Jerrold I. Hirsch, *The History of the National Training Laboratories, 1947–1986: Social Equality Through Education and Training* (New York: Peter Lang, 1987), ix, 31, 35, 41–45, 46, 72–73; David R. Segal, *Recruiting for Uncle Sam: Citizenship and Military Manpower Policy* (Lawrence: Univ. Press of Kansas, 1989), 50–53; Richard Beckhard, *Organization Development: Strategies and Models* (Reading, Mass.: Addison-Wesley, 1969), v, 2, 9, 13. Brinsfield, *Encouraging Faith,* pt. 1, 36–37, notes that the name of the OD program was changed in 1974 to "Parish Development" so that "a theological underpinning and perspective" could be incorporated. "It had been felt by some that the program up to that point had been too secular."

39. "Information Letter" quoted in Office of the Chief of Chaplains, "Annual Report of Major Activities: Historical Review 1 July 1974–30 June 1975," 41–42, Department of Army, Washington, D.C., [1975], typescript, U.S. Army Center of Military History, Washington, D.C.; Hyatt paraphrased in D. Mallicoat, "The Chaplain Today," *Soldiers* 29 (Apr. 1974): 8; "Interview with Major General Gerhardt Hyatt," July 19, 1976, 60; Hyatt, "Army Chaplaincy Today and Tomorrow," 6.

40. Gerhardt W. Hyatt, "A Pastoral-Prophetic Ministry to Leaders," *Chaplain* 30 (Fall Quarter 1973): 28, 25, 23, 24; Hyatt, "Men of Faith," 2, 3; "Keynote Address," *MC* 47 (May–June 1974): 8.

41. "Interview with Major General Gerhardt Hyatt," Sept. 22, 1976, 12, 13; Office of the Chief of Chaplains, Department of the Army, "1 July 1975 to 30 September 1976," 60, Chief of Chaplains, Department of the Army, Washington, D.C., 1976, typescript, U.S. Army Center of Military History, Washington, D.C.. See also, for other chaplains' views on the use of behavioral sciences methods, OCCH, "Historical Review," 1971–72, 14; OCCH, "Annual Report," 1972–73, 91–92; Chaplain (Maj.) Billy W. Libby and Capt. Michael Rohrbaugh, "Talking It Out," *Army* 22 (Dec. 1972): 39–41; H. Newton Maloney, "Toward a Theology for Organization Development," *Chaplain* 32 (Third Quarter 1975): 31–42. For a study of the use of behavioral science methods in industry, including the human relations approach to management, which chaplains adopted, see Loren Baritz, *The Servants of Power: A History of the Use of Social Science in American Industry* (Middletown, Conn.: Wesleyan Univ. Press, 1960).

42. Venzke, *Confidence in Battle,* 162. See also Ackerman, *He Was Always There,* 205–8.

43. OCCH, "Annual Report," 1973–74, 34; Hyatt, "Army Chaplaincy, Today and Tomorrow," 6–7; Brinsfield, *Encouraging Faith,* pt. 1, 32–33. The OCCH Human Relations Ministry paralleled programs instituted by the Defense Race Relations Institute (DRRI) and the Race Relations Education Board established by the

Department of Defense in 1971; see Richard O. Hope, *Racial Strife in the U.S. Military: Toward the Elimination of Discrimination* (New York: Praeger, 1979).

44. Newsletter, Mar. 1, 1974, quoted in OCCH, "Annual Report," 1973–74, 38, 39.

45. OCCH, "Historical Review," 1970–71, 51; OCCH, "Historical Review," 1971–72, 53; OCCH, "Annual Report," 1972–73, 66; OCCH, "Annual Report," 1973–74, 38.

46. "Conclusions to the Conference," *MCR* (Jan. 1972): 61–62.

47. OCCH, "Annual Report," 1972–73, 66, and see also Chaplain (MAJ) John C. Pearson, "The Black Experience in the Military Chapel," *MCR* (Winter 1975): 23–33.

48. OCCH, "Historical Review," 1970–71, 52; OCCH, "Annual Report," 1972–73, 63. The black denominations represented at the conference were the African Methodist Episcopal Church, the Christian Methodist Episcopal Church, the National Baptist Convention of America, the Progressive National Baptist Convention, and the National Baptist Convention, USA.

49. OCCH, "Annual Report," 1972–73, 64–65; Brinsfield, *Encouraging Faith,* pt. 1, 195.

50. OCCH, "Annual Report," 1973–74, 45–48.

51. My discussion of PET is based on the following: Brinsfield, *Encouraging Faith,* pt. 1, 34–35; OCCH, "Annual Report," 1973–74, 44–47; "Interview with Chaplain (Major General) Orris Kelly," 4–10; COL Homer W. Kiefer Jr. and Chaplain (MAJ) James D. Bruns, "PET You Bet!" *Field Artillery Journal* 12 (Sept.–Oct. 1974): 27–28; Chaplain (MAJ) Paul E. Phelps, "'Heavy P.E.T.ing' in Marne Land," *MCR* (Winter 1976): 69–74; Chaplain (CPT) Charles Daniel Witmer, "Supervisory Effectiveness Training: A New P.E.T. Model for Ministry to the 'System,'" *MCR* (Winter 1979): 95–105; LTC Don W. Barber, C.E., "The Impact of 'Personal Effectiveness Training' for Supervisors (S.E.T.) In One Battalion," *MCR* (Winter 1979): 107–12.

52. Office of Chief of Chaplains, "Historical Review 1 October 1976–30 September 1977," 27, Department of the Army, Washington, D.C., [1977], typescript, U.S. Army Center of Military History, Washington, D.C.; and see also OCCH, "Annual Report," 1974–75, 29.

53. The *Military Chaplains' Review* took note of the reaction in its spring 1978 issue by publishing two opposing views: "Why Christians Cannot Minister to Systems," by Chaplain (COL) Mark M. McCullough Jr., and "Why Christians Must Minister to Systems," by Chaplain (MAJ) Carl R. Stephens. The two articles generated so much interest that the editor decided to publish a second "point-counter-point" the following year, by the same two chaplains, on the related topic of prophetic ministry. See Chaplain (COL) Mark M. McCullough Jr., "Why Christians Cannot Minister to Systems," and Chaplain (MAJ) Carl R. Stephens, "Why Christian Must Minister to Systems," *MCR* (Spring 1978): 9–22; and McCullough, "Chaplains as Prophets—Innocents, Martyrs and Moralizers," and Stephens, "The Ageless Need for a Prophetic Ministry," *MCR* (Winter 1979): 19–36.

54. Dennis C. Kinlaw, "The Secularization of Religion and the Management of Chaplains," *Chaplain* 28 (Nov.–Dec. 1971): 9–11, 13; Stephens, "Ageless Need for

a Prophetic Ministry," 30–32; Stephens, "Why Christians Must Minister to Systems," 20.

55. Reverend Thomas A. Harris, "The Chaplain: Prophet, Jester, or Jerk," *MCR* (Fall 1983): 82, 83, 84; Stephens, "Why Christians Must Minister to Systems," 19.

56. Kermit D. Johnson, "How Army Chaplains View Themselves, Their Work and Their Organization," *Chaplain* 34 (Second Quarter 1977): 30. In the army, the chaplain's role in substance abuse prevention and treatment programs seemed especially contentious. Hicks, "Chaplain and the Army's Drug and Alcohol Abuse Program," 28, insisted that chaplain participation in such programs was a "*prophetic* responsibility" and that chaplains were in "a most advantageous position" to engage in them. However Venzke, *Confidence in Battle,* 162, points to "frustration" and negative opinions chaplains expressed in evaluations of the army's drug abuse programs. For one chaplain's very negative view of the army's drug and alcohol abuse rehabilitation program, in which he himself served, see John W. Schumacher, *A Soldier of God Remembers: Memoir Highlights of a Career Army Chaplain* (Winona Lake, Ind.: Grace Brethren North American Missions, 2000), 161–62.

57. OCCH, "Annual Report," 1973–74, 65. See also Chaplain Leonard Stegman, quoted in Brinsfield, *Encouraging Faith,* pt. 1, 8.

58. "Interview with Major General Gerhardt Hyatt," July 19, 1976, 58.

59. On commanders' positive view of chaplains' institutional ministry, see, for example, General Bernard W. Rogers, Chief of Staff, United States Army, "The Challenges of the Chaplaincy," *MCR* (Fall 1977): 1–3; OCCH, "Annual Report," 1973–74, 63–68; OCCH, "Annual Report," 1972–73, 30–32.

60. In June 1971, when the army discontinued mandatory character education in OMH for personnel other than those in basic (BCT) and advanced (AIT) training, OCCH promptly began work on what became the new HSD program. OCCH, "Historical Review," 1971–72, 64, citing DA Message R 301320Z, June 1971.

61. Headquarters, Department of the Army, *Human Self-Development Discussion Topics/Our Moral Heritage* (Pamphlet 165–10; Washington, D.C., May 15, 1972), 1; OCCH, "Historical Review," 1971–72, 63; Headquarters, Department of the Army, "Personnel-General, Human Self Development Program" (Army Regulation 600–630; Washington, D.C., Oct. 29, 1971), 2–3.

62. OCCH, "Historical Review," 1971–72, 63, 64; Department of the Army, "Personnel-General, Human Self Development Program" (AR 600-30, Oct. 29, 1971), 1–3.

63. OCCH, "Historical Review," 1971–72, 62; OCCH, "Annual Report," 1972–73, 103.

64. Department of the Army, "Personnel-General, Human Self Development Program" (AR 600-30, Oct. 29, 1971), 1.

65. OCCH, "Annual Report," 1972–73, 103.

66. OCCH, "Historical Review," 1971–72, 62–63; Department of the Army, "Personnel-General" (AR 600-30, Oct. 29, 1971), 1; OCCH, "Annual Report," 1973–74,

77; OCCH, "Annual Report," 1972–73, 102–3. The idea of using character education to remedy the crisis of the late 1960s and 1970s originated with OMH. In the "Historical Review" for 1970–71, OCCH argued that OMH was vital to the effort to "humanize" the military environment and uniquely suited to addressing not just the symptoms but "the causes of human turbulence and bad behavior." The subjects discussed in the OMH classes, it noted, "are at the very heart of contemporary issues and are most timely in a time of social turbulence, dissent, racial tensions, changing life styles, and other 'people problems.'" OCCH, "Historical Review," 1970–71, 89, 90.

67. Headquarters, Department of the Army, *Human Self-Development Discussion Topics* (Pamphlet 165-11; Washington, D.C., June 30, 1972), n.p.; Headquarters, Department of the Army, *Human Self-Development Discussion Topics* (Pamphlet 165-8-1; Washington, D.C., Feb. 28, 1974), n.p.

68. OCCH, "Historical Review," 1971–72, 63; Headquarters, Department of the Army, *Human Self-Development Discussion Topics* (Pamphlet 165-10; Washington, D.C.: 15 May 1972), iii; OCCH, "Annual Report," 1972–73, 108; Department of the Army, *Human Self-Development Discussion Topics* (Pamphlet 165-11, 30 June 1972), III-B-1.

69. Department of the Army, *Human Self-Development Discussion Topics* (Pamphlet 165-10, May 15, 1972), ii, iii; Department of the Army, *Human Self-Development Discussion Topics* (Pamphlet 165-8-1, Feb. 28, 1974), ii, iv; Department of the Army, "Personnel-General" (AR 600-30, Oct. 29, 1971), 3.

70. Titles of topics from "Tab F, The American Moral Heritage Series," typescript, CG Ad Hoc Committee, file 201-05 (67) Committee, AD HOC, Instructional Files, Acc. No. 71-A-3095, RG 247, Washington National Records Center, Suitland, Md.

71. OCCH, "Annual Report," 1973–74, 77.

72. My discussion of the values clarification approach is based on the following: Howard Kirschenbaum and Sidney B. Simon, eds., *Readings in Values Clarification* (Minneapolis: Winston Press, 1973), 2–30; Richard D. Hersh, John P. Miller, and Glen D. Fielding, *Models of Moral Education: An Appraisal* (New York: Longman, 1980), 9–10, 74–98; David Purpel and Kevin Ryan, eds., *Moral Education . . . It Comes with the Territory* (Berkeley, Calif.: McCutchan, 1976), 73, 152–69.

73. OCCH, "Historical Review," 1971–72, 62; OCCH, "Annual Report," 1973–74, 76.

74. "Morality and the Conscience," in Department of the Army, *Human Self-Development Discussion Topics* (Pamphlet 165-11, June 30, 1972), III-B-6; Department of the Army, *Human Self-Development Discussion Topics* (Pamphlet 165-10, May 15, 1972), v.

75. Chaplain (MAJ) Hugh J. Bickley and Chaplain (CPT) Ford F. G'Segner, "Games and Values Clarification: Aids in Human Self-Development," *MCR* (Fall 1975): 43–44, 46, 51. The two chaplains were very much involved in institutional ministry. Bickley was the director of the PET program at Redstone Arsenal, while G'Segner was a Race Relations/Equal Opportunity Training Program instructor and

a consultant in Human Relations. They practiced team teaching in both the HSD and Race Relations Training programs.

76. Robert D. Crick and Douglas J. Groen, "The Fitzsimmons Model for the Human Self-Development Program," *MCR* (Apr. 1972): 53, 57.

77. Jack S. Boozer, "The Military Chaplaincy: One Calling, Two Roles," *Chaplain* 27 (Dec. 1970): 4, 5, 8. An occasional contributor to the *Chaplain*, Boozer had been an army chaplain in World War II, after which he served in various academic capacities at Emory University for nearly forty years. Beth Bassett, "Jack S. Boozer 1918–1989," *Emory Magazine* (Oct. 1989): 2.

78. OCCH, "Annual Report," 1973–74, 103–4; Brinsfield, *Encouraging Faith,* pt. 1, 68.

79. Chaplain (COL) Bertram C. Gilbert, "Value Education," *MCR* (Apr. 1972): 49, 51.

80. It was replaced by a new, even more flexible program called Chaplain Support Activities, which was only "recommended," not mandatory, even for BCT/AIT personnel. Headquarters, Department of the Army, "Chaplain Support Activities" (Army Regulation 600-30; Washington, D.C.: Jan. 14, 1977); OCCH, "Annual Report," 1974–75, 26, 92–93; OCCH, "1 July 1975 to 30 September 1976," 20–22.

81. "The Special Ministry of the Chief of Chaplains, an Interview with Chaplain (MG) Gerhardt W. Hyatt, US Army Ret.," *MCR* (Summer 1978): 4, 5.

6. "The Conscience of the Army"

1. King (Lt. Col., USA, Ret.) quoted in Captain James Jay Carafano, U.S. Army, "Officership, 1966–1971," *Military Review* 60 (Jan. 1989): 45. In March 1970, Maj. Gen. Samuel W. Koster, commander of the Americal Division in Vietnam (who had become Superintendent of West Point in June 1968), was formally charged with failing to obey lawful regulations and dereliction of duty in the My Lai cover-up; his deputy in Vietnam, Brig. Gen. George Young, was also censured. Mylander, *Generals,* 11.

2. Cincinnatus [Cecil B. Currey], *Self-Destruction: The Disintegration and Decay of the United States Army during the Vietnam Era* (New York: W. W. Norton, 1981), 68–69; Spector, *After Tet,* 265, 266; Mylander, *Generals,* 10; James Kitfield, *Prodigal Soldiers: How the Generation of Officers Born of Vietnam Revolutionized the American Style of War* (New York: Simon and Schuster, 1995), 108.

3. Cincinnatus, *Self-Destruction,* 174.

4. Riker-Coleman, "Reflection and Reform," 23; Loory, *Defeated,* 27–28. See also Cincinnatus, *Self-Destruction,* 129–30.

5. W. C. Westmoreland, General, United States Army, Chief of Staff, to Commandant, United States Army War College, Apr. 18, 1970, Inclosure [*sic*] 1 in U.S. Army War College, *Study on Military Professionalism* (Carlisle Barracks, Pa.: U.S. Army War College, June 30, 1970), 53.

6. Cincinnatus, *Self-Destruction,* 130; Mylander, *Generals,* 301. In his keynote address to the November 1970 Army Commanders' Conference, Westmoreland adverted to the *Study,* saying, "Without question, the integrity of the Officer Corps has declined. You will be shocked later today when you hear a report by the Commandant of the Army War College on this subject." "Edited Transcript: Keynote Address by General W. C. Westmoreland." For a fascinating description of the way Westmoreland and other generals reacted to a briefing on the *Study* by its authors, Lieutenant Colonels Walt Ulmer and Mike Malone, see Kitfield, *Prodigal Soldiers,* 107–12.

7. U.S. Army War College, *Study on Military Professionalism,* i, iii–iv, 2. See also Riker-Coleman, "Reflection and Reform," 4–5, 24–25, 33–34.

8. See, for example, Lt. Col. Ernest L. Webb, "When Ethic Codes Clash: Absolute vs. Situational," *Army* 28 (Mar. 1978): 31–33.

9. U.S. Army War College, *Study on Military Professionalism,* v, vi, B2; Gabriel and Savage, *Crisis in Command,* 85.

10. As Anthony E. Hartle has pointed out, "The American professional military ethic has not been formally and systematically codified. The formal aspects of the code are found primarily in the oath of enlistment and the oath of commissioning (the wording of the commission actually awarded to officers), and the codified laws of war, though a variety of official publications contribute to the accepted guidelines for conduct." Hartle, *Moral Issues in Military Decision Making,* 44.

11. U.S. Army War College, *Study on Military Professionalism,* vi.

12. "Interview with Major General Gerhardt Hyatt," Sept. 22, 1976, 24–26; OCCH, "Annual Report," 1972–73, 43, 104; OCCH, "Annual Report," 1974–75, 40, 55. See also Brinsfield, *Encouraging Faith,* pt. 1, 41–43.

13. Hyatt, "Pastoral-Prophetic Ministry to Leaders," 23, 26; Hyatt, "Army Chaplaincy, Today and Tomorrow," 4.

14. John W. Brinsfield, "Army Values and Ethics: A Search for Consistency and Relevance," *Parameters* 27 (Autumn 1998): 69; Chaplain (COL) John W. Brinsfield, "Alexander's Challenge: Issues in Teaching Leadership," *Army Chaplaincy* (Winter 1998): 13.

15. In 1973, the U.S. Army Chaplain School was officially redesignated as the U.S. Army Chaplain Center and School. The creation of TRADOC was part of a "sweeping reorganization of the U.S. Army." See Brinsfield, *Encouraging Faith,* pt. 1, 395–97; OCCH, "Annual Report," 1974–75, 39; and OCCH, "Annual Report," 1973–74, 48.

16. OCCH, "Annual Report," 1974–75, 39.

17. *Program of Instruction, 5-16-C20-56A, Chaplain Officer Basic* (Fort Monmouth, N.J.: U.S. Army Chaplain Center and School, July 1, 1984), 12, 18, 33, 34, 35.

18. Chaplain (MAJ) Benjamin C. Manning, *U.S. Army Chaplain Officer Basic Course, Chaplain Values Training 1, Subcourse No. CH 0555, Edition 8* (Fort

Monmouth, N.J.: U.S. Army Chaplain Center and School/U.S. Army Training Support Center, Aug. 1988), pp. 1.1, 1.2, 1.3.

19. "Workshops Moving Forward," *MC* 49 (Nov.–Dec. 1976): 9.

20. Office of Chief of Chaplains, "Annual Historical Review 1 October 1978–30 September 1979," 20, 24, Department of the Army, Washington, D.C., [1979], typescript, U.S. Army Center of Military History, Washington, D.C.; OCCH, "Annual Report," 1973–74, 48; OCCH, "Annual Report," 1974–75, 27, 40, 55; Office of the Chief of Chaplains, "Annual Historical Review 1 October 1982–30 September 1983," 6, Department of the Army, Washington, D.C., [1983], typescript, U.S. Army Center of Military History, Washington, D.C.; Brinsfield, *Encouraging Faith,* pt. 1, 43, 136.

21. Brinsfield, *Encouraging Faith,* pt. 1, 67; Office of the Chief of Chaplains, "Historical Review 1 October 1977–30 September 1978," 20, Department of the Army, Washington, D.C., [1978], typescript, U.S. Army Center of Military History, Washington, D.C.

22. Stover, "Army Sponsors Ethics Workshop," 900–4.

23. "A Proposal for a Course in Ethics" (1973), quoted in Joseph Hodgin Beasley, "Implications of Teaching Ethics: The West Point Experience" (Ph.D. diss.: Univ. of North Carolina, 1985), 82. West Point had no philosophy department, but the English Department offered a course, Philosophy 201, which Beasley described as "a smorgasbord of philosophy and religion" (42–44).

24. Beasley, "Implications of Teaching Ethics," 42–44, 82–83, 92–94.

25. Ibid., 156–58, 168, 203. See also Brinsfield, *Encouraging Faith,* pt. 1, 67, 134.

26. John Wesley Brinsfield Jr., "Developing a Ministry of Teaching the History of Ethics and World Religions at the United States Military Academy, West Point, New York" (D.Min. thesis, Drew Univ., 1983), 58, 72, 78. Prior to his assignment at West Point, Brinsfield had taught ethics and leadership at the U.S. Army Aviation Center and School, where he and other chaplains had used the case study approach with success. See Chaplain (CPT) John W. Brinsfield, "Ethics and the Angry Young Man: An Instructor's View," *MCR* (Summer 1980): 56. Brinsfield and other chaplains were much influenced by the moral development theories of the Harvard University psychologist Lawrence Kohlberg, as were many civilian educators. During the 1970s and 1980s, the *Military Chaplains' Review* published a number of articles on Kohlberg and his theories.

27. Brinsfield, "Developing a Ministry," 94, 96, 184, 187, 189, 193, 200–201. For two articles that provide insight into Brinsfield's approach to his course, see Chaplain (MAJ) John W. Brinsfield, "By the Book: The Military Ethics of General William T. Sherman," *MCR* (Spring 1982): 41–58; and "From Plato to NATO; the Ethics of Warfare; Reflections on the Just War Theory," *MCR* (Spring 1991): 21–36. Another army chaplain who taught the ethics of warfare in the mid-1980s was a chaplain instructor at the U.S. Army War College, Donald L. Davidson, whose academic credentials included a Th.M. in ethics, with emphasis on war and morality,

from Harvard University. Chaplain (MAJ-P) Donald L. Davidson, "Religious Strategists: The Churches and Nuclear Weapons," *MCR* (Spring 1984): 5.

28. OCCH, "Annual Report," 1972–73, 42–43. See also OCCH, "Historical Review," 1971–72, 32.

29. Peter L. Stromberg, Malham M. Wakin, Daniel Callahan, *The Teaching of Ethics in the Military* (Hastings-on-Hudson, N.Y.: Hastings Center, 1982), 59–60, 64, regarded the teaching of military ethics in military science and leadership courses and through "chaplains' programs" as a means of emphasizing the importance of specific military virtues but cautioned that it "will probably not enhance moral reasoning or aid in the search for the ultimate foundations or justifications of moral values; nor will [it] acquaint students with the views of the classic moral philosophers and the complex moral concerns involved in just war issues."

30. *Department of the Army Historical Summary, Fiscal Year 1975,* comp. Karl E. Cocke (Washington, D.C.: Center of Military History, United States Army, 1978), 92–93, http://www.history.army.mil/books/DAHSUM/1975/ch09.htm/ (accessed Nov. 7, 2009).

31. Besides Earl Stover's report on the Ethics Workshop, for additional evidence of service school students' negative view of ethics as a "'soft,' untestable subject" that dealt with the "ideal" rather than the "real" world, see Brinsfield, "Ethics and the Angry Young Man," 55.

32. Major Robert A. Fitton, U.S. Army, "Leadership Doctrine and Training: A Status Report," *Military Review* 65 (May 1985): 39–40. See also, for the thinking of Army Chief of Staff General John A. Wickham Jr. regarding what he called "the ethical foundation of military leadership," Loveland, *American Evangelicals, 277–79.* On the limitations of the programs the Army instituted in the early 1970s to address the erosion of the professional ethic, see Riker-Coleman, "Reflection and Reform," 33–34, 48.

33. "The Chaplain Professional Development Plan," U.S. Army, 1979, quoted in Chaplain (Major) Robert Vickers, "The Military Chaplaincy: A Study in Role Conflict," *MCR* (Spring 1986): 83.

34. Brinsfield, *Encouraging Faith,* pt. 1, 169–70.

35. Ronald Reagan, "Remarks at the Annual Convention of the National Association of Evangelicals in Orlando, Florida," Mar. 8, 1983, in *Public Papers of the Presidents of the United States: Ronald Reagan: 1983* (Washington, D.C.: U.S. Government Printing Office, 1982–91), bk. 1, 363, 364.

36. For comments on Reagan's national defense policies, see, for example, Steven E. Miller, "Nuclear Arms Control: The Freeze Debate Heats Up," *New Leader,* Mar. 22, 1982, 3; "Living with Mega-Death," *Time,* Mar. 29, 1982, 19, 25; "Freeze Vote: Body Blow to Arms Strategy?" *U.S. News & World Report,* Nov. 15, 1982, 32.

37. Riley quoted in Kenneth L. Woodward, "Churchmen Vs the Bomb," *Newsweek,* Jan. 11, 1982, 70 (LexisNexis; hereafter LN).

38. Kenneth A. Briggs, "Religious Leaders Objecting to Nuclear Arms," *NYT*, Sept. 8, 1981, pp. A1, A20; Fox Butterfield, "Anatomy of the Nuclear Protest," *NYT Magazine*, July 11, 1982, 14; "On the March—U.S. Version of Peace Crusade," *U.S. News & World Report*, Mar. 22, 1982, 25; L. Bruce van Voorst, "The Churches and Nuclear Deterrence," in *The Political Role of Religion in the United States*, ed. Stephen D. Johnson and Joseph B. Tamney (Boulder, Colo.: Westview Press, 1986), 280.

39. Donald L. Davidson, *Nuclear Weapons and the American Churches: Ethical Positions on Modern Warfare* (Boulder, Colo.: Westview Press, 1983), 180–81, Appendix B.

40. The signers included Reverend Dwain Epps, director for international affairs of the National Council of Churches, Avery Post, president of the United Church of Christ, and Catholic bishops Thomas Gumbleton of Detroit and Walter Sullivan of Richmond. "Group Opposes 'Star Wars' Funding," *WP*, May 14, 1985, p. A8; A. James Reichley, *Religion in American Public Life* (Washington, D.C.: Brookings Institution, 1985), 353.

41. See, for example, "Weaving a Seamless Garment Out of Peace, Freedom, and Security," *CT* (May 15, 1987): 42–44, and, for a longer discussion of both conservative and radical evangelicals, Loveland, *American Evangelicals*, chaps. 15–18.

42. Jim Wallis, *Agenda for Biblical People* (San Francisco: Harper & Row, 1984), xiii.

43. Gene Preston, "Coffin Confronts the Cadets," *CC* 98 (July 15–22, 1981): 730–32; Gene Preston, "Religion at West Point," *CT* (Nov. 6, 1981): 77; Noel Gayler, "Nuclear Deterrence—Its Moral and Political Implications" (address given at the U.S. Air Force Academy, May, 1983), in *Military Ethics* (Washington, D.C.: National Defense Univ. Press, 1987), 165; Richard Halloran, *To Arm a Nation: Rebuilding America's Endangered Defenses* (New York: Macmillan, 1986), 44.

44. See, for example, Ethics and Professionalism Committee, USMA, *War and Morality*, technical report distributed by Defense Technical Information Center, Defense Logistics Agency, Cameron Station, Alexandria, Va., 1978; and *Armed Forces and Society* and *Parameters*, 1980–1983. For a firsthand account of how questions of national defense policy were debated within the military, see Admiral Noel Gayler, "Opposition to Nuclear Armament" ("Questions and Answers" segment), *Annals of the American Academy of Political and Social Science* 469 (Sept. 1983): 22. On the opinions and attitudes of officers of all the service branches regarding national defense policy and nuclear weapons in the mid-1980s, see Loveland, *American Evangelicals*, 262–65, 272–73.

45. Kermit D. Johnson, *Realism and Hope in a Nuclear Age* (Atlanta: John Knox Press, 1988), 12; "Of Peace and Policy" (interview with Johnson) *Sojourners* 12 (Oct. 1983), 26.

46. "Pax Christi Concerns," *CC* 102 (June 19–26, 1985): 608.

47. "Challenge to Chaplains," *CC* 103 (Mar. 5, 1986): 234. The signers of the letter included Lutheran bishops, theologians, seminary professors, reserve chaplains, and members of the Commission for a New Lutheran Church. For a more detailed discussion of the "Open Letter" and the response it provoked, as well as other LPF efforts to engage in dialogue with chaplains, see Steven Schroeder, *A Community and a Perspective: Lutheran Peace Fellowship and the Edge of the Church, 1941–1991* (Lanham, Md.: Univ. Press of America, 1993), chap. 6.

48. Schroeder, *Community and a Perspective,* 104–5.

49. Gordon C. Zahn, "Military Chaplains: Defining their Ministry," *America* 147 (Aug. 7–14, 1982): 67, 69, 70.

50. The NCCB committee was made up of five bishops serving on the NCCB's Committee on War and Peace. Allen E. Carrier, "'Softer' Anti-Nuke Letter Relieves Chaplains," *Army Times,* Apr. 18, 1983, p. 4.

51. See, for example, Lieutenant Commander John N. Petrie, U.S. Navy, "Exodus or Entrenchment: The Catholic Dilemma of Duty," *Naval War College Review* 35 (July–Aug. 1982): 69–76; Mary Ellen Ruff, "Catholic and Military," *MCR* (Fall 1982): 95–101.

52. *NYT,* Dec. 19, 1982, p. 46, and see also Dec. 15, 1981, pp. 1, B16.

53. Allen E. Carrier, "Bishops Soften Stance on Nuclear Arms in Letter," *Air Force Times,* Apr. 18, 1983, p. 2; Carrier, "'Softer' Anti-Nuke Letter," 4.

54. The vote on the complete text was 238 in favor and 9 opposed. The official title of the pastoral letter was "The Challenge of Peace: God's Promise and Our Response." My summary of the main points of the pastoral letter is based on the following: Philip J. Murnion, *Catholics and Nuclear War: A Commentary on The Challenge of Peace: The U.S. Catholic Bishops' Pastoral Letter on War and Peace* (New York: Crossroad, 1983), which contains the text of the letter; George W. Cornell, "Bishops Strengthen Anti-Nuke Stand in Letter," *Nashville Banner,* May 3, 1983, p. A-6; *NYT,* May 5, 1983, p. B16; Kenneth L. Woodward with Donna Foote, "The Bishops Call a 'Halt,'" *Newsweek,* May 16, 1983, https://www -lexisnexis-com. libezp.lib.lsu.edu/ (accessed Dec. 2, 2009).

55. Cooke's letter quoted in *NYT,* June 8, 1983, p. 12. For the section of the pastoral letter addressed to the men and women in military service, which also commented on such matters as the conduct of war, disobedience of unlawful orders, dehumanizing practices in military training, and respect for basic human rights of "our own forces" as well as of "adversaries," see Murnion, *Catholics and Nuclear War,* 330–32. Cooke's emphasis on the binding character of the "moral principles of the church" may have suggested to readers of his letter that the whole of the bishops' pastoral was morally binding on Catholics. However, John W. Coffey, "The American Bishops on War and Peace," *Parameters* 13 (Dec. 1983): 30–31, pointed out that "a bishops' conference as such has no teaching authority. Only the Pope or the whole College of Bishops with the Pope can proclaim morally binding principles for Catholics. Thus the pastoral letter carries moral authority only when it reiterates the

formal teaching of the universal Church or when it reaffirms natural law principles of the just-war theory." Coffey quoted the following statement in the pastoral letter: "The applications of principles in this pastoral letter do not carry the same authority as our statements of universal moral principles and formal church teaching."

56. Brinsfield, *Encouraging Faith,* pt. 1, 171; Chaplain (Colonel, USAF Reserve) John E. Groh, *Facilitators of the Free Exercise of Religion: Air Force Chaplains, 1981–1990* (Washington, D.C.: Office of the Chief of Chaplains, USAF, 1991), 268.

57. Davidson, "Religious Strategists," 20, 22–25. This article was reprinted, under the same title, in *MCR* (Spring 1984): 5–21.

58. On the discussion of whether the chaplaincies should define their positions in the nuclear debate, see Groh, *Facilitators of the Free Exercise of Religion,* 171, 269, 271, 272. The statement by Army Chief of Chaplains Johnson was in response to a proposal from the army chief of staff that OCCH issue a white paper on the subject of "just wars." See Memorandum for Chief of Staff, Army, from Kermit D. Johnson, Chaplain (Major General), USA, Chief of Chaplains (signed "for the Chief of Chaplains" by Executive Officer Paul O. Forsberg, Chaplain [Colonel], USA, and with a notation "recommendations approved, CSA"), Aug. 1, 1980, File 701-01 Chaplain Instruction Files (82), Acc. No. 247-88-001, RG 247, Washington National Records Center, Suitland, Md.

59. Hessian quoted in Kenneth Briggs, "Service Chaplains Ponder War Letter," *NYT,* Aug. 8, 1983, p. B7.

60. Briggs, "Service Chaplains Ponder War Letter," B7.

61. Davidson, "Religious Strategists," 20.

62. Brinsfield, *Encouraging Faith,* pt. 1, 171.

63. Davidson, "Religious Strategists," 29n16.

64. Johnson, *Realism and Hope,* 13; Chaplain (MG) Kermit D. Johnson, "A New Stage: Beyond 'In-House' Ethical Issues," *MCR* (Spring 1982): v.

65. Johnson, *Realism and Hope,* 36, 67; "Of Peace and Policy," 25–27.

66. Memorandum for Chief of Staff, Army, from Kermit D. Johnson, Chaplain (Major General), USA, Chief of Chaplains, Mar. 31, 1982, in File 701-01, Chaplain Instruction Files (82), Acc. No. 247-88-001, RG 247, Washington National Records Center, Suitland, Md.; Johnson, *Realism and Hope,* 12–14, 108–14; Brinsfield, *Encouraging Faith,* pt. 1, 144–48.

67. Brinsfield, *Encouraging Faith,* pt. 1, 147. Upon retirement from the military, Johnson served as Associate Director of the nongovernmental Center for Defense Information in Washington from 1983 to 1986, and thereafter as a member of the Central America Working Group in the Washington Office of the Presbyterian Church, USA. For his postretirement publications, see in addition to *Realism and Hope* (1988), Kermit D. Johnson, "The Nuclear Reality: Beyond Niebuhr and the Just War," *CC* 99 (Oct. 13, 1982): 1014–17; Kermit D. Johnson, "The Morality of Nuclear Deterrence," in *The Choice: Nuclear Weapons Versus Security,* ed. Gwyn Prins (London: Chatto & Windus–Hogarth Press, 1984); Kermit D. Johnson, "Just

War and Nuclear Deterrence," in *The Peacemaking Struggle: Militarism and Resistance: Essays Prepared for the Advisory Council on Church and Society of the Presbyterian Church (U.S.A.),* ed. Ronald H. Stone and Dana W. Wilbanks (New York: Univ. Press of America, 1985).

68. Chaplain (COL) John W. Brinsfield, "The Chaplaincy and Moral Leadership," *Army Chaplaincy* (Summer 1995): 6, 8, 25, 26.

7. Ministering on the Battlefield

1. Brinsfield, *Encouraging Faith,* pt. 1, 163. On "doctrine and the Army," see Headquarters, Department of the Army, *Operations* (Field Manual 3-0; Washington, D.C., June 14, 2001), p. 1.14. For the historical and cultural context in which AirLand Battle doctrine was formulated, see Roger Spiller, "Doctrine, Military," in *The Oxford Companion to American Military History,* ed. John Whiteclay Chambers II (New York: Oxford Univ. Press, 1999), 234.

2. *Chaplain and Chaplain Assistant in Combat Operations* (FM 16-5, Dec. 10, 1984), 38.

3. Headquarters, Department of the Army, *Religious Support Doctrine: The Chaplain and Chaplain Assistant* (Field Manual 16-1; Washington, D.C., Nov. 27, 1989), p. 1.9.

4. Headquarters, Department of the Army, *Religious Support* (Field Manual 16-1; Washington, D.C., May 26, 1995), p. 2.1; Headquarters, Department of the Army, *Religious Support* (Field Manual 1-05; Washington, D.C., Apr. 18, 2003), pp. 1.9, 2.1, 2.2, 2.5.

5. Brinsfield, *Encouraging Faith,* pt. 1, 159, 175–76, 182–85, 198–99.

6. *Religious Support Doctrine* (FM 16-1, Nov. 27, 1989), p. 1.12; *Chaplain and Chaplain Assistant in Combat Operations* (FM 16-5, Dec. 10, 1984), 4. In other chaplain field manuals and army regulations, the usual phrasing of the prohibition was "will not bear arms," but other phrasing also appeared, including, for example, "the policy of the Chief of Chaplains forbids chaplains to bear arms" and "shall not bear arms." In 2003–2006, the noncombatant status of chaplains became a "key policy issue" in the army, when some chaplains who had served in Iraq and Afghanistan challenged the policy. See Dr. John W. Brinsfield Jr. and Chaplain (Lieutenant Colonel) Kenneth E. Lawson, *A History of the United States Army Chaplaincy: Taking Spiritual Leadership to the Next Level: The Hicks Years, 2003–2007* (Washington, D.C.: Office of the Chief of Chaplains, Department of the Army, 2007), pt. 1, 139, 197, 199.

7. Brinsfield, *Encouraging Faith,* pt. 1, 64, 75–76, 175, 183.

8. *Chaplain and Chaplain Assistant in Combat Operations* (FM 16-5, Dec. 10, 1984), 5, 37; *Religious Support Doctrine* (FM 16-1, Nov. 27, 1989), p. 1.8. See also Jacqueline Earline Whitt, "Conflict and Compromise: American Military Chaplains and the Vietnam War" (Ph.D. diss., Univ. of North Carolina at Chapel Hill, 2008), 80–82.

9. Chaplain (COL) Wayne E. Kuehne, "Faith and the Soldier: Religious Support on the Airland Battlefield," 5, U.S. Army War College Military Studies Program paper, Carlisle Barracks, Pa., 1988. According to Brinsfield, *Encouraging Faith,* pt. 1, 175, Kuehne was the OCCH Force Structure Officer. On the UMT, see Brinsfield, *Encouraging Faith,* pt. 1, 75–76, 175, 179–81, 183. For a comment on how "Forward Thrust" revolutionized the chaplain's role on the battlefield, see Chaplain (Colonel) Harold D. Roller, in "Transcript of Oral History of Roller, Harold D.," Dec. 11, 2003, in Special Collections, Oral Histories, William Madison Randall Library, University of North Carolina at Wilmington, Wilmington, N.C., http://library.uncw.edu/web/collections/oralhistories/transcripts/369.html/ (accessed May 19, 2007).

10. *Chaplain and Chaplain Assistant in Combat Operations* (FM 16-5, Dec. 10, 1984), 4, 36, 37; *Religious Support Doctrine* (FM 16-1, Nov. 27, 1989), p. 1.8, chap. 5.

11. Brinsfield, *Encouraging Faith,* pt. 1, 197–98, 405, 420; Chaplain (Lieutenant Colonel) Herbert J. McChrystal III, "Spiritual Fitness: An Imperative for the Army Chaplaincy," 15, USAWC Strategy Research Project, U.S. Army War College, Carlisle Barracks, Pa., 1998. The entire issue of the Summer 1986 *Military Chaplains' Review* was devoted to the topic "Training for Combat Ministry: The Chaplain at the National Training Center." On the continued emphasis on chaplains' "technical and tactical training for deployment and combat," see, for example, *Religious Support* (FM 16-1, May 26, 1995), p. 4.1; *Religious Support* (FM 1-05, Apr. 18, 2003), pp. 3.9–3.10.

12. *Chaplain and Chaplain Assistant in Combat Operations* (FM 16-5, Dec. 10, 1984), 7.

13. Ibid., 49, 51–53, 56–58.

14. *Religious Support Doctrine* (FM 16-1, Nov. 27, 1989), pp. 5.30–5.31; Headquarters, Department of the Army, *Religious Support Handbook for the Unit Ministry Team* (TC [Training Circular] 1-05; Washington, D.C., May 10, 2005), p. C.1.

15. Brinsfield, *Encouraging Faith,* pt. 1, 66–67, 113, 133.

16. Rogers, "Challenges of the Chaplaincy," 3. In a 2003 interview, Chaplain (COL) Harold D. Roller, Commandant of the U.S. Army Chaplain Center and School, talked about how the role of the chaplain had been significantly "enhanced" during the past thirty years. "Chaplains now are very highly regarded by commanders," he observed, adding that "the commander today leans on the chaplain far more than the commander did say 20 to 30 years ago." They were valued for assisting the commander not only with "taking care of the soldiers" but also with "ethical issues within the command." According to Roller, "The chaplain can come up to the commander, is expected to come up to the commander and give him the bad news. Maybe he doesn't want to hear it, but the expectation is that the chaplain is not only allowed to do it but encouraged to do it. I think that's a large difference." Roller attributed the enhancement to the chaplains' new role as teachers of ethics. "When that happened[,] more and more the role of the chaplain, even the battalion chaplain,

would be an ethical advisor to the commander and [it] was enhanced and took on a broader meaning and gave the chaplain more latitude in being able to advise the commander in that regard." Roller, "Transcript of Oral History of Roller."

17. Headquarters, Department of the Army, *The Chaplain* (Field Manual 16-5; Washington, D.C., July 8, 1977), p. 1.3.

18. *Chaplain and Chaplain Assistant in Combat Operations* (FM 16-5, Dec. 10, 1984), 21; *Religious Support Doctrine* (FM 16-1, Nov. 27, 1989), pp. 3.7, 3.8.

19. *Religious Support Doctrine* (FM 16-1, Nov. 27, 1989), p. 3.8; *Religious Support* (FM 16-1, May 26, 1995), p. 3.8; *Religious Support* (FM 1-05, Apr. 18, 2003), p. 3.25.

20. *Religious Support* (FM 1-05, Apr. 18, 2003), 1.5; Headquarters, Joint Chiefs of Staff, *Religious Support in Joint Operations* (Joint Publication 1-05; [Washington, D.C.], June 9, 2004), p. I.1; *Religious Support Handbook* (TC 1-05, May 10, 2005), p. 2.1.

21. *Religious Support Handbook* (FM 1-05, Apr. 18, 2003), pp. 1.5, 3.13, 3.5, 3.8; *Religious Support Handbook* (TC 1-05, May 10, 2005), pp. 4.4, 4.6; Headquarters, Joint Chiefs of Staff, *Religious Support in Joint Operations* (JP 1-05, June 9, 2004), p. II.3.

22. *Chaplain and Chaplain Assistant in Combat Operations* (FM 16-5, Dec. 10, 1984), 12, 21; *Religious Support Doctrine* (FM 16-1, Nov. 27, 1989), p. 3.9. During the Vietnam War, the Army Chaplain School had introduced classes dealing with illegal orders and the procedure for reporting actual or alleged atrocities (Brinsfield, *Encouraging Faith,* pt. 1, 393). In the 1970s, the army developed new, more rigorous and specific training for soldiers in the laws of war. See Guenter Lewy, "The Punishment of War Crimes: Have We Learned the Lessons of Vietnam?" *Parameters* 9 (Dec. 1979): 12–19. The same was true for officer candidates. After My Lai, the army strengthened their instruction in the laws of war. The previous one hour of instruction was replaced by a two-hour session, and "the content was substantially upgraded" (Riker-Colman, "Reflection and Reform," 39–40). By 2009, according to army officials, soldiers (and their commanders) received much more extensive training regarding war crimes, "throughout their careers," with particular emphasis placed on the treatment of civilians (Mahr, "Army Makes Adjustments").

23. Headquarters, Department of the Army, *Moral Leadership/Values: Responsibility and Loyalty* (Pamphlet 165-15; [Washington, D.C.]: U.S. Government Printing Office, Dec. 15, 1986), p. 1.1.

24. *Chaplain and Chaplain Assistant in Combat Operations* (FM 16-5, Dec. 10, 1984), 11–12; Headquarters, Department of the Army, "Religious Activities: Chaplain Activities in the United States Army" (Army Regulation 165-1; Washington, D.C., Mar. 25, 2004), 21, http://www.army.mil/usapa/epubs/pdf/r165_1.pdf/ (accessed Nov. 1, 2009); *Religious Support Handbook* (TC 1-05, May 10, 2005), Appendix E. On the implementation of Moral Leadership Training, see Brinsfield, *Encouraging Faith,* pt. 1, 196.

25. Lee Lawrence, "Military Chaplains: A Presbyterian Pastor Patrols with his Flock of Soldiers in Iraq," *Christian Science Monitor,* Nov. 27, 2007, http://www.printthis.clickability.com/ (accessed Dec. 12, 2008). See also Sharon Cohen, "Chaplains Preserve Stories of the Fallen," Mar. 23, 2008, http://www.msnbc.msn.com/ (accessed July 1, 2008); Kristin Henderson, "'In the Hands of God,'" *WP Magazine,* Apr. 30, 2006, W08 (LN).

26. Dan Elliott, "Chaplain First Killed in Action since 1970," *Baton Rouge Advocate,* Sept. 3, 2010, p. 3A. Goetz was the first army chaplain killed in action since the Vietnam War, according to the spokesman.

27. "Profile: Iraq Military Chaplain, October 22, 2004," *Religion and Ethics Newsweekly,* http://www.pbs.org/wnet/religionandethics/week808/p-profile.html/ (accessed Nov. 4, 2009). See also Kent Annan, "For God & Country: Chaplains Who Serve the U.S. Armed Forces," *InSpire* (Spring 2002): 13.

28. "The Soul of War," transcript of radio program, Speaking of Faith from American Public Media, http://being.publicradio.org/programs/soulofwar/transcript.shtml/ (accessed July 17, 2008).

29. Glenn Palmer, "Mainline Chaplains," *CC,* Nov. 14, 2006, http://findarticles.com/ (accessed Oct. 30, 2009).

30. Lawrence, "Military Chaplains," *Christian Science Monitor,* Nov. 27, 2007.

31. Rod Dreher, "Ministers of War," *National Review,* Mar. 10, 2003, http://www.nationalreview.com/ (accessed Nov. 25, 2008).

32. "Profile: Iraq Military Chaplain," Oct. 22, 2004.

33. Cohen, "Chaplains Preserve Stories of the Fallen," Mar. 23, 2008.

34. Troy Moon, "Faith in the Face of War: Chaplains Share Stories of Sacrifice on the Front Lines," *Pensacola News Journal,* n.d., http://www.png.com/ (accessed Nov. 19, 2008).

35. Chaplain (Captain) Glenn Palmer, "Unloading Soldiers' Burdens," *ArmyTimes,* n.d., http://www.armytimes.com/ (accessed Oct. 30, 2009). On Critical Event Debriefings, see *Religious Support Handbook* (TC 1-05, May 20, 2005), pp. C-11 - C-16.

36. *Chaplain* (FM 16-5, Aug. 27, 1964), 42; *Chaplain* (FM 16-5, Dec. 26, 1967), 41.

37. Ackerman, *He Was Always There,* 25.

38. *Chaplain* (FM 16-5, Dec. 26, 1967), 41.

39. John W. Brinsfield, "The Army Chaplaincy and World Religions: From Individual Ministries to Chaplain Corps Doctrine," *Army Chaplaincy* (Winter–Spring 2009): 14, 15.

40. Headquarters, Department of the Army, "Religious Support: Army Chaplain Corps Activities" (Army Regulation 165-1; Washington, D.C., Dec. 3, 2009), 21.

41. Brinsfield, "Army Chaplaincy and World Religions," 15.

42. Chaplain (LTC) Ira Houck, "The U.S. Army Chaplaincy's Involvement in Strategic Religious Engagement," *Army Chaplaincy* (Winter–Spring 2009): 53;

Chaplain (MAJ) Brian P. Crane, "Locating Christendom's Center of Gravity," *Army Chaplaincy* (Winter–Spring 2009): 82; Chaplain (Major) Timothy K. Bedsole, "Religion: The Missing Dimension in Mission Planning," *Special Warfare* (Nov.–Dec. 2006): 14; Chris Seiple, "Ready . . . Or Not? Equipping the U.S. Military Chaplain for Inter-Religious Liaison," *Review of Faith and International Affairs* 7 (Winter 2009), http://rfiaonline.org/ (accessed Feb. 10, 2010).

43. Dr. Pauletta Otis, "Chaplains Advising Commanders in a post-9/11 World and Beyond," *Army Chaplaincy* (Winter–Spring, 2009): 33.

44. Headquarters, Joint Chiefs of Staff, *Religious Support in Joint Operations* (JP 1-05, June 9, 2004), pp. viii, 1.2, 1.3, II.7. See also Raymond L. Bingham, "Bridging the Religious Divide," *Parameters* 36 (Autumn 2006): 50–66; Bedsole, "Religion," 9–10. For an earlier statement on the need to recognize the influence of "religions and systems of belief" in responding to "worldwide strategic challenges," see *Religious Support* (FM 16-1, May 26, 1995), p. 2.1.

45. Headquarters, Department of the Army, *Chaplain and Chaplain Assistant in Combat Operations* (FM 16-5, Dec. 10, 1984), 15–16; *Religious Support Doctrine* (FM 16-1, Nov. 27, 1989), pp. 1.11, 3.10; *Religious Support* (FM 1-05, Apr. 18, 2003), p. A.2. The chaplain worked with the Civil-Military Operations (CMO) officer in such training. Headquarters, Joint Chiefs of Staff, *Religious Support in Joint Operations* (JP 1-05, June 9, 2004), GL-2, defines civil-military operations (CMO) as follows: "The activities of a commander that establish, maintain, influence, or exploit relations between military forces, governmental and nongovernmental civilian organizations and authorities, and the civilian populace in a friendly, neutral, or hostile operational area in order to facilitate military operations, to consolidate and achieve operational US objectives."

46. *Chaplain* (FM 16-5, July 8, 1977), p. 1.3; *Chaplain and Chaplain Assistant in Combat Operations* (FM 16-5, Dec. 10, 1984), 15, Appendix B; *Religious Support Doctrine* (FM 16-1, Nov. 27, 1989), pp. 1.11, 3.8; *Religious Support* (FM 16-1, May 26, 1995), pp. 3.7–3.8; *Religious Support* (FM 1-05, Apr. 18, 2003), pp. 3.4, 3.13, 3.25, Appendix A; Headquarters, Joint Chiefs of Staff, *Religious Support in Joint Operations* (JP 1-05, June 9, 2004), p. viii, II-2; *Religious Support Handbook* (TC 1-05, May 10, 2005), pp. 2.5, 3.9, 3.10, 4.5, Appendix D; "Religious Support" (AR 165-1, Dec. 3, 2009), 10, 12.

47. *Religious Support Handbook* (TC 1-05, May 10, 2005), p. 4.8; Headquarters, Joint Chiefs of Staff, *Religious Support in Joint Operations* (JP 1-05, June 9, 2004), pp. I.1, ix, II.2, II.4.

48. During the Vietnam War, the 1967 chaplain field manual described a "staff function" that had not been listed in the 1964 manual. It involved "stability operations," in which the army focused on "internal defense and internal development." The manual noted that "this situation provides special responsibilities for the chaplain; it also provides greater opportunities for service. The liaison and rapport that the chaplain is able to effect with indigenous religious groups and leaders may be of

inestimable value to the people and to the national interests of the United States. Certainly the command will depend heavily upon this aspect of the chaplain's work." This same statement appeared in the 1977 chaplain manual. In the 1984 field manual, the chaplain was directed to consult with the Civil-Military Operations Office and supervisory chaplain on ways of developing "friendly relations with local civilians" and meeting "the human welfare needs produced by combat." When called upon to assist the civil affairs program, one of the actions the chaplain could take was to "establish liaisons with local missionaries, clergy, and religious leaders." In the 1989 field manual, the phrasing was "establishes liaison with local missionaries and religious leaders." *Chaplain* (FM 16-5, Dec. 26, 1967), 42; *Chaplain* (FM 16-5, July 8, 1977), p. 4.3; *Chaplain and Chaplain Assistant in Combat Operations* (FM 16-5, Dec. 10, 1984), 15; *Religious Support Doctrine* (FM 16-1, Nov. 27, 1989), p. 3.10.

49. CH (LTC) Kenneth E. Lawson, "The United States Army Chaplaincy in the Balkans, 1995–2005," in Brinsfield and Lawson, *History of the United States Army Chaplaincy,* pt. 2, 17, 17, 22.

50. Ibid., 166, 140, 161; and for other mention by Lawson of chaplains' religious leader engagement and related activities in the Balkans, see 94, 100, 105, 126, 144, 147, 159, 161, 168, 169. Other descriptions of religious leader engagement in Bosnia and Kosovo appear in Peter Ephross, "Chaplain Envisions Role for Clergy in Rebuilding Kosovo," *Jewish Telegraphic Agency Daily News Bulletin,* June 1999, http://www.resnicoff.net/kosovo.html/ (accessed June 19, 2010); Timothy Bedsole, "The World Religions Chaplain: A Practitioner's Perspective," *Review of Faith and International Affairs* 7 (Winter 2009), http://rfiaonline.org/ (accessed Feb. 10, 2010); and "Transcript of Oral History of Baktis, Peter A.," Chaplain (Major) Peter A. Baktis interviewed by Paul Zarbock and John Brinsfield, Fort Jackson, S.C., Dec. 11, 2003, Archives, Oral Histories, William Madison Randall Library, University of North Carolina, Wilmington, http://library.uncw.edu/web/collections/oral histories/transcripts/382.html/ (accessed Sept. 1, 2010).

51. Lawson, "United States Army Chaplaincy in the Balkans," 165, 167, 168.

52. Navy Chaplain (CPT) George Adams, "Chaplains as Liaisons with Religious Leaders," *Army Chaplaincy* (Winter–Spring 2009): 41.

53. Brinsfield and Lawson, *History of the United States Army Chaplaincy,* pt. 1, 135; Chaplain (Lieutenant Colonel) Kenneth Lawson, "Doctrinal Tension: The Chaplain and Information Operations," *Military Intelligence Professional Bulletin* (Apr.–June 2009): 29; William Sean Lee, Chaplain (Colonel), ARNG, Christopher J. Burke, PHD, Lieutenant Colonel, USAF, Zonna M. Crayne, Lieutenant Colonel, ANG, "Military Chaplains as Peace Builders: Embracing Indigenous Religions in Stability Operations," 3, 10, 17, research report, Maxwell Air Force Base, Ala., Apr. 2004; Chaplain (Lieutenant Colonel) Scottie Lloyd, United States Army, "Chaplain Contact with Local Religious Leaders: A Strategic Support," 1, 5, 6, 10, USAWC Strategy Research Report, U.S. Army War College, Carlisle Barracks, Pa., Mar. 18, 2005.

54. Adams, "Chaplains as Liaisons with Religious Leaders," 42.

55. Jacqueline E. Whitt, "Dangerous Liaisons: The Context and Consequences of Operationalizing Military Chaplains," *Military Review* (Mar.–Apr. 2012): 60.

56. Michael J. Carden, "Muslim Chaplains Serve Various Roles During Deployment," *American Forces Press Service,* Aug. 29, 2008, http://www.defense.gov/ (accessed June 25, 2010).

57. Chaplain (COL) LaMar Griffin, "RLL and the Emerging Role for Chaplains in Shaping Full Spectrum Operations," *Army Chaplaincy* (Winter–Spring 2009): 46, 47.

58. *Religious Support* (FM 1-05, Apr. 18, 2003), pp. iii, 1.5, 1.7, 3.3, 3.4, 3.5, 3.8, 3.25, A.1.

59. *Religious Support Handbook* (TC 1-05, May 10, 2005), p. 3.10. According to the handbook (3.17), the G5 was "the principal staff officer for the commander in all matters pertaining to the civilian impact on military operations and the political, economic, and social effects of military operations on civilian personnel in the area of operations." The Joint Chiefs of Staff recognized religious leader engagement in a 2004 publication as follows: "The JFCH [joint forces chaplain], after careful consideration and only with the JFC's [joint force commander's] approval, may serve as a point of contact to HN [host nation] civilian and military religious leaders, institutions, and organizations, including established and emerging military chaplaincies, through the CMOC [civil-military operations center]." Headquarters, Joint Chiefs of Staff, *Religious Support in Joint Operations* (JP 1-05, June 9, 2004), p. II.3.

60. Griffin, "RLL and the Emerging Role for Chaplains," 45.

61. The term "force protection," which was changed to "protection warfighting function" in 2008, refers to "the related tasks and systems that preserve the force so the commander can apply maximum combat power." Headquarters, Department of the Army, *Operations* (Field Manual 3-0; Washington, D.C., Feb. 27, 2008), Glossary-12.

62. Chaplain (MAJ) Jonathan Etterbeek quoted in Lloyd, "Chaplain Contact with Local Religious Leaders," 9, and see also 13.

63. Lee, Burke, and Crayne, "Military Chaplains as Peace Builders," 17; Carden, "Muslim Chaplains Serve Various Roles."

64. Chaplain (COL) William Sean Lee, "Blessed are the Peacemakers: The Emerging Role of Army Chaplains as Religious Liaisons," *Army Chaplaincy* (Winter–Spring 2005): 58.

65. Expressions of concern about the military's use of information chaplains provided in their assessments of indigenous religions or as a result of religious leader engagement were evident in the 1980s. A preliminary draft of the 1989 chaplain field manual included the following cautionary statement: "The chaplain's function is to further religious and humanitarian concerns, not to enhance combat effectiveness through the use of religious information." The statement was revised somewhat for the final version of the manual, issued in November 1989. It stated that "one of the

chaplain's staff functions is to support religious humanitarian concerns, through the use of religious information." Headquarters, Department of the Army, *Religious Support Doctrine: The Chaplain and Chaplain Assistant* (Field Manual 16-1, Final Coordinating Draft; Washington, D.C., Apr. 1989), p. 1.14; Headquarters, Department of the Army, *Religious Support Doctrine* (FM 16-1, Nov. 27, 1989), p. 1.11. Nor was concern about chaplains and IO limited to the Army. In May 2002, the United States Central Command Air Forces (CENTAF) and Air Intelligence Agency Headquarters sponsored a symposium on the "Role of Religion in Information Operations" at Lackland Air Force Base, San Antonio, Texas. One of its purposes was to "examine appropriate roles for chaplains in . . . the operational environment." See Chaplain (COL) Greg W. Hill and Chaplain (MAJ) Rob Meyer, "The Role of Religion in Information Operations," *Army Chaplaincy* (Summer–Fall 2002), http://www.usachcs.army.mil/TACarchive/ACsumfal02/HillMeyer.htm/ (accessed Nov. 5, 2009).

66. Lawson, "Doctrinal Tension," 25–26, 27, 29, 30. See also Hill and Meyer, "Role of Religion in Information Operations"; and Chaplain (Colonel) David E. Smith, "The Implications Of Chaplaincy Involvement Within Information Operations," *Iosphere* (Fall 2006): 43–50.

67. Lawson, "Doctrinal Tension," 25, 28.

68. Lee, "Blessed are the Peacemakers," 61; Lawson, "Doctrinal Tension," 30; John W. Brinsfield and Eric Wester, "Ethical Challenges for Commanders and Their Chaplains," *Joint Force Quarterly* 54 (July 2009): 20–21; Brinsfield, "Army Chaplaincy and World Religions," 17; Chaplain (Colonel) Thomas C. Vail, "Religious Engagement and Diplomacy: Training the 21st Century U.S. Military Chaplaincy," *Military Intelligence Professional Bulletin* 31 (Jan.–Mar. 2011): 40–41.

69. *Religious Support* (FM 1-05, Apr. 18, 2003), p. A.1.

70. *Operations* (FM 3-0, Feb. 27, 2008), Glossary-3, defined "center of gravity" as "the source of power that provides moral or physical strength, freedom of action, or will to act."

71. Department of the Army, Office of the Chief of Chaplains, "Religious Leader Liaison Policy Letter," Sept. 30, 2008, *Army Chaplaincy* (Winter–Spring 2009): 36–39.

72. Chaplain (MG) Douglas L. Carver, Chief of Chaplains, "From the Chief," *Army Chaplaincy* (Winter–Spring 2009): 1–2. For another statement of the purpose of the CWR, see Chaplain (COL) Chet Lanious, "Introduction: Framing the World Religions Discussion," ibid., 10.

73. "Religious Support" (AR 165-1, Dec. 3, 2009), 12.

74. Chaplain (COL) Mike Hoyt, "Religion Counts: Planning Considerations for Religious Liaisons," *Army Chaplaincy* (Winter–Spring 2009): 73; Houck, "U.S. Army Chaplaincy's Involvement in Strategic Religious Engagement," 49, 50; Navy Chaplain (CPT) George Adams, "Chaplains as Liaisons with Religious Leaders," *Army Chaplaincy* (Winter–Spring 2009): 43; Brinsfield and Wester, "Ethical Challenges for Commanders and their Chaplains," 21. The discussion that took place between two

chaplains in Iraq, recounted by Kristin Henderson in the *Washington Post Magazine,* illustrates the problem a chaplain's religious orientation might pose for RLL. They were talking about joining an officer and soldiers of their unit in helping Muslims renovate a mosque that happened to be located on the base. The one chaplain regarded the effort as a means of "bridge-building," as well as the military's responsibility, in keeping with Army Regulation 12-4. The other chaplain was unwilling to participate, because he believed doing so would constitute an endorsement of the Muslim religion, something he, as a Southern Baptist, was not supposed to give. He told Henderson, "I am a Baptist, nothing more, nothing less, and I'm not comfortable with going outside my religion. If it was up to me, I'd have it [the mosque] shut down." Henderson, "'In the Hands of God,'" W08 (LN). Chaplain Timothy Bedsole noted that some chaplains' "theological prejudices" also limited their ability to advise the commander on indigenous religions. Bedsole, "Religion," 12.

75. Griffin, "RLL and the Emerging Role for Chaplains," 47–48.

76. Lee, "Blessed are the Peacemakers," 59, 60; Lawson, "Doctrinal Tension," 29; Vail, "Religious Engagement and Diplomacy," 38. See also Hoyt, "Religion Counts: Planning Considerations for Religious Liaisons," 73; Chaplain (Lieutenant Colonel) Ira C. Houck III, "Strategic Religious Engagement for Peacebuilding," Civilian Research Project, U.S. Army War College, Carlisle Barracks, Pa., Mar. 17, 2009, 6.

77. Adams, "Chaplains as Liaisons with Religious Leaders," 42–43. See also Brinsfield and Wester, "Ethical Challenges for Commanders and their Chaplains," 20–21.

78. Houck, "U.S. Army Chaplaincy's Involvement in Strategic Religious Engagement," 50, citing George Adams, *Chaplains as Liaisons with Religious Leaders: Lessons from Iraq and Afghanistan* (Washington, D.C.: United States Institute of Peace, 2006), 4.

79. Adams, "Chaplains as Liaisons with Religious Leaders," 43.

80. The Joint Chiefs of Staff recognized religious leader engagement in a 2004 publication as follows: "The JFCH [Joint Forces Chaplain], after careful consideration and only with the JFC's [joint force commander's] approval, may serve as a point of contact to HN [host nation] civilian and military religious leaders, institutions, and organizations, including established and emerging military chaplaincies, through the CMOC [civil-military operations center]." Headquarters, Joint Chiefs of Staff, *Religious Support in Joint Operations* (JP 1-05, June 9, 2004), p. II.3.

81. For a valuable assessment of the promise and problems of RLL, as of 2009, by a civilian religious leader who was very interested and involved with the U.S. military in matters relating to religion and security, see Seiple, "Ready . . . Or Not?"

82. In a research paper written for the Army War College in 2009, Chaplain (Lieutenant Colonel) Ira C. Houck III described a new version of religious leader engagement, which he called "Strategic Religious Engagement" (SRE), that would expand the chaplain's IR and RLL roles to include involvement in "peacebuilding processes," using conflict resolution analysis and strategies. In performing SRE,

chaplains would not only engage "moderate clerics and political-religious leaders" but also "reach out to personally enlist radical and militant religious leaders." Their peace-building effort would involve communication and collaboration with a host of individuals and groups: "minority religious groups as well as institutionally affiliated religious actors, lay and clerical human rights advocates, development and relief workers, missioners, denominational structures, and international and multi-religious bodies"; the commander's staff; individuals in "the Joint, Interagency, Intergovernmental, and Multinational environment"; and the chaplaincies of allied and multinational forces. Houck, "Strategic Religious Engagement for Peacebuilding," 1, 4, 5, 10–11, 13, 22, 24. In 2005, in a U.S. Army War College research paper, Chaplain (Colonel) Stephen L. Cook had proposed another new role for the chaplain: serving as "a religious professional" on a U.S. ambassador's "country team." He included a detailed description of Chaplain (Colonel) Douglas Carver's experience in Iraq, serving a "diplomatic role" as a "religious representative" of Ambassador Paul Bremer. Chaplain (Colonel) Stephen L. Cook, "U.S. Military Chaplains on the Ambassador's Country Team," Strategy Research Project, U.S. Army War College, Carlisle Barracks, Pa., Mar. 28, 2005.

83. Houck, "Strategic Religious Engagement for Peacebuilding," 3, 26n9.

84. *Operations* (FM 3-0, Feb. 27, 2008), "Foreword."

8. Building Soldier Morale

1. Douglas Johnston, "U.S. Military Chaplains: Redirecting a Critical Asset," *Review of Faith and International Affairs* 7 (Winter 2009), http://rfiaonline.org/ (accessed Feb. 10, 2010).

2. The Chaplain School, *The Chaplain and Military Morale* (ST 16-155; Carlisle Barracks, Pa.: Aug. 1950), 1–2.

3. Ibid., 6–7, 100–102, 130–32 (emphasis in original).

4. *The Chaplain* (Field Manual 16-5; Washington, D.C., Jan. 1952), 13; *Chaplain* (FM 16-5; Washington, D.C., Apr. 15, 1958), 3, 7, 52.

5. *Character Guidance Manual* (FM 16-100, Mar. 1961), 2. See also *Character Guidance Manual* (FM 16-100, June 26, 1968), 12.

6. OCCH, *Summary of Major Events and Problems,* 26.

7. Chaplain (Colonel) Richard B. Cheatham, U.S. Army, Director of Plans, Programs and Policies, wrote the OCCH response: "Material for Senate Preparedness Investigating Subcommittee," Aug. 14, 1962, in File 721-01 Stennis Subcommittee Study, Acc.No. 68-A-3353, RG 247, Washington National Records Center, Suitland, Md. See also Brady, "Narrative Description of the Character Guidance Program," 6.

8. *Chaplain* (FM 16-5, Aug. 27, 1964), 1, 41, 42; *Chaplain* (FM 16-5, Dec. 26, 1967), 3, 40, 41.

9. Klitgaard, "Onward Christian Soldiers," 1378–79; United Church of Christ, "Ministries to Military Personnel," 81, 82; Zahn, "Scandal of the Military

Chaplaincy," 313, 314, 315; Zahn, "What Did You Do during the War, Father?" 196, 197, 198.

10. *The Chaplain* (Field Manual 16-5; Washington, D.C., May 11, 1970), 1 (emphasis added).

11. The 1970 memorandum appears to be an early draft of a pamphlet the Command Chaplain Agency issued in April 1972, *The Chaplain's Role as Related to Soldier Motivation, Final Study*. Based on behavioral research, it discussed the chaplain's role in "soldier motivation" in the light of communist indoctrination of the Korean War POWs, the current threat of war with the Soviet Union, the new approach the Army was developing to deal with the "new breed" of soldiers coming into the all-volunteer Army, and the growing emphasis among segments of the civilian religious community on "[social] involvement-centered ethics," including antimilitarism. Some of its statements suggest that OCCH's commentary had made an impact. The pamphlet described the chaplain as a "priceless commodity." Religion was declared to be one of several supports of morale. The pamphlet noted that "a person who has matured in our society usually has acquired certain spiritual values. These values not only strengthen an individual's character, but also provide him with a source of inner strength and stability." The pamphlet emphasized that the chaplain's function was one of motivation rather than indoctrination: "The chaplain definitely is not a political officer! Whereas the political officer conducts indoctrination through dialectical distortion," the chaplain "is the motivator of the free mind for humane, and above all, theocentric spiritual ends." United States Army Combat Developments Command Chaplain Agency, *The Chaplain's Role as Related to Soldier Motivation, Final Study* (Fort Lee, Va., Apr. 1972), viii, 16, 19, 20, 24, 28, 32. The mission of the U.S. Army Combat Developments Command Chaplain Agency was "to develop and recommend current and future chaplain concepts and objectives, organizational and operational doctrine, materiel requirements, and field test requirements." *Chaplain* (FM 16-5, Dec. 26, 1967), 74.

12. OCCH, "Historical Review," 1970–71, 62–63. Compare Jack S. Boozer's disparagement of the chaplain's role as "a morale officer of the command," in "The Military Chaplaincy: One Calling, Two Roles," 4.

13. Headquarters, United States Army Training and Doctrine Command, *Military Operations: US Operational Concept for Religious Support in Combat Areas* (TRADOC Pamphlet 525-26; Fort Monroe, Va.: Mar. 15, 1983), 2 (emphasis added).

14. *Chaplain and Chaplain Assistant in Combat Operations* (FM 16-5, Dec. 10, 1984), 11, 3, 51, 58; *Chaplain* (FM 16-5, Aug. 27, 1964), 41.

15. Chaplain (COL) Jay H. Ellens, "Preparation for Combat: Emotional and Spiritual," *MCR* (Spring 1984): 31, 32, 33.

16. In the 1980s, Chaplain Kuehne served as force structure plan officer and, subsequently, director of plans, policy, development and training in the Army Office of the Chief of Chaplains. He was also director of the Combat Developments Directorate at

the Army Chaplain School. Brinsfield, *Encouraging Faith,* pt. 1, 64, 76, 167, 175, 177, 179, 180, 182, 184, 198, 270, 276, 304.

17. Kuehne, "Faith and the Soldier," 8.

18. Ibid., 59n68.

19. Although Kuehne quoted Clausewitz as saying that "all military action is intertwined with psychological ["von geistigen"] forces and effects," he disagreed with translators of Clausewitz who substituted "psychology" or "mind" for the German words "Geist" and "geistig." Kuehne translated them as "spirit" and "spiritual." Ibid., 7, 53–54n30, 54n33.

20. Ibid., 3–4. Kuehne cited Stouffer's *American Soldier: Combat and Its Aftermath.*

21. Marshall quoted in Kuehne, "Faith and the Soldier," 6. For other quotations of the statement, see Brinsfield, "Army Values and Ethics," 81; John W. Brinsfield and Peter A. Baktis, "The Human, Spiritual, and Ethical Dimensions of Leadership in Preparation for Combat," in *The Future of the Army Profession,* ed. Don M. Snider and Lloyd J. Matthews, 2nd ed., rev. and expanded (New York: McGraw-Hill, 2005), 472; *Chaplain and Chaplain Assistant in Combat Operations* (FM 16-5, Dec. 10, 1984), 6; *Military Operations* (TRADOC Pamphlet 525–26, Mar. 15, 1983), inside cover; *Fit to Win: Spiritual Fitness* (Pamphlet 600-63-12; Washington, D.C.: Sept. 1, 1987), foreword; *Religious Support Handbook* (TC 1-05, May 10, 2005), B-3.

22. Kuehne, "Faith and the Soldier," 49n7.

23. Ibid., 52n22, 38, 47, and see also 43.

24. *Religious Support Doctrine* (FM 16-1, Nov. 27, 1989), p. 5.1. The preface to the manual (vii) noted that "this field manual sets forth doctrine and guidance for commanders, staffs, chaplains, and chaplain assistants," and that "doctrinal principles," which were "prominently outlined" throughout the manual, were "statements of policy, broad in scope, directed toward the successful accomplishment of a religious support mission."

25. Ibid., pp. 5.34, 5.2; *Religious Support* (FM 16-1, May 26, 1995), p. 1.5.

26. *Religious Support* (FM 16-1, May 26, 1995), p. F.1.

27. *Religious Support Doctrine* (FM 16-1, Nov. 27, 1989), p. 5.31; see also *Religious Support* (FM 16-1, May 26, 1995), p. F.1.

28. *Religious Support Doctrine* (FM 16-1, Nov. 27, 1989), pp. 5.31–5.33; *Religious Support* (FM 16-1, May 26, 1995), pp. 2.1, F.2.

29. *Religious Support* (FM 16-1, May 26, 1995), p. F.1.

30. *Fit to Win* (Pamphlet 600-63-12, Sept. 1, 1987), 1. The pamphlet was part of the army's Health Promotion Program.

31. *Religious Support* (FM 16-1, May 26, 1995), p. F.2. For guidance on chaplains engaging in spiritual fitness training at Army installations to ensure units were "spiritually prepared for deployment and combat," see also *Religious Support* (FM 1-05, Apr. 18, 2003), pp. 1.6, 7.9; *Religious Support Handbook* (TC 1-05, May 10, 2005),

p. B.2; "Chaplain Activities in the United States Army" (AR 165-1, Mar. 25, 2004), 21, 22; "Religious Support: Army Chaplain Corps Activities" (AR 165-1, Dec. 3, 2009), 31.

32. *Fit to Win* (Pamphlet 600-63-12, Sept. 1, 1987), 6–8.

33. *Religious Support* (FM 1-05, Apr. 18, 2003), pp. 7.4, 7.7, 7.8.

34. *Religious Support Handbook* (TC 1-05, May 20, 2005), pp. J.1, J.3.

35. Ibid., pp. J.3, J.4, C.7, C.8.

36. Ibid., pp. J.1, F.3, F.1.

37. Ibid., pp. J.2–J.3.

38. Ibid., p. J.6.

39. John W. Brinsfield, "Reality Check: The Human and Spiritual Needs of Soldiers and How to Prepare Them for Combat," in *The Future of the Army Profession,* ed. Lloyd J. Matthews (Boston: McGraw-Hill, 2002), 398, 400; Brinsfield, "Army Values and Ethics," 70, 81. The Autumn 1998 issue of *Parameters* (vol. 27) provided biographical information on Brinsfield on pages 70 and 84n21. It gave his current title, Director of Ethical Program Development at the U.S. Army War College, and described him as a graduate of Vanderbilt University and the U.S. Army War College, with an M.Div. from Yale Divinity School, a Ph.D. in history from Emory University, and a D.Min. in ethics from Drew University. It noted that he had taught military ethics in the Army since 1976, and from December 1990 to May 1991, had served as the ARCENT PERSCOM Chaplain during Operations Desert Shield and Desert Storm.

40. Brinsfield, "Reality Check," 401, 406.

41. Charles H. Lippy, *Pluralism Comes of Age: American Religious Culture in the Twentieth Century,* ebook (Armonk, N.Y.: M. E. Sharpe, 2000), 160.

42. Jerry Adler. "In Search of the Spiritual," *Newsweek,* Aug. 29, 2005, https://www-lexisnexis-com.libezp.lib.lsu.edu/ (accessed Dec. 2, 2009).

43. Pew Forum on Religion & Public Life, "Eastern, New Age Beliefs Widespread: Many Americans Mix Multiple Faiths," Dec. 2009, 2, 7, 12, http://www.pewforum.org/uploadedfiles/Topics/Beliefs_and_Practices/Other_Beliefs_and_Practices/multiplefaiths.pdf/ (accessed Sept. 7, 2012); Pew Forum on Religion & Public Life, "'Nones' on the Rise: One-in-Five Adults Have No Religious Affiliation" (Oct. 9, 2012), 9–10, 22, 24, 41, 43–45, 48, 53–54, 56, http://www.pewforum.org/Unaffiliated/nones-on-the-rise.aspx/ (accessed Oct. 15, 2012).

44. The chaplaincy's six key values (SACRED) were formulated in November 1999 at a Board of Directors workshop. Chaplain (Colonel) Douglas B. McCullough, Chaplain (Colonel) John W. Brinsfield, and Chaplain (Major) Kenneth E. Lawson, *Courageous in Spirit, Compassionate in Service: The Gunhus Years, 1999–2003* (Washington, D.C.: Office of the Chief of Chaplains, Department of the Army, 2003), 9, 10, 175. For Gunhus's definition of "Spirituality," "spiritual," and his making "Spiritual Leadership" the "centerpiece" of the chaplaincy's strategic plan, see ibid., 10, 194.

45. Chaplain (MG) Gaylord T. Gunhus, Chief of Chaplains, "From the Chief," *Army Chaplaincy* (Summer–Fall, 2000), http://www.usachcs.army.mil/ (accessed July 30, 2010). Another of the Chaplain Corps SACRED values was "Religious Leadership: Model spiritual truths wisely and courageously." McCullough, Brinsfield, and Lawson, *Courageous in Spirit,* 175.

46. Chaplain (MG) David H. Hicks, Chief of Chaplains, "From the Chief," and Chaplain (MAJ) Maury Stout, "Taking Spiritual Leadership to the Next Level: A White Paper for The United States Army Chaplain Corps," *Army Chaplaincy* (Winter–Spring 2005): 1, 6–7.

47. Chaplain (COL) J. Gordon Harris, "Spiritual Well-Being and Maturity in the Bible," and Chaplain (MAJ) Ronald Thomas, "Pastoral Counseling is Spiritual Leadership," *Army Chaplaincy* (Summer–Fall 2000), http://webharvest.gov/peth04/20041019110609/http://www.usachcs.army.mil/ (accessed July 30, 2010).

48. *Religious Support Doctrine* (FM 16-1, Nov. 27, 1989), pp. 1.1, 1.8, 5.1, 5.2, 5.31, 5.33, 5.34.

49. *Religious Support* (FM 16-1, May 26, 1995), F.2.

50. Gallup Report cited in Groh, *Facilitators of the Free Exercise of Religion,* 36; "Religious Preferences of the U.S. Population and Military Personnel, 201," in David R. Segal and Mady Wechsler Segal, "America's Military Population," *Population Bulletin* 59 (Dec. 2004): 25, http://www.maaf.info/resources/PopRef Bureau2004.pdf/ (accessed Nov. 9, 2009); "Religion of Active Duty Personnel By Service (no Coast Guard). As of March 31, 2009" (produced by the Defense Manpower Data Center, May 1, 2009). I wish to thank Bob Smietana, a reporter for the *Tennessean,* for sending me the Defense Manpower Data Center document, September 2009.

51. Brinsfield, "Reality Check," 403.

52. Brady, "Narrative Description of the Character Guidance Program," 6.

53. *Military Operations* (TRADOC Pamphlet 525-26, Mar. 15, 1983), 4; *Fit to Win* (Pamphlet 600-63-12, Sept. 1, 1987), 1.

54. Army Well-Being Strategic Plan of 2001, quoted in Brinsfield, "Reality Check," 404; "Chaplain Activities in the United States Army" (AR 165-1, Mar. 25, 2004), 52.

55. Brinsfield, "Army Values and Ethics," 83, 82. The second statement was quoted, without attribution, in the *Religious Support Handbook* (TC 1-05, May 10, 2005), B-5.

56. Brinsfield, "Reality Check," 416, 418; emphasis added.

57. "Religious Support" (AR 165-1, Dec. 3, 2009), 1, 9; emphasis added. For statements about chaplains facilitating the free exercise of religion and avoiding "even the appearance of any establishment of religion," see "Chaplain Activities in the United States Army" (AR 165-1, Mar. 25, 2004), 1, 5, 6.

58. *Fit to Win* (Pamphlet 600-63-12, Sept. 1, 1987), 1, cited "resilience" as one of the "personal qualities," along with "endurance and emotional and moral stability,"

nurtured by spiritual fitness training. For references to soldier "resilience" and "spiritual resiliency," see also "Religious Support" (AR 165-1, Dec. 3, 2009), 11, 12.

59. Chaplain (MG) Douglas L. Carver, Chief of Chaplains, "From the Chief," *Army Chaplaincy* (Summer–Fall 2009), 1, 2 (emphasis in original). Carver quoted the popular evangelist Ruth Haley Barton, well known in evangelical circles as a practitioner and teacher of "spiritual transformation" and "spiritual leadership" (2). She, herself, had a lengthy article in the Summer–Fall 2009 *Army Chaplaincy*, pp. 41–54, based on a talk she had delivered at a Chaplain Corps Strategic Leader Development Training at Hilton Head, South Carolina, and her best-selling book, *Strengthening the Soul of Your Leadership*. "Ruth Haley Barton, Founder of the Transforming Center," http://www.transformingcenter.org/in/about/ruth-bio.shtml/ (accessed Sept. 23, 2012).

60. Chaplain (LTC) Abdul-Rasheed Muhammad, "TAQWAH and SABR: The Foundation of Spiritual Resilience in Islam," and Chaplain (CPT) Henry Soussan, "A Jewish Perspective on Spiritual Resilience," *Army Chaplaincy* (Summer–Fall 2009): 55–60 and 69–73, respectively.

61. Chaplain (COL) Mike Dugal, "Spiritual Resiliency and the Senior Chaplain's Role," *Army Chaplaincy* (Summer–Fall 2009): 8. In "From the Chief," 3, Chief of Chaplains Carver described the mission of the Center for Spiritual Leadership (CSL) as follows: "The CSL team has already begun to supply the Chaplain Corps with tailored spiritual development programs and products to advance life-long learning and to enhance the comprehensive spiritual fitness of our branch. The CSL is a growing resource to support Chaplain Corps' ability to remain spiritually resilient and its ability to empower the soul of our Army's Soldiers."

62. Dr. John W. Brinsfield and Chaplain (LTC) Peter A. Baktis, "The Human, Spiritual and Ethical Dimensions of Leadership in Preparation for Combat," BG Rhonda Cornum, "Why Comprehensive Soldier Fitness? A Personal and Organizational Perspective," Chaplain (LTC) Dean Bonura, "Combat Trauma, Resiliency and Spirituality," Chaplain (MAJ) William Scritchfield, "A Question of Spiritual Resiliency in Context," all in *Army Chaplaincy* (Summer–Fall 2009): 23–24, 15, 16, 76, 87, respectively.

63. Soussan, "Jewish Perspective on Spiritual Resilience," 73; Brinsfield and Baktis, "Human, Spiritual and Ethical Dimensions of Leadership," 25–30, 32–33, 35–36; Scritchfield, "Question of Spiritual Resiliency in Context," 87; Brinsfield, "Army Values and Ethics," 82; *Religious Support Handbook* (TC 1-05, May 10, 2005), B-5.

9. Addressing Religious Pluralism

1. Israel Drazin and Cecil B. Currey, *For God and Country: The History of a Constitutional Challenge to the Army Chaplaincy* (Hoboken, N.J.: KTAV Publishing House, 1995), 1–3, 163–64, 198, 199.

2. Brinsfield, *Encouraging Faith,* pt. 1, 121; Drazin and Currey, *For God and Country,* 205. See also Gregory J. Darr, "For God & Country: The Constitutional Question of the U.S. Army Chaplaincy," *MCR* (Winter 1992): 101.

3. Barry A. Kosmin, Egon Mayer, Ariela Keysar, "American Identification Survey, 2001" (New York: Graduate Center of the City Univ. of New York, 2001), http://www.gc.cuny.edu/faculty/research_briefs/aris.pdf/ (accessed Jan. 5, 2011); Barry A. Kosmin and Ariela Keysar, "American Religious Identification Survey Summary Report March 2009" (Hartford, Conn.: Trinity College, 2009), http://www.americanreligionsurvey-aris.org/reports/ARIS_Report_2008.pdf/ (accessed Jan. 5, 2011) (hereafter cited as ARIS 2008). The data provided in the surveys came from interviews with the adult population of the United States. The 2001 survey included data from a 1990 survey; the 2008 survey included data from the 1990 and 2001 surveys. The "total numbers" reported in the surveys are rounded-off extrapolations from the data.

4. ARIS 2008, 3, 4, 5.

5. Ibid., 5, 6.

6. Ibid., 5.

7. Ibid., n.p.

8. Ibid., 9.

9. Ibid., n.p., 3.

10. Ibid., 3.

11. Ibid., 4. Muslim, Hindu and Buddhist leaders contended that the number of Americans affiliated with their faith groups was much larger than various surveys showed. See, e.g., David Briggs, "Islamic Faiths Change Face of U.S. Religion," *Baton Rouge Advocate,* Aug. 31, 1993, p. 6A.

12. ARIS 2008, 3.

13. Ibid., 7.

14. Ibid., 4.

15. Ibid., n.p.

16. Ibid., n.p.

17. Gallup Report cited in Groh, *Facilitators of the Free Exercise of Religion,* 36.

18. "Religious Preferences of the U.S. Population and Military Personnel, 2001," in Segal and Segal, "America's Military Population," 25.

19. "Religion of Active Duty Personnel By Service (no Coast Guard). As of March 31, 2009."

20. "Department of Defense Directive: Accommodation of Religious Practices Within the Military Services" (Number 1300.17, June 18, 1985, Feb. 3, 1988, certified current as of Nov. 21, 2003), http://www.dtic.mil/whs/directives/crresp/pdg/130017p.pdf/ (accessed Jan. 21, 2009). In 1999, the DoD revised and updated the guidelines for religious accommodation, in response to the increasing diversity of religious beliefs in the military, especially the increase in the number of military

personnel listing no religious preference. See Glen Elsasser, "Religious Pluralism Is Newest Theater for Military Action," *Chicago Tribune,* July 6, 1999, sec. 1, p. 4. For a good discussion of Army regulations and policy statements regarding religious accommodation, see Major Michael J. Benjamin, "Justice, Justice Shall You Pursue: Legal Analysis of Religion Issues in the Army," *Army Lawyer* (Nov. 1998), 1–18. For recent versions of DoD directives and service branch regulations, see http://www.deomi.org/DiversityMgmt/RelAccomMilitary.cfm/ (accessed Dec. 14, 2010).

21. Brinsfield, *Encouraging Faith,* pt. 1, 243; Groh, *Facilitators of the Free Exercise of Religion,* 277. For a comprehensive listing of religious practices to be accommodated, see, for example, Headquarters, Department of the Army, "Army Command Policy" (Army Regulation 600-20; Washington, D.C., June 13, 2002), 32–34.

22. On the commander's responsibility regarding free exercise, see, for example, Headquarters, Department of the Army, "Chaplain Activities in the United States Army" (Army Regulation 165-1; Washington, D.C., June 26, 2000), 3, 5; on the responsibility to decide on requests for accommodation, and for guidelines and procedural rules, see Headquarters, Department of the Army, "Army Command Policy" (2002), 32–35.

23. "Chaplain Activities in the United States Army" (AR 165-1, June 26, 2000), 1, 6, 7.

24. Benjamin, "Justice, Justice Shall You Pursue," 11, 13, 14. For a much publicized instance of a chaplain's advocacy in behalf of Jewish personnel who sought to wear the yarmulke while in military uniform, which influenced the U.S. House of Representatives to pass a religious apparel amendment allowing it, see http://enwikipedia.org/wiki/Arnold_Resnicoff/ (accessed Dec. 6, 2010); "The Religious Apparel Amendment," *Congressional Record,* May 11, 1987, http://www.resnicoff.net/CongRec87.html/ (accessed Jan. 21, 2010); Brinsfield, *Encouraging Faith,* pt. 1, 243; Groh, *Facilitators of the Free Exercise of Religion,* 21.

25. 1988 DoD directive quoted in Groh, *Facilitators of the Free Exercise of Religion,* 107, and see also 11–12. Even before the implementation of religious accommodation, DoD had accepted as official policy a late 1970s report by the Conference of Ecclesiastical Endorsing Agents that set forth requirements endorsing agents must meet in order to be accepted by the Armed Forces Chaplains Board, one of which read as follows: "Should be expected to provide chaplains who are willing to respect the integrity of, and whenever possible, to work cooperatively with other religious groups and faiths." Paul J. Weber, "The First Amendment and the Military Chaplaincy: The Process of Reform," *Journal of Church and State* 22 (Autumn 1980): 468.

26. Department of Defense Directive 1304.19, *Appointment of Chaplains for the Military Departments,* June 11, 2004, quoted in CDR William A. Wildhack III, CHC, USNR, "Navy Chaplains at the Crossroads: Navigating the Intersection of Free Speech, Free Exercise, Establishment, and Equal Protection," *Naval Law Review* 51 (2005): 229, 229n, 230.

27. See, for example, Chaplain School of the United States Army, *Program of Instruction* (Chaplain Officer Basic, 5-16-C20-56A; Fort Monmouth, N.J.: U.S. Army Chaplain and School [*sic*], 1 July 1984), 13–14; "Army Chaplain Boot Camp," *Religion & Ethics,* Apr. 4, 2008, http://www.pbs.org/ (accessed Sept. 30, 2012). According to Army Regulation 600-20, the Commanding General, TRADOC, was to ensure that training in the accommodation of religious practices was "provided for commanders, chaplains, and judge advocates." Headquarters, Department of the Army, "Army Command Policy" (2002), 32.

28. Brinsfield, *Encouraging Faith,* pt. 1, 243.

29. See, for example, Brinsfield, *Encouraging Faith,* pt. 1, 32–34, 138–39, 190, 196–97, 206; Chaplain (LTC-P, United States Army) Jerry L. Robinson, "A Chronological Record of Historical Events Relating to Diversity in the U.S. Army Chaplaincy as Viewed by Chaplain (Major General) (Retired) Matthew A. Zimmerman, Jr.," 7–10, Strategy Research Project, U.S. Army War College, Carlisle Barracks, Pa., 1997; Paul Otterstein, "Theological Pluralism in the Air Force Chaplaincy," *MCR* (Fall 1987): 119–20; Chaplain (Brigadier General) Charles McDonnell, "Multi-Cultural Ministry and The Unit Ministry Team," *MCR* (Fall 1987): 80.

30. Brinsfield, *Encouraging Faith,* pt. 1, 195; Susanne Kappler, "Chaplain Recalls Path to Making History," *Fort Jackson Leader,* June 12, 2009, http://www.army.mil/ (accessed Aug. 15, 2012); Bob Smietana, "Buddhist Chaplain Is Army First," *USA Today,* Sept. 8, 2009, http://www.usatoday.com/ (accessed Aug. 15, 2012); Gary Sheftick, "Chaplain Corps Turns 236 with New Strength," Army News Service, July 28, 2011, http://www.army.mil/ (accessed Aug. 15, 2011); Ari L. Goldman, "Religion Notes," *NYT,* Jan. 30, 1993, p. 28; Alan Cooperman, "For Gods and Country," Feb. 19, 2007, *WP,* http://www.washingtonpost.com/ (accessed June 25, 2010); James Dao, "Atheists Seek Chaplain Role in the Military," *NYT,* Apr. 26, 2011, http://www.nytimes.com/ (accessed Aug. 15, 2012); Billy Hallowell, "Atheists Continue Their Major Push for Non-Believing Military Chaplains," Nov. 11, 2011, and Billy Hallowell, "Non-Theist Soldier Wants U.S. Army to Official Recognize Humanism as a 'Faith Group," Feb. 8, 2012, http://www.theblaze.com/ (accessed Oct. 8, 2012). Besides certification by the Armed Forces Chaplains Board, other requirements for appointment as an army chaplain include the following: meeting the same qualifications as other officers for regular army commissioning, having an ecclesiastical endorsement from one's faith group, and completion of a baccalaureate degree of not less than 120 semester hours, and of a graduate degree in theological or religious studies, with at least 72 semester hours in graduate work in those two fields. Additional requirements include being at least twenty-one years of age but younger than forty-two at the time of commissioning and having a minimum of two years of full-time professional experience, validated by the endorsing agency. "Army Chaplain Corps: Chaplain Requirements," http://www.goarmy.com/chaplain/about/requirements.html/ (accessed Sept. 30, 2012); "Military Chaplain," 2, *Wikipedia,* http://en.wikipedia.org/ (accessed Sept. 30, 2012).

31. Loveland, *American Evangelicals,* 22–27.

32. "Study of Representation of Religious Faiths in the Armed Forces, Section 513, DOD Defense Authorization Act, 1987," Jan. 1987, cited in Groh, *Facilitators of the Free Exercise of Religion,* 36, 559n6. Mandated by Congress in connection with the 1987 Defense Authorization Act, the study apparently included reserve as well as active-duty chaplains.

33. "Religion of Active Duty Chaplains by Service (no Coast Guard). As of March 31, 2009."

34. The Reverend Dr. Bertram C. Gilbert, "The Scandal and the Glory," *MCR* (Spring 1984): 23.

35. Retired U.S. Navy Chaplain S. David Chambers referred to this as "The Protestant Problem," *MCR* (Fall 1987): 81–88.

36. Loveland, *American Evangelicals,* 92–94, 93n18, 303.

37. *Chaplain* (FM 16-5, Apr. 15, 1958), 13; *Chaplain* (FM 16-5, Dec. 26, 1967), 13–14.

38. OCCH Newsletter, Jan. 1, 1988, p. 5, quoted in Brinsfield, *Encouraging Faith,* pt. 1, 246; Darr, "For God & Country," 102–3.

39. "Diversity" was one of the Army Chaplain Corps' "core values." It committed chaplains to "a Chaplaincy that honors and enables the practices of religious expressions of an extremely diverse constituency," a chaplaincy that "is not merely 'politically correct' but is generally sensitive to the diverseness of race, gender and culture." Quoted in Chaplain (Col.) Douglas B. McCullough, U.S. Army Reserve, "Religious Leadership for the Army," in McCullough, Brinsfield, and Lawson, *Courageous in Spirit,* 11. In a 1994 "Statement on Equal Opportunity," Chief of Chaplains Matthew A. Zimmerman described the Army chaplaincy as "multi-faith, ethnically and religiously diverse, and supportive of the soldier's right to free exercise of religion." He went on to declare, "No chaplain serving in the Army is expected to compromise the tenets of his or her religion, nor will the free exercise of any chaplain's faith be inhibited." Brinsfield, *Encouraging Faith,* pt. 1, 338. The U.S. District Court decision, *Rigdon v. Perry* (1997), specifically declared that chaplains had the right of free exercise, as well as free speech. Kim A. Lawton, "Military Chaplains Sue Over 'Project Life' Ban," *CT* 40 (Dec. 9, 1996): 74; "In the Federal Courts: District Judge Rules Military Chaplains May Advocate Against Abortion," *Reproductive Freedom News* 6 (Apr. 18, 1997): 2–3 (LN).

40. On the shift away from evangelistic activity and the reasons for it, see Berg, "'Proclaiming Together'?" 61, 63; Hutcheson, *Mainline Churches and the Evangelicals,* 117, 138, 177.

41. "Monthly newsletter, OCofCh, CHPE (219)," June 1, 1969, 2, in OCCH, *Historical Review,* 1968–69, 33. OCCH had been requested to issue the statement by Chaplain (Colonel) Joseph B. Messing, director of administration and management.

42. OCCH, *Historical Review,* 1968–69, 32–33.

43. Loveland, *American Evangelicals,* 173–80.

44. “Commission Aids Evangelical Chaplains, *United Evangelical Action* (Jan.–Feb. 1987): 11; James H. Young, “Is Religious Proselytism in the Military Legitimate?” *Faith for the Family* (Nov. 1979): 6–7; Barry C. Black, “The Seventh-Day Adventist Military Chaplain: A Study of Beliefs and Functions in Tension with Military Life” (D.Min. diss., Eastern Baptist Theological Seminary, 1982), 122. Black was identified in the dissertation as an SDA chaplain on active duty with the U.S. military.

45. “Military Chaplains Manual, The Christian and Missionary Alliance” (Mar. 16, 1983), 7 (emphasis added).

46. To compare evangelical chaplains’ views of evangelism with those of civilian evangelicals, see “Where Has Evangelism Been? Where is it Going?” *United Evangelical Action* (Nov.–Dec. 1991): 6–9

47. Chaplain (MAJ) Jerry E. Malone, “The Chaplain as an Advocate of Religious Freedom,” *MCR* (Fall 1983): 57; Mathis quoted in Gil A. Stricklin, “Evangelism in the Ranks,” *MCR* (Fall 1991): 54; Robert G. Leroe, “Faith Sharing in the Military,” *MCR* (Fall 1991): 71.

48. National Guard chaplain and command chaplain quoted in Stricklin, “Evangelism in the Ranks,” 54–57.

49. National Guard chaplain quoted in Stricklin, “Evangelism in the Ranks,” 54.

50. Leroe, “Faith Sharing in the Military,” 74.

51. Malone, “Chaplain as an Advocate of Religious Freedom,” 52, 57–58.

52. Chaplain (CPT) Rick D. Mathis, “Constitutional Guidelines for the Military Chaplain Evangelist and Chaplaincy,” *MCR* (Fall 1991): 33–34, 43; “Constitutional Guidelines for Chaplains,” *Centurion* (Dec. 1992): 3–5.

53. Mathis, “Constitutional Guidelines,” 37–38; Roy Mathis quoted in Stricklin, “Evangelism in the Ranks,” 55.

54. On the spate of Supreme Court decisions, see Laurence H. Tribe, “The Cross and the Sword: Separating the Realms of Authority,” in *Government Intervention in Religious Affairs,* ed. Dean M. Kelley (New York: Pilgrim Press, 1986), 14. On chaplains’ discussion of public prayer, see Chaplain (COL) Bertram Gilbert, USA, Ret., “On Prayers in Jesus’ Name,” *MCR* (Fall 1987): 123.

55. LCDR A. E. Resnicoff, CHC, USN, “Prayers that Hurt: Public Prayer in Interfaith Settings,” *MCR* (Winter 1987): 31, 33.

56. Gilbert, “On Prayers in Jesus’ Name,” 123–25.

57. Robert G. Leroe, “Public Prayer,” *MCR* (Spring 1992): 50, 53.

58. In 1982, the commission changed its name to National Conference on Ministry to the Armed Forces and began expanding its membership beyond the Protestant faith to include Roman Catholic, Jewish, Orthodox, Buddhist, and Islamic faith groups. By 2009, it represented more than two hundred religious denominations and faith groups. “National Conference on Ministry to the Armed Forces,” http://www.ncmaf.org/ (accessed Jan. 20, 2009). According to the NCMAF’s coordinator, Clifford T. Weathers, the “Code” had been formulated by a

"representative group of endorsers." It was presented to a plenary session of the NCMAF for discussion and went through several revisions before receiving final NCMAF approval. It "represents the NCMAF standard for chaplains who serve in the military chaplaincy." Clifford T. Weathers, Coordinator, NCMAF, to author, July 22, 1992.

59. National Conference on Ministry to the Armed Forces, "The Covenant and The Code of Ethics for Chaplains of the Armed Forces," http://www.ncmaf.org/policies/PDFs/CodeofEthics-NCMAF.pdf/ (accessed Mar. 16, 2011).

60. Commission Resource Board, NAE Chaplains Commission, "Guidelines for Cooperation with Chaplains of Other Faiths," [Nov. 1992], quoted in Loveland, *American Evangelicals,* 307–8.

61. [Commission Resource Board, NAE Chaplains Commission], "Freedom of Expression in Public Prayer: Guidelines for Evangelical Chaplains," [Nov. 1992], quoted in Loveland, *American Evangelicals,* 308–9.

62. Captain (U.S. Army Reserve) Matthew Zimmerman, Administrator, NAE Commission on Chaplains and Military Personnel, telephone conversation with author, Jan. 15, 2008.

10. The 2005–2006 Culture War

1. Portions of this chapter first appeared in Anne C. Loveland, "The God Squadron," *Religion in the News* 8 (Fall 2005): 2–5, 25, and Anne C. Loveland, "Evangelical Proselytizing at the U.S. Air Force Academy: The Civilian-Military Controversy, 2004–2006," *Journal of Ecumenical Studies* 44 (Winter 2009): 11–25.

2. Mike Soraghan, "AFA Religion Debate Erupts in D.C.," *Denver Post,* June 21, 2005, p. A05 (LN); M. E. Sprengelmeyer, "A Holy War in D.C.," *Rocky Mountain News* (Denver, Colo.), June 21, 2005, p. A05 (LN).

3. Charlie Brennan, "Leader of AFA Panel Is 'Born-Again.'" *Rocky Mountain News,* June 25, 2005, p. A22 (LN).

4. Laurie Goodstein, "Air Force Bans Leaders' Promotion of Religion," *NYT,* Aug. 30, 2005, p. A10 (LN); Alan Cooperman, "A Noisy Takeoff for Air Force Guidelines on Religion," *WP,* Oct. 31, 2005, p. A20 (LN).

5. "Interim Guidelines concerning Free Exercise of Religion in the Air Force," Aug. 29, 2005, http://www.religioustolerance.org/relintolafal.htm/ (accessed Sept. 20, 2005).

6. "Air Force Action: New Religion Guidelines Issued," Aug. 31, 2005, http://blog.au.org/; Laura A. Colarusso, "New Religious Tolerance Guidelines Issued," *Air Force Times,* Aug. 30, 2005, http://www.airforcetimes.com/ (both accessed Sept. 20, 2005). According to information on the ADL and AU websites, http://www.adl.org/ and http://www.au.org/ (Mar. 2, 2011), the ADL, founded in 1913, was well known for fighting against anti-Semitism, and Americans United, founded in 1947, described itself as a "religious liberty watchdog group." Both organizations spon-

sored programs and services to counteract religious bigotry and to promote church-state separation as a means of safeguarding religious freedom.

7. Eric Gorski, "Air Force Tries to Walk Thin Line on Faith," *Denver Post,* Nov. 9, 2005, p. A01 (LN); Alan Cooperman, "Air Force Eases Rules on Religion," *WP,* Feb. 10, 2006, p. A05 (LN). On FOF and the Christian Coalition, see Devlin Buckley, "Evangelicals Exploit Air Force Academy," *Online Journal,* Dec. 23, 2005, http://onlinejournal.com/artman/publish/printer_363.shtml/ (accessed Dec. 10, 2007). On the ACLJ, see "American Center for Law and Justice," http://www.aclj.org/Default.aspx/ (accessed Apr. 4, 2011.

8. Jim Belshaw, "Cadet in Man Still Battles," *Albuquerque Journal,* May 6, 2005, p. B1 (LN).

9. Mikey Weinstein, "Commentary: Religious Freedom Has Gone AWOL in the Air Force," n.d., *Religion News Service,* Aug. 22, 2005, e-mail to author from Mark Silk, Aug. 29, 2005.

10. Laurie Goodstein, "Evangelicals Are Growing Force in the Military Chaplain Corps," *NYT,* July 12, 2005, pp. A1, A20.

11. Minnery quoted in Alan Cooperman, "Air Force Withdraws Paper for Chaplains," *WP,* Oct. 11, 2005, p. A03 (LN).

12. During the 2005–2006 culture war, three global Christian organizations, the World Council of Churches, the Vatican, and the World Evangelical Alliance, engaged in formulating a "code of conduct to guide activities seeking converts to Christianity," which would, among other things, be based on "a distinction between aggressive proselytizing and evangelism." They concluded their work in June 2011. See "Largest Global Christian Groups Appear to Agree on Conversion 'Code of Conduct,'" World Council of Churches media release, Aug. 16, 2007, and Martin Revis, "Kobia: Church Must Promote 'Non-Aggressive' Evangelism," *Ecumenical News International,* May 1, 2007, http://www.ucc.org/ (accessed Dec. 6, 2007); and "WEA, WCC and Vatican Launch Historic Joint Document on Ethics of Christian Mission," June 28, 2011, http://www.worldevangelicals.org/ (accessed July 2, 2011).

13. "Chaplain (Major General) Charles C. Baldwin," *Air Force Link,* http://www.af.mil/ (accessed Dec. 10, 2007); Cooperman, "Noisy Takeoff for Air Force Guidelines," A20 (LN); Gorski, "Air Force Tries to Walk Thin Line," A01 (LN)

14. Eric Gorski, "Air Force Academy Seeks Thin Line Between Religious Freedom, Tolerance," *Denver Post,* June 26, 2005, p. A01 (LN).

15. Lawsuit paraphrased and quoted in Tim Korte, "Air Force Sued Over Religious Intolerance," *Yahoo! News,* http://news.yahoo.com/ (accessed Oct. 6, 2005). See also Laurie Goodstein, "National Briefing Rockies: Colorado: Air Force Is Sued Over Religion," *NYT,* Oct. 7, 2005, p. A21 (LN).

16. Reverend Ted Haggard, "Declaration of Reverend Ted Haggard," Feb. 7, 2006, http://www.mca-usa.org/ (accessed Jan. 20, 2009). Haggard's declaration appears to apply to the plaintiffs' "amended complaint," for which see "In the

United States District Court for the District of New Mexico, Michael L. Weinstein, et al., Plaintiffs, vs. United States Air Force and Michael Wynne, Acting Secretary of the Air Force, Defendants. Memorandum Opinion and Order, Civ. No. 05-1064 JP/ LAM," http://www.nmcourt.fed.us/ (accessed Feb. 12, 2011).

17. Haggard, "Declaration of Reverend Ted Haggard." See also "President of National Association of Evangelicals Expresses Concern over Erosion of Religious Freedom (United States)," *Religion News Service,* Sept. 9, 2006, http://pluralism.org/ (accessed Feb. 15, 2011).

18. "President of National Association of Evangelicals Expresses Concern over Erosion of Religious Freedom (United States)," Sept. 9, 2006.

19. Christopher Hitchens, "Fighting Words: GI Jesus," Oct. 2, 2006, http:// www.slate.com/; Weinstein quoted in "Analysis: Evangelizing the Troops," NPR All Things Considered, Nov. 25, 2005, http://nl.newsbank.com/ (both accessed Dec. 8, 2006); Weinstein quoted in "Religion in the Air Force," Feb. 22, 2006, in *NewsHour,* PBS, http://www.pbs.org/newshour/bb/religion/jan-june06/air_02-22.html/ (accessed Mar. 11, 2006). The phrase, "ambassadors of Christ in uniform" was from the mission statement of Officers Christian Fellowship, an evangelical parachurch organization in the armed forces.

20. Tom Minnery quoted in Julia Duin, "Air Force Sets Revised Rules for Prayers by Its Chaplains," *Washington Times,* Feb. 10, 2006, p. A03 (LN); and see also Walter Jones quoted in Neela Banerjee, "Proposal on Military Chaplains and Prayer Holds Up Bill," *NYT,* Sept. 19, 2006, p. A19.

21. Letter published in "Lawmakers Join Jones' Call for President to Protect Military Chaplains' First Amendment Rights," press release, Oct. 25, 2005, http:// jones.house.gov/ (accessed Dec. 10, 2007). See also Rick Maze, "Lawmakers Seek Free-Speech Guarantee for Chaplains," *Air Force Times,* Oct. 26, 2005, http://www. airforcetimes.com/ (accessed Nov. 22, 2005). In "Praying in Jesus' Name," *MC* 79 (Jan.–Feb. 2006): 11, Chaplain (Colonel) Herman Keizer Jr., USA (Ret.), chaplain-endorser of the Christian Reformed Church, explained why "many evangelical Christian chaplains believe it is necessary to end every one of their prayers in Jesus' name no matter what the setting" as follows: "For these chaplains, the only proper prayer is one that is prayed to God the Father through and in the name of Jesus, the mediator. This stems from a particular understanding about the teachings of Jesus on prayer as recorded in the New Testament. These chaplains address their prayers to a personal God, through Jesus, and in the mediating power of his name. For them, offering a non-sectarian or generic prayer is not seen as keeping fidelity with their theologies and endorsing faith groups."

22. Amy Fagan, "Executive Order Sought on Prayers," *Washington Times,* Oct. 20, 2005, p. A04 (LN).

23. See, for example, "Thousands Petition President in Support of Jones' Effort to Protect Military Chaplains' First Amendment Rights," press release, Nov. 4, 2005;

"More than 100,000 Citizens Petition President in Support of Jones' Effort to Protest Military Chaplains' First Amendment Rights," press release, Nov. 17, 2005; "Legislators Join Jones to Unveil 150,000 Signatures Urging President to Protect Military Chaplains' First Amendment Rights," press release, Dec. 14, 2005, http://jones.house.gov/ (accessed Dec. 10, 2007).

24. Adelle Banks, "Public Prayer in Jesus' Name Debated," *Religion News Service,* Apr. 28, 2006, http://www.baptiststandard.com/; Ken Walker, "Prayer in Jesus' Name Remains an Issue in Military Chaplaincy," *Baptist Press* online news service, Feb. 20, 2006, http://www.bpnews.net/ (both accessed Dec. 10, 2006). For the full text of the letter to President Bush and the entire list of organizations represented by the evangelical leaders who signed it, see "29 Organizations Join Jones' Call to Protect Religious Freedom for Military Chaplains," Jan. 4, 2006, http://jones.house.gov/ (accessed Dec. 10, 2007).

25. "Revised Interim Guidelines concerning Free Exercise of Religion in the Air Force," Feb. 9, 2007, http://www.af.mil/library/guidelines.pdf/ (accessed Feb. 23, 2009).

26. "Chief of Navy Chaplains Official Statement on Public Prayer in the Navy," [Jan. 19, 2006?], http://www.mca-usa.org/wp-content/uploads/2009/08/Prayer_Policy_Statement.pdf/ (accessed Feb. 24, 2011); Department of the Navy, Office of the Secretary, "SECNAV INSTRUCTION 1730.7C," Subject: Religious Instruction Within the Department of the Navy, Washington, D.C., Feb. 21, 2006, http://www.persuade.tv/frenzy2/SECNAVINST17307C.pdf/ (accessed Feb. 24, 2011). For statements Chief of Chaplains Iasiello made earlier to a *Washington Post* reporter, see Alan Cooperman, "Military Wrestles with Disharmony Among Chaplains," *WP,* Aug. 30, 2005, p. A01 (LN).

27. Foxman quoted in E. J. Kessler, "Campaign Confidential," Feb. 17, 2006, http://www.forward.com/ (accessed Jan. 6, 2008); Lynn quoted in "Air Force Issues Troubling Guidelines On Religion, Says Americans United," Americans United for Separation of Church and State, press release, Feb. 9, 2006, http://www.au.org/ (accessed Sept. 22, 2007).

28. Weinstein quoted in Erin Curry Roach, "Air Force Religion Guidelines Garner Both Praise and Criticism," *Baptist Press,* Feb. 10, 2006, http://www.bpnews.net/ (accessed Feb. 21, 2009).

29. Mikey Weinstein, "Military Religious Freedom Foundation Denounces New U.S. Navy Guidelines on Religion," press release, Military Religious Freedom Foundation, Mar. 23, 2006, http://www.jewsonfirst.org/06a/mrff_navy_instruc.pdf/ (accessed Feb. 26, 2011). Weinstein had by this time founded the Military Religious Freedom Foundation (MRFF), which, according to its website, was "dedicated to ensuring that all members of the United States Armed Forces fully receive the Constitutional guarantees of religious freedom to which they and all Americans are entitled by virtue of the Establishment Clause of the First Amendment." Ibid.

30. "U.S. Air Force Commended for its Guidelines on Religious Exercise," *Religious Action Center of Reform Judaism,* press release, Feb. 9, 2006, http://rac.org/ (accessed Jan. 5, 2008). See also Kessler, "Campaign Confidential."

31. Minnery quoted in Kessler, "Campaign Confidential."

32. Haggard paraphrased in Cooperman, "Air Force Eases Rules on Religion," A05 (LN).

33. "The National Association of Evangelicals Statement on Religious Freedom for Soldiers and Military Chaplains," Feb. 7, 2006, http://mca-usa.org/wp-content/uploads/2009/09/NAEStatement.pdf/ (accessed Feb. 21, 2011).

34. Jones quoted in Cooperman, "Air Force Eases Rules on Religion," A05 (LN); and in Duin, "Air Force Sets Revised Rules for Prayers," A03 (LN).

35. There was an effort, in early February, to persuade Bush to direct Defense Secretary Donald H. Rumsfeld to "order the military to relax its stance," but it failed. Walker, "Prayer in Jesus' Name Remains an Issue."

36. Alan Cooperman and Ann Scott Tyson, "House Injects Prayer into Defense Bill," *WP,* May 12, 2006, p. A05 (LN); Amy Fagan, "Warner Holds Up Chaplain Freedom," *Washington Times,* Sept. 20, 2006, p. A04 (LN).

37. Cooperman and Tyson, "House Injects Prayer into Defense Bill." The so-called prayer amendment appeared as Sec. 590 in the House Resolution 5122, National Defense Authorization Act for FY 2007, passed on May 11, 2006. See "House Resolution 5122, National Defense Authorization Act for FY 2007," *MC* 79 (May 2006): 5.

38. Jones quoted in Cooperman and Tyson, "House Injects Prayer into Defense Bill."

39. Rev. Herman Keizer Jr., NCMAF Chairman and Director, Chaplaincy Ministries Christian Reformed Church, to Congressman Duncan Hunter, Congressman Ike Skelton, Congressman Dennis Hastert, Congresswoman Nancy Pelosi, Congressman Vern Ehlers, May 5, 2006; and L. V. Iasiello, Rear Admiral, CHC, U.S. Navy, Chief of Navy Chaplains, to the Honorable Steve Israel, U.S. House of Representatives, Washington D.C., May 9, 2006, http://www.washingtonpost.com (accessed Jan. 5, 2008).

40. Comment on the chiefs of chaplains and quotation of the DoD statement in "Sen. John Warner Statement on House Chaplain's Provision," Sept. 19, 2006, http://www.govtrack.us/congress/record.xpd/ (accessed Jan. 7, 2008); and in Banerjee, "Proposal on Military Chaplains and Prayer," A19 (LN).

41. Haggard quoted in "Unneeded and Divisive: Let Us Pray that Congress Stops Meddling with Military Chaplains," editorial, *WP,* Sept. 21, 2006, p. A24 (LN).

42. "Baptist Joint Committee, Other Religious Groups, Oppose Military Prayer Legislation," Blog from the Capital, Sept. 8, 2006, http://www.bjconline.org/cgi-bin/2006/09/baptist_joint_committee_other.html/ (accessed Dec. 10, 2006). The Baptist Joint Committee for Religious Liberty comprised the following: Alliance of Baptists; American Baptist Churches USA; Baptist General Association

of Virginia; Baptist General Conference; Baptist General Convention of Missouri; Baptist General Convention of Texas; Baptist State Convention of North Carolina; Cooperative Baptist Fellowship; National Baptist Convention of America; National Baptist Convention, USA, Inc.; National Missionary Baptist Convention; North American Baptist Conference; Progressive National Baptist Convention; Religious Liberty Council; and Seventh Day Baptist General Conference. The other religious groups that signed the statement were the following: American Jewish Committee; Anti-Defamation League; The Episcopal Church; Evangelical Lutheran Church in America, Washington Office; Friends Committee on National Legislation; General Conference of the Seventh-day Adventist Church; Jewish Council for Public Affairs; National Council of Jewish Women; Presbyterian Church (USA), Washington Office; Sikh American Legal Defense and Education Fund; the Interfaith Alliance; Union for Reform Judaism; United Methodist Church, General Board of Church and Society; and Unitarian Universalist Association of Congregations.

43. Jennifer Siegel, "GOP Push for Sectarian Public Prayers in Military Opposed by Orthodox Group," Sept. 26, 2006, http://www.forward.com/ (accessed Jan. 6, 2008).

44. On the founding, purposes, and membership of the Congressional Prayer Caucus, and the role it played in holding up the final passage of the defense authorization bill, see Congressional Prayer Caucus website, http://www.house.gov/forbes/ (accessed June 21, 2011); and Chris Rodda, "A Little History Lesson For Rep. Walter B. Jones About Military Chaplains," Jan. 11, 2009, http://www.huffingtonpost.com/ (accessed Jan. 10, 2011).

45. Forbes quoted in Kate Wiltrout, "Prayer Debate Puts Military Chaplains on the Spot," *Virginian-Pilot,* Sept. 23, 2006, http://www.jewsonfirst.org/06cprint/sep128p.html/ (accessed Feb. 23, 2009).

46. Laura M. Colarusso, "Defense Authorization Bill Upends Religious Guidelines," *AirForceTimes,* Oct. 16, 2006, http://www.airforcetimes.com/ (accessed June 13, 2011); "Congress Passes Defense Bill, Jones Sees Progress for Rights of Military Chaplains," press release, Oct. 2, 2006, http://jones.house.gov/ (accessed Dec. 10, 2007).

11. Developing a Culture of Pluralism

1. Fagan, "Executive Order Sought on Prayers," A04 (LN).

2. Julia Duin, "White House to Push Military on Jesus Prayer," *Washington Times,* Jan. 23, 2006, p. A03 (LN).

3. On the Associated Gospel Churches, one of the oldest chaplain-endorsing bodies, see "The History of Associated Gospel Churches," http://www.agcchaplains.com/history.html/ (accessed June 16, 2011). On the ICECE, see Bryant Jordan, "Seminaries Threaten to Stop Sending Chaplains to Military," *ArmyTimes,* Mar. 6, 2006, http://www.armytimes.com/legacy/new/1-292925-1579644.php/ (accessed

Feb. 16, 2011); and Alan Cooperman, "Chaplains Group Opposes Prayer Order," *WP*, Mar. 30, 2006, p. A04 (LN).

4. See, for example, Baugham paraphrased and quoted in Chad Groening, "Former Army Chaplain Calls for Executive Order to End USAF's Religious Persecution," *Agape Press*, Jan. 6, 2006, http://www.persecution.org/ (accessed Jan. 6, 2008); Baugham quoted in Adelle M. Banks, "Air Force Issues Revised Guidelines on Religion," *CT*, Feb. 9, 2006, http://www.christianitytoday.com/ (accessed June 15, 2010); Baugham quoted in Duin, "Air Force Sets Revised Rules for Prayers," A03 (LN).

5. Duin, "White House to Push Military on Jesus Prayer," A03 (LN). This was the same story in which Duin quoted army spokesperson Martha Rudd saying that "sectarian prayer" had not been "an issue" among army chaplains. Apparently Rudd did not know about Stertzbach when she talked with Duin, and Duin did not see fit to apprise her of his complaints, which is not surprising since the *Times* was generally sympathetic to the conservative evangelical point of view in the culture war.

6. Walter B. Jones, "Jones Requests Investigation of Chaplain's Removal," press release, Feb. 7, 2006, http://jones.house.gov/ (accessed Dec. 10, 2007).

7. Ibid.

8. Ibid.

9. Chad Groening, "Chaplain Rebels at Prayer Censorship, Then Removed from Assignment," *AgapePress*, Feb. 23, 2006, http://headlines.agape.press.org/ (accessed Jan. 6, 2008). The constituency of the Associated Gospel Churches included "Independent Baptists, Bible Churches and other Bible-believing churches." See "The History of Associated Gospel Churches," http://www.agcchaplains.com/history.html/ (accessed May 20, 2011).

10. Jones, "Jones Requests Investigation."

11. Walker, "Prayer in Jesus' Name Remains an Issue."

12. Groening, "Chaplain Rebels at Prayer Censorship."

13. "Military Chaplains Should Be Able to Pray According to Their Faith," United States House of Representatives, Feb. 8, 2006, Sec. 51, http://www.govtrack.us/congress/record.xpd?id=109-h20060208-51/ (accessed Oct. 19, 2009).

14. On the contentiousness in the Navy Chaplain Corps, see Ward Sanderson, "War in the Chaplain Corps," and "War in the Chaplain Corps: Naples at the Center of Battle," *Stars and Stripes Sunday Magazine*, Nov. 23 and Nov. 30, 2003, http://www.stripes.com/ (accessed Sept. 18, 2009).

15. On the origin of the motto, in 1943, see Wildhack, "Navy Chaplains at the Crossroads," 237n125.

16. Brinsfield, *Encouraging Faith*, pt. 1, 243; Groh, *Facilitators of the Free Exercise of Religion*, 105, 106.

17. Hutcheson, *Churches and the Chaplaincy*, 125.

18. National Conference on Ministry to the Armed Forces, *The Covenant and the Code of Ethics for Chaplains of the Armed Forces: A Project of the National Conference on Ministry to the Armed Forces* (n.p., 1990).

19. Cooperman, "Air Force Withdraws Paper for Chaplains," Oct. 11, 2005, p. 3A (LN); Laurie Goodstein, "Air Force Rule on Chaplains Was Revoked," *NYT,* Oct. 12, 2005, p. 16A (LN).

20. *Chaplain* (FM 16-5, Aug. 27, 1964), 5.

21. *Chaplain* (FM 16-5, July 8, 1977), ii, p. 1.1.

22. Ibid., p. 1.1. For slightly different wording, see *Chaplain* (FM 16-5, Apr. 15, 1958), 16.

23. *Chaplain* (FM 16-5, Apr. 15, 1958), 16.

24. *Chaplain and Chaplain Assistant in Combat Operations* (FM 16-5, Dec. 10, 1984), 11; *Religious Support Doctrine* (FM 16-1, Nov. 27, 1989), p. 1.2; *Religious Support* (FM 1-05, Apr. 18, 2003), p. 1.9.

25. *Religious Support Doctrine* (FM 16-1, Nov. 27, 1989), p. 1.2. For similarly worded statements, see *Religious Support* (FM 16-1, May 26, 1995), p. 1.2; *Religious Support* (FM 1-05, Apr. 18, 2003), pp. 1.9, 1.10.

26. *Religious Support Doctrine* (FM 16-1, Nov. 27, 1989), p. 1.11.

27. *Religious Support* (FM 16-1, May 26, 1995), p. 1.2.

28. See, for example, *Religious Support Doctrine* (FM 16-1, Nov. 27, 1989), p. 2.4; "Chaplain Activities in the United States Army" (AR 165-1, June 26, 2000), 6.

29. The 1989 field manual provided definitions: "As used in this manual, 'to perform' means to do the activity or task personally. 'To provide for' means to do what is necessary to ensure that the activity or task is accomplished." *Religious Support Doctrine* (FM 16-1, Nov. 27, 1989), p. 1.1. A 1985 DoD directive included the perform/provide guidance, in the following statement: "Persons appointed to the chaplaincy shall be able to perform a ministry for their own specific faith groups, and provide for ministries appropriate to the rights and needs of persons of other faith groups." DoD directive 1304.19 (1984) quoted in Groh, *Facilitators of the Free Exercise of Religion,* 12.

30. *Religious Support* (FM 16-1, May 26, 1995), p. 1.2. See also *Religious Support* (FM 1-05, Apr. 18, 2003, pp. 1.1, 1.9, 1.10.

31. *Religious Support* (FM 16-1, May 26, 1995), p. 1.2.

32. *Religious Support* (FM 1-05, Apr. 18, 2003), p. 1.1.

33. Ibid., p. 1.9, and see also p. 1.5.

34. *Religious Support Handbook for the Unit Ministry Team* (TC 1-05, May 10, 2005), pp. 1.2, 2.1, 2.4, 2.6.

35. Ibid., p. 1.2.

36. See, for example, *Chaplain* (FM 16-5, Apr. 15, 1958), 13: "Protestant chaplains are required to conduct a general service of worship which is acceptable and meaningful to the maximum number of Protestant personnel in the command. . . . Protestant denominational services are a secondary responsibility of chaplains but are encouraged for those denominations which require them by Church law to be conducted by their chaplains. See also *Chaplain* (FM 16-5, Dec. 26, 1967), 13–14.

37. OCCH Newsletter, Jan. 1, 1988, 5, quoted in Brinsfield, *Encouraging Faith,* pt. 1, 246; "Chaplain Activities in the United States Army" (AR 165-1, Mar. 25, 2004), 5.

38. *Chaplain* (FM 16-5, Apr. 15, 1958), 17 (emphasis added). For a briefer statement, which also declared that patriotic ceremonies "are not religious services . . . but may contain religious elements such as an invocation, prayer or benediction," see Department of the Army, *Chaplain* (FM 16-5, Jan. 1952), 8.

39. Headquarters, Department of the Army, "Religious Activities: Duties of Chaplains and Commanders' Responsibilities" (Army Regulation 165-20; Washington, D.C., Oct. 15, 1979), 2.

40. *Religious Support Doctrine* (FM 16-1, Nov. 27, 1989), p. 2.3.

41. Department of the Army, "Chaplain Activities in the United States Army" (AR 165-1, June 26, 2000), 6.7. The "no compromise" rule regarding religious services read as follows: "Chaplains are authorized to conduct rites, sacraments, and services as required by their respective denomination. Chaplains will not be required to take part in worship when such participation is at variance with the tenets of their faith." Compare the statement in *Chaplain* (FM 16-5, Apr. 15, 1958), 12: "The chaplain is required by law to conduct appropriate public religious services for the command to which he is assigned. . . . No chaplain is required to conduct or participate in any service, rite, or sacrament contrary to the requirements of his denomination."

42. *Religious Support* (FM 16-1, May 26, 1995), p. 1.9.

43. *Religious Support* (FM 16-1, Nov. 27, 1989), p. 1.2. The reference to "an establishment of any one religious viewpoint or doctrine" may have been a response to a 1986 memorandum from the Office of the Judge Advocate General (OJAG) alerting OCCH to questions raised by Julie B. Kaplan in an article in the *Yale Law Journal* that considered the chaplaincy in the light of the establishment clause. Paraphrasing Kaplan, the author of the memorandum summarized her argument in one sentence: "The Government may not provide chaplain services for any other purpose . . . than to preserve the right of service members to the free exercise of religion, and especially not to implement a military vision of religion that enhances secular military values such as morale, patriotism, and the national interest." Memorandum quoted in Brinsfield, *Encouraging Faith,* pt. 1, 241–42. The phrase "to implement a military vision of religion that enhances secular military values such as morale, patriotism, and the national interest" was a direct quote from Julie B. Kaplan, "Military Mirrors on the Wall: Nonestablishment and the Military Chaplaincy," *Yale Law Journal* 95 (May 1986): 1212.

44. *Religious Support* (FM 16-1, May 26, 1995), p. 1.2.

45. "Chaplain Activities in the United States Army" (AR 165-1, June 26, 2000), 6; "Chaplain Activities in the United States Army" (AR 165-1, Mar. 25, 2004), 6.

46. "Chaplain Activities in the United States Army" (AR 165-1, June 26, 2000), 1, 5; "Chaplain Activities in the United States Army" (AR 165-1, Mar. 25, 2004), 1, 5.

47. Zimmerman, "Statement of Equal Opportunity," May 12, 1994, quoted in Brinsfield, *Encouraging Faith,* pt. 1, 338.

48. Brinsfield, *Encouraging Faith,* pt. 1, 331.

49. Shea quoted in Brinsfield, *Encouraging Faith,* pt. 1, 372.

50. Ibid., 377.

51. "Chaplain Gunhus' Address to the Senior Leader Training Conference," Chief of Chaplains Senior Leader Training Conference, Hilton Head, S.C., Jan. 25, 2000, in McCullough, Brinsfield, and Lawson, *Courageous in Spirit,* 10–11. On the origin, listing, and definitions of the Chaplain Corps' key values—Spirituality, Accountability, Compassion, Religious Leadership, Excellence, and Diversity, that is, SACRED—see 175. Two chaplain manuals, *Religious Support* (FM 1-05, Apr. 18, 2003), p. 1.4, and *Religious Support Handbook* (TC 1-05, May 10, 2005), p. 2.6, listed the "Army Chaplaincy Values," defining "Diversity" as "Being respectful of different views and ideas that are not like our own."

52. "Chaplain Gunhus' Address to the Senior Leader Training Conference," 11.

53. For Gunhus's views on prophetic leadership, see McCullough, Brinsfield, and Lawson, *Courageous in Spirit,* 190, 196.

54. Chaplain (Major) Maury Stout, Action Officer, Directorate of Ministry Initiatives, Office of the Chief of Chaplains, "Taking Spiritual Leadership to the Next Level: A White Paper for the United States Army Chaplain Corps," June 2004, in Brinsfield and Lawson, *History of the United States Army Chaplaincy,* pt. 1, 233, 238. According to Chaplain Glenn Bloomstrom (207), Hicks assigned the task of writing the white paper to Chaplain (Major) Maury Stout, Action Officer in the OCCH Directorate of Ministry Initiatives, with instructions to elaborate the concept of spiritual leadership from a "religiously pluralistic" perspective.

55. *Religious Support Handbook* (TC 1-05, May 10, 2005), p. 1.2.

56. Chaplain (Colonel) Glenn Bloomstrom, "Office of the Chief of Chaplains: Directorate Functions, Challenges and Accomplishments, 2003–2006," in Brinsfield and Lawson, *History of the United States Army Chaplaincy,* pt. 1, 189, 197, 198.

57. Army Chief of Chaplains (Major General) David H. Hicks, newsletter, Dec. 2005, quoted by James Gordon Klingenschmitt in "Chief of Navy Chaplains[,] Official Statement on Public Prayer in the Navy," distributed Jan. 19, 2006, http://www.persuade.tv/frenzy/NewNavyPrayerPolicy1.pdf/ (accessed Jan. 1, 2011).

58. Hicks quoted in Russell D. Moore, "Uniform Prayers," *Touchstone: A Journal of Mere Christianity,* July/Aug. 2006, http://www.touchstonemag.com/ (accessed May 17, 2011). Moore referred to the advisory as "a March communiqué to army chaplains."

59. Gaylord T. Gunhus, Chaplain (Major General) USA RET, "Letter to the American Center for Law and Justice," *MC* 79 (Jan.–Feb. 2006): 6–7.

Epilogue

1. Drazin and Currey, *For God and Country,* 205.

2. Ibid., 205, 229n14.

3. Kaplan, "Military Mirrors on the Wall," 1210, 1212.

4. Benjamin, "Justice, Justice Shall You Pursue," 6, 16, 17.

5. Chaplain (COL) Terry A. Dempsey, "Asymmetric Threats to the United States Army Chaplaincy in the 21st Century," 5, 6, 13, 14–15, Strategy Research Project, U.S. Army War College, Carlisle Barracks, Pa., 2000.

6. Rodda, "Little History Lesson For Rep. Walter B. Jones"; David Waters, "Amendment Would Let Military Chaplains Pray as They Wish," *WP,* May 26, 2010, http://onfaith.washingtonpost.com/ (accessed May 28, 2012).

7. Robert Hodierne, "Poll: More Troops Unhappy with Bush's Course in Iraq," *Army Times,* Dec. 29, 2006, http://www.armytimes.com/ (accessed Apr. 6, 2011).

8. Nicole Neroulias (Religion News Service), "New Accusations Surface Against U. S. Military Proselytizing," May 5, 2009, http://pewforum.org/ (accessed Feb. 25, 2011); "Proselytizing in the Military Likely to Continue Under Obama," *Public Record,* Dec. 27, 2008, http://pubrecord.org/ (accessed Nov. 8, 2009).

9. Neela Banerjee, "Religion and Its Role Are in Dispute at the Service Academies," *NYT,* June 25, 2008, http://www.nytimes.com/ (accessed June 25, 2010); Anna Mulrine, "Too Much Religion at Military Academies? West Point Cadet Revives Charge," *Christian Science Monitor,* Dec. 7, 2012, http://www.csmonitor.com/USA/Military/2012/1207 (accessed Dec. 14, 2012); Michael Stone, "West Point Cadet Resigns, Protests Christian Fundamentalism in Military," *Examiner.com,* Dec. 6, 2012, http://www.examiner.com (accessed Dec. 14, 2012); Blake Page, "Why I Don't Want to be a West Point Graduate," *Huff Post Politics,* Dec. 3, 2012, http://www.huffingtonpost.com/blake-page/west-point-religious-freedom_b_2232279.html (accessed Dec. 18, 2012).

10. Brinsfield and Lawson, *History of the United States Army Chaplaincy,* pt. 1, 207. Rick and Kay Warren donated 375,000 copies of *The Purpose Driven Life,* "the largest single private contribution" the army had ever received, "valued at more than ten million dollars."

11. Lisa Miller, with Larry Kaplow, "A Good Book in Camouflage," *Newsweek,* Apr. 7, 2008, 17 (LN). Henry and Richard Blackaby, the authors, used donations from "friends in the business world" to finance the printing and distribution of one hundred thousand copies of the book (with a camouflage cover) free of charge to the army. Miller and Kaplow quoted journalist Jeff Sharlet on the implications of the army distribution: "'The military stands for our democratic nation, not for any religion,' he says. The ubiquity of this devotional 'creates the appearance that this is an approved religion, that it's favored by the state.'"

12. Rick Warren, *The Purpose Driven Life: What on Earth Am I Here For?* (Grand Rapids, Mich.: Zondervan, 2002), 9, 11–12, 20, back cover.

13. Americans United for Separation of Church and State, "Hunting Souls In Afghanistan: Video Exposes Military Proselytizing Plan," May 5, 2009, http://blog.au.org/ (accessed Nov. 7, 2009); Peter Graff, "US Denies Letting Troops Convert Afghans," May 4, 2009, http://www.commondreams.org/ (accessed Nov. 8, 2009).

14. Kathryn Joyce, "Christian Soldiers: The Growing Controversy Over Military Chaplains Using the Armed Forces to Spread the Word," *Newsweek Web Exclusive*

Homepage Story, June 19, 2009, http://www.militaryreligiousfreedom.org/press-releases/newsweek_christian.html/ (accessed Nov. 19, 2009); Headquarters, Multi-National Corps—Iraq, Baghdad, Iraq, "General Order Number 1 (GO-1)," [Feb. 14, 2008], signed by Lloyd J. Austin III, Lieutenant General, USA Commanding, http://www.jrtc-polk.army.mil/115th_BDE/generalorder1.pdf/ (accessed May 26, 2012).

15. For examples of the work of the Secular Coalition for America and the Military Association of Atheists and Freethinkers, see "Atheists in Foxholes: Preventing Religious Discrimination in the Military under an Obama Administration," Nov. 10, 2008, http://secular.org/news/Obama_military_accommodation.html/ (accessed May 31, 2012); "Re: Secular Coalition Urges Changes to Religious Accommodation Practices in the Military," Nov. 10, 2008, http://secular.org/news/Secular_Coalition_MAAF_Obama.pdf/ (accessed Nov. 9, 2009); Leo Shane III, "Military Atheists Want New Rules on Prayer," *Stars and Stripes,* Nov. 12, 2008, http://www.stripes.com/ (accessed May 22, 2012); Greg M. Epstein, "Military Needs Chaplains for Humanists, Atheists," July 25, 2008, http://newsweek.washingtonpost.com/ (accessed Nov. 8, 2009). Stefani E. Barner wrote a book titled *Faith and Magick in the Armed Forces: A Handbook for Pagans in the Military,* published by Llewellyn, in which, among other things, she listed "the potential roadblocks thrown up by military chaplains and shows how best to overcome them without risk of punishment." "Faith and Magick in the Armed Forces: A Handbook for Pagans in the Military," *Publishers Weekly,* Mar. 24, 2008 (LN).

16. "Proselytizing in the Military Likely to Continue Under Obama," Dec. 27, 2008. *NYT* reporter Neela Banerjee quoted Mikey Weinstein saying that since 2004, he had been contacted by "more than 5,500 service members and, occasionally military families about incidents of religious discrimination," and that "96 percent of the complainants were Christians, and the majority of those were Protestants." Neela Banerjee, "Soldier Sues Army, Saying His Atheism Led to Threats," *NYT,* Apr. 26, 2008, http://www.nytimes.com/ (accessed May 28, 2012).

17. Banerjee, "Soldier Sues Army"; Jane Lampman, "Are U.S. Troops Being Force-fed Christianity?" *Christian Science Monitor,* Oct. 4, 2007, http://www.csmonitor.com/2007/1004/p13s02-lire.html/ (accessed May 31, 2012).

18. "Proselytizing in the Military Likely to Continue Under Obama," Dec. 27, 2008.

19. Klitgaard, "Onward Christian Soldiers," 1378–79; United Church of Christ, "Ministries to Military Personnel," 81, 82; Zahn, "Scandal of the Military Chaplaincy," 313, 314, 315; Zahn, "What Did You Do during the War, Father?" 196, 197, 198.

20. Dr. Scott R. Borderud, "Introduction: The Attraction of Chaplain Ministry," paper presented at a symposium of the International Society for Military Ethics (ISME), 2007, http://www.usafa.edu/isme/ISME07/Borderud07.html/ (accessed Sept. 14, 2009).

21. Pauletta Otis, "An Overview of the U.S. Military Chaplaincy: A Ministry of Presence and Practice," *Review of Faith and International Affairs* 7 (Winter 2009), http://rfiaonline.org/ (accessed Feb. 10, 2010).

22. Andrew Todd, "Reflecting Ethically with British Army Chaplains," *Review of Faith and International Affairs* 7 (Winter 2009), http://rfiaonline.org/ (accessed Feb. 10, 2010).

23. For chaplains' difficulty with such words and their implications, see Master Sgt. Eric B. Pilgrim, "Spiritual Fitness: What is it, can we train it and if so, how?" 2008, http://www.hooah4health.com/spirit/FHPspirit.htm/ (accessed Jan. 10, 2011).

24. Dr. Erik Wingrove-Haugland, U.S. Coast Guard Academy, "A Pluralistic Approach to Religion in the Military: Accommodating Diversity, Utilizing Consensus, Motivating Sacrifice, and Encouraging Growth," paper presented at a symposium of the International Society for Military Ethics (ISME), 2007, http://www.usafa.edu/isme/ISME07/Wingrove-Haugland07.html/ (accessed Sept. 14, 2009).

25. Martin L. Cook, Professor and Deputy Department Head, Department of Philosophy, U.S. Air Force Academy, "Religion and the US Military," paper presented at a symposium of the International Society for Military Ethics (ISME), 2007, http://www.usafa.edu/isme/ISME07/Cook07.html/ (accessed Sept. 24, 2009).

26. Chaplain William McCoy, *Under Orders: A Spiritual Handbook for Military Personnel* (Ozark, Ala.: ACW Press, 2005), 14, 18, 19. See also Chaplain (MAJ) William McCoy, "The Main Thing Is the Only Thing," *Army Chaplaincy* (Winter–Spring 2006): 26–30.

27. Petraeus quoted in Bryant Jordan, "Petraeus Book 'Endorsement' Draws Fire," Aug. 20, 2008, http://www.military.com/news/article/petraeus-bok-endorsement-draws-fire.html/ (accessed Nov. 8, 2009).

28. Chris Rodda, "Petraeus Endorses 'Spiritual Handbook,' Betrays 21% of Our Troops," Aug. 17, 2008, http://www.huffingtonpost.com/chris-rodda/petraeus-endorses-spiritu_b_119242.html/ (accessed Nov. 8, 2009). For more on the incident, see Chris Rodda, "Removal of Petraeus Endorsement from 'Spiritual Handbook' Not Enough," Aug. 21, 2008, http://www.huffingtonpost.com/chris-rodda/removal-of-petraeus-endor_b_120095.html/ (accessed Nov. 8, 2009); Eric Young, "Petraeus Endorsement of Faith-Based Military Book Draws Fire," Aug. 22, 2008, http://www.christianpost.com/ (accessed Nov. 8, 2009).

29. Stephen Mansfield, *The Faith of the American Soldier* (New York: Jeremy P. Tarcher/Penguin, 2005), 4–5, 12–13, 126, 130. Mansfield said that he based the "code" on recent speeches delivered by General William Boykin.

30. George W. Casey Jr., "Comprehensive Soldier Fitness: A Vision for Psychological Resilience in the Army," *American Psychologist* 66 (Jan. 2011): 1. Two groups of psychologists took sides on CSF. The American Psychological Association (APA) supported it and even published a special issue of the *American Psychologist* describing and praising the program. Roy Eidelson, president of Psychologists for Social Responsibility, published an article in *Psychology Today*, in which he

described the program as a "$125 million resilience training initiative" based on "U.S. Army–APA collaboration." He contended that despite its "worthy aspirations," that is, "to reduce and prevent the adverse psychological consequences of combat for our soldiers and veterans," the program raised serious concerns. One was the fact that although it was described as a "training program," it was also a "research program of enormous size and scope," designed to test a "mere hypothesis," and did not offer soldiers, who were required to participate in it, the "protections routinely granted to those who participate in research studies." In Eidelson's opinion, "Respect for informed consent is more, not less, important in total environments like the military where individual dissent is often severely discouraged and often punished." Eidelson also pointed out "potential ethical concerns related to the uncertain effects of the CSF training itself," such as whether it might "actually cause harm" and whether "soldiers who have been trained to resiliently view combat as a growth opportunity [would] be more likely to ignore or under-estimate real dangers, thereby placing themselves, their comrades, or civilians at heightened risk of harm." Or in another scenario, whether CSF training, "by increasing perseverance in the face of adversity," might "lead soldiers to engage in actions that may later cause regret (e.g., the shooting of civilians at a roadblock in an ambiguous situation) thereby increasing the potential for PTSD or other post-combat psychological difficulties." Eidelson's concerns were in part a reaction to what he saw as CSF's strong reliance on research in the field of "positive psychology," whose advocates he described as focusing not on "distress and pathology" but emphasizing "human strengths and virtues, happiness, and the potential to derive positive meaning from stressful circumstances." Roy Eidelson, "Dangerous Ideas: The Dark Side of 'Comprehensive Soldier Fitness,'" *Psychology Today,* Mar. 25, 2011, http://www.psychologytoday.com/ (accessed July 25, 2012).

31. "New Comprehensive Soldier Fitness," Oct. 12, 2009, http://www.military.com/military-report/new-comprehensive-soldier-fitness/ (accessed July 25, 2012); "Comprehensive Soldier Fitness," n.d., http://www.acsim.army.mil/readyarmy/ra_csf.htm/ (accessed Aug. 17, 2012); "What is Comprehensive Soldier Fitness?" n.d., http://csf.army.mil/ (accessed Aug. 17, 2012); "Comprehensive Soldier Fitness Execution Order," Sept. 7, 2010, http://csf.army.mil/resilience/supportdocs/ALARACT-097-2010-FINAL.pdf/ (accessed Aug. 17, 2012); Casey, "Comprehensive Soldier Fitness," 1; "Comprehensive Soldier Fitness," Wikipedia, http://en.wikipedia.org/wiki/Comprehensive_Soldier_Fitness/ (accessed May 22, 2012). In the 1987 *Fit to Win: Spiritual Fitness* pamphlet, the army admitted the difficulty of assessing spiritual fitness. "Care must be taken when we talk about assessing or measuring the spiritual," the pamphlet stated. "There are no universally accepted instruments available to determine spirituality or to measure spiritual well-being." *Fit to Win* (Pamphlet 600-63-12, Sept. 1, 1987), 1. Twelve years later the army had changed its thinking on the matter.

32. Casey, "Comprehensive Soldier Fitness," 1.

33. Ibid., 2. Between January 1 and November 17, 2009, there were 19 suicides by Army soldiers in Iraq; another 140 active duty and 71 reserve component soldiers also committed suicide. These figures exceeded the number for 2008. See Sgt. Lindsey Bradford, "Army Develops New Program as 2009 Suicide Numbers Remain High," Nov. 21, 2009, http://www.army.mil/ (accessed Nov. 25, 2009). In 2010 and 2011, the suicide rate "leveled off," but as of June 2012, the Pentagon reported 154 suicides for active-duty troops during the first 155 days of that year, an 18 percent increase over the same period of 2011. By the end of the year, the number for the army had increased to 182 (Robert Burns, "Troops' Suicides Nearly One a Day," *Baton Rouge Advocate,* June 7, 2012, p. 1A; Burns, "2012 Suicides Hit Record High of 349," *Baton Rouge Advocate,* Jan. 15, 2012, p. 3A). See also Patrik Jonsson, "Troubled Soldiers Turn to Chaplains for Help," *Christian Science Monitor,* Mar. 8, 2006, p. 1 (LN). For a chaplain's commentary on soldier suicide and its causes, see Rev. Frank E. Wismer III, *War in the Garden of Eden: A Military Chaplain's Memoir from Baghdad* (New York: Seabury Books, 2008), 76–81.

34. Cornum, "Why Comprehensive Soldier Fitness?" 15–16.

35. See, for example, on the "chaplain-led Strong Bonds program," "Clergy Reflect on Spiritual Resiliency at Fort Drum since 9/11," http://www.army.mil/ (accessed July 26, 2012); "Stronger Relationships Mean a Stronger Army," http://www.strongbonds.org/skins/strongbonds/home.aspx/; and "Army Chaplain Corps 235th Anniversary," http://www.army.mil/ (both accessed July 31, 2012).

36. Cornum, "Why Comprehensive Soldier Fitness?" 13.

37. "Comprehensive Soldier Fitness," http://csf.army.mil/whatiscsf.html/ (accessed July 25, 2012).

38. Kenneth I. Pargament and Patrick J. Sweeney, "Building Spiritual Fitness in the Army: An Innovative Approach to a Vital Aspect of Human Development," *American Psychologist* 66 (Jan. 2011): 58. This appeared in the special *American Psychologist* issue devoted to CSF.

39. Colonel Patrick J. Sweeney, USA, USMA, Dr. Jeffrey E. Rhodes, Master Sergeant Bruce Boling, USAF, "Spiritual Fitness: A Key Component of Total Force Fitness," *Joint Force Quarterly* 66 (3rd Quarter 2012): 36–37.

40. See, for example, Sgt. Mary Katzenberger, "'Vanguards' Begin Bible Study Event Open to All" and "'Vanguards' Host Prayer Breakfast to Strengthen Spiritual Fitness," http://www.dvidshub.net/; and Elvia Kelly, "Luncheon Enlightens Community of Spiritual Resiliency," Oct. 21, 2011, http://www.army.mil/ (all accessed July 26, 2012); and "Chaplain Hosts First 205th Infantry Brigade Prayer Luncheon," Dec. 15, 2010, http://www.dvidshub.net/ (accessed Jan. 10, 2011).

41. Rob Schuette, "Spiritual Resiliency Helps Soldiers Weather Life's Traumas," Dec. 11, 2009, http://www.army.mil/; Tom Michele, "Woolridge Discusses Spiritual Resiliency with Fort McCoy Senior Leadership," Jan. 14, 2011, http://www.mccoy.army.mil/ (both accessed July 26, 2012).

42. Nancy Rasmussen, "Armywide Program Advocates Resilience through Spirituality," Jan. 13, 2011, http://www.army.mil/ (accessed July 26, 2012).

43. Steve Szkotak, "Army Probes Treatment of Soldiers Skipping Event," *Baton Rouge Advocate,* Aug. 21, 2010, p. 9A.

44. "Army Chaplain Holds Christian Prayer During Suicide Prevention Class, Soldiers Say," Oct. 2, 2012, http://www.huffingtonpost.com/ (accessed Oct. 8, 2012).

45. Military Association of Atheists and Freethinkers, "Military Needlessly Excludes Atheists with Pervasive Spirituality Approach," Jan. 7, 2010, http://www.maaf.info/spirituality.html/ (accessed Jan. 20, 2011); Secular Coalition for America, "End Religious Discrimination in the Military," n.d., http://secular.org/ (accessed May 31, 2012).

46. The Public Record Staff, "Army Manual Promotes Christianity to Combat Epidemic of Suicides," Jan. 3, 2009, http://pubrecord.org/ (accessed Oct. 8, 2012).

47. "Army 'Stand Down' Focuses Day on Suicide Training," http://www.boston.com/; Billy Hallowell, "Atheists Demand That 'Personal Religious Expressions' Be Removed from Military's Suicide Prevention Training," Oct. 4, 2012, http://www.theblaze.com/ (both accessed Oct. 8, 2012).

48. Crews quoted in "Atheists Ask Military to Drop 'Religious Expressions' in Suicide Prevention," Oct. 4, 2012, http://www.christianpost.com/ (accessed Oct. 8, 2012). On the Chaplain Alliance for Religious Liberty, the Alliance Defense Fund, and the Speak Up Movement, see various Chaplain Alliance press releases in "In the News," http://www.chaplainalliance.org/; "The Debate over Chaplains' Religious Liberty Heats Up," http://blogs.christianpost.com/; "Faith Under Fire: When Religious Freedom Becomes the Target," http://www.speakupmovement.org/; "Alliance Defense Fund," http://www.rightwingwatch.org/ (all accessed Oct. 8, 2012).

49. Ira C. Lupu and Robert W. Tuttle, "Instruments of Accommodation: The Military Chaplaincy and the Constitution," *West Virginia Law Review* 110 (Fall 2007): 94, 118, 122, 123, 136, 139, 140, 153–59, 162–64; Richard D. Rosen, "*Katcoff v. Marsh* at Twenty-Two: The Military Chaplaincy and the Separation of Church and State," *University of Toledo Law Review* 38 (Summer 2007): 1142, 1161, 1174, 1176, 1177.

50. For two recent statements regarding the "unique" role of chaplains and the Chaplain Corps, see the following: Chaplain (COL) Gilbert H. Pingel, "Commandant's Notes," *Army Chaplaincy* (Summer–Fall 2000): 2: "The Army looks to us for spiritual leadership and direction. No other branch is charged with that responsibility. It is our turf; it is our piece of the action." Chaplain (CPT) Daniel W. Hardin, "Theology of Suffering and Evil," *Army Chaplaincy* (Summer–Fall 2009): 98–99: "Every chaplain should have a strategy for addressing inner woundedness and be comfortable working cooperatively with other healing professionals, valuing their contribution to the hurting individual. I believe the Chaplain Corps has an opportunity of a lifetime before it—finding its definition in something that is

squarely in its lane—connecting hurting people to God. No other institution can do that; no other is commissioned and expected to do that. We have the possibility to see the greatest spiritual revival among service members because the need is great. Christian Chaplains are spiritual leaders charged with bringing the good, healing news of Jesus Christ into the suffering of our contemporary operating environment. Let us be faithful to the call."

51. Pew Forum on Religion & Public Life, "'Nones' on the Rise: One-in-Five Adults Have No Religious Affiliation," Oct. 9, 2012, 9–10, 13, 23.

Selected Bibliography

This bibliography lists sources that pertain to the main events and major figures involved in U.S. Army chaplains' cultural transition during the period 1945 through 2012. Sources that relate specifically to the subject matter of the individual chapters may be found, with full bibliographic citation, in the endnotes to those chapters.

Archival and Manuscript Sources

Clay, Lucius. Papers. Manuscript Department, U.S. Army Military History Institute. Carlisle Barracks, Pa.

Davis, Franklin M. Papers. Manuscript Department, U.S. Army Military History Institute. Carlisle Barracks, Pa.

Hyatt, Gerhardt W. Papers. Manuscript Department, U.S. Army Military History Institute. Carlisle Barracks, Pa.

Memorandum for Chief of Staff, Army, from Kermit D. Johnson, Chaplain (Major General), USA, Chief of Chaplains. Aug. 1, 1980. File 701-01, Chaplain Instruction Files (82). Acc. No. 247-88-001. RG 247. Washington National Records Center, Suitland, Md.

"Transcript of Oral History of Roller, Harold D." Dec. 11, 2003. Special Collections, Oral Histories. William Madison Randall Library, University of North Carolina, Wilmington. http://library.uncw.edu/web/collections/oralhistories/transcripts/369.html/. Accessed May 19, 2007.

Chaplain Periodicals

Army and Navy Chaplain. Washington, D.C.: Chaplains' Association of the Army and Navy of the United States, 1945–48.

Army Chaplaincy: Professional Bulletin of the Unit Ministry Team. Fort Monmouth, N.J.: U.S. Army Chaplain Center and School, 1993–2012.

Chaplain. Washington, D.C.: General Commission on Chaplains and Armed Forces Personnel, 1944–78.

Military Chaplain. Washington, D.C.: Military Chaplains Association of the United States, 1948–2010.

Military Chaplains' Review. [Fort Wadsworth, Staten Island, N.Y.]: Chaplains, U.S. Army, 1972–92.

Government Publications and Serials

The Army Chaplain School

The Chaplain School. *The Chaplain and Military Morale.* ST 16-155; Carlisle Barracks, Pa., Aug. 1950.

The Chaplain School, Carlisle Barracks, Pennsylvania. *The Army Character Guidance Program.* ST 16-151, Mar. 1, 1950.

Manning, Chaplain (MAJ) Benjamin C. *U.S. Army Chaplain Officer Basic Course, Chaplain Values Training 1, Subcourse No. CH 0555, Edition 8.* Fort Monmouth, N.J.: U.S. Army Chaplain Center and School/U.S. Army Training Support Center, Aug. 1988.

[U.S. Army Chaplain Center and School]. *Program of Instruction, 5-16-C20-56A, Chaplain Officer Basic.* Fort Monmouth, N.J.: U.S. Army Chaplain Center and School/U.S. Army Training Support Center, Aug. 1988.

U.S. Army Chaplain School, Fort Hamilton, New York. *Character Guidance Common Subjects: Instructional Guide for U.S. Army Service Schools.* [Fort Hamilton, N.Y.?], Dec. 1968.

Defense Manpower Data Center

"Religion of Active Duty Personnel By Service (no Coast Guard). As of March 31, 2009." Produced by the Defense Manpower Data Center, May 1, 2009. Copy provided to author by *Nashville Tennessean* reporter Bob Smietana.

Department of Defense

"Department of Defense Directive: Accommodation of Religious Practices Within the Military Services." Directive 1300.17. June 18, 1985. http://www.dtic.mil/whs/directives/corresp/pdg/130017.pdf/. Accessed Jan. 21, 2009.

Department of Defense, Office of Armed Forces Information and Education. *The U.S. Fighting Man's Code.* [Washington, D.C.], Nov. 1955.

———. *The U.S. Fighting Man's Code.* [Washington, D. C., 1959].

Serials

Department of the Army, Office of the Chief of Chaplains. *Historical Review.* Title varies: *Historical Review, Summary of Major Events, Annual Report,* and *Annual Historical Review.* Washington, D.C., 1959–93.

Chaplain Field Manuals

Department of the Army. *The Chaplain.* Field Manual 16-5; Washington, D.C., Jan. 1952.

Headquarters, Department of the Army. *The Chaplain.* Field Manual 16-5; Washington, D.C., Apr. 15, 1958.

——. *The Chaplain.* Field Manual 16-5; Washington, D.C., Aug. 27, 1964.

——. *The Chaplain.* Field Manual 16-5; Washington, D.C., Dec. 26, 1967.

——. *The Chaplain.* Field Manual 16-5; Washington, D.C., May 11, 1970.

——. *The Chaplain.* Field Manual 16-5; Washington, D.C., July 8, 1977.

——. *The Chaplain and Chaplain Assistant in Combat Operations.* Field Manual 16-5; Washington, D.C., Dec. 10, 1984.

——. *Religious Support.* Field Manual 16-1; Washington, D.C., May 26, 1995.

——. *Religious Support.* Field Manual 1-05; Washington, D.C., Apr. 18, 2003.

——. *Religious Support Doctrine: The Chaplain and Chaplain Assistant.* Field Manual 16-1, Final Coordinating Draft; Washington, D.C., Apr. 1989.

——. *Religious Support Doctrine: The Chaplain and Chaplain Assistant.* Field Manual 16-1; Washington, D.C., Nov. 27, 1989.

——. *Religious Support Handbook for the Unit Ministry Team.* TC [Training Circular] 1-05; Washington, D.C., May 10, 2005.

Headquarters, Joint Chiefs of Staff. *Religious Support in Joint Operations.* Joint Publication 1-05; [Washington, D.C.], June 9, 2004.

Army Regulations

Headquarters, Department of the Army. "Chaplain Activities in the United States Army." Army Regulation 165-1; Washington, D.C., June 26, 2000.

——. "Chaplain Activities in the United States Army." Army Regulation 165-1; Washington, D.C., Mar. 25, 2004.

——. "Chaplain Support Activities." Army Regulation 600-30; Washington, D.C., Jan. 14, 1977.

——. "Religious Activities: Duties of Chaplains and Commanders' Responsibilities." Army Regulation 165-20; Washington, D.C., Oct. 15, 1979.

——. "Religious Support: Army Chaplain Corps Activities." Army Regulation 165-1; Washington, D.C., Dec. 3, 2009.

Articles

Ahlstrom, Sydney E. "The Radical Turn in Theology and Ethics: Why It Occurred in the 1960's." *Annals of the American Academy of Political and Social Science* 387 (Jan. 1970).

[Appelquist, A. Ray.] "Editor's Notes: Dissent Over Vietnam." *Chaplain* 25 (May–June 1968).

Benjamin, Michael J. "Justice, Justice Shall You Pursue: Legal Analysis of Religion Issues in the Army." *Army Lawyer* (Nov. 1998).

Berg, Thomas C. "'Proclaiming Together'? Convergence and Divergence in Mainline and Evangelical Evangelism, 1945–1967." *Religion and American Culture* 5 (Winter 1995).

Boozer, Jack S. "The Military Chaplaincy: One Calling, Two Roles." *Chaplain* 27 (Dec. 1970).

Brinsfield, John W. "Alexander's Challenge: Issues in Teaching Leadership." *Army Chaplaincy* (Winter 1998).

——. "The Army Chaplaincy and World Religions: From Individual Ministries to Chaplain Corps Doctrine." *Army Chaplaincy* (Winter–Spring 2009).

——. "Army Values and Ethics: A Search for Consistency and Relevance." *Parameters* 27 (Autumn 1998).

——. "By the Book: The Military Ethics of General William T. Sherman." *Military Chaplains' Review* (Spring 1982).

——. "The Chaplaincy and Moral Leadership." *Army Chaplaincy* (Summer 1995).

——. "Ethics and the Angry Young Man: An Instructor's View." *Military Chaplains' Review* (Summer 1980).

——. "From Plato to NATO; the Ethics of Warfare; Reflections on the Just War Theory." *Military Chaplains' Review* (Spring 1991).

Brinsfield, John W. , and Peter A. Baktis. "The Human, Spiritual, and Ethical Dimensions of Leadership in Preparation for Combat." In *The Future of the Army Profession,* edited by Don M. Snider and Lloyd J. Matthews. 2d ed., rev. and expanded. New York: McGraw-Hill, 2005.

Burchard, Waldo W. "Role Conflicts of Military Chaplains." *American Sociological Review* 19 (Oct. 1954).

Cook, Martin L., Professor and Deputy Department Head, Department of Philosophy, U.S. Air Force Academy. "Religion and the US Military." Paper presented at a meeting of the International Society for Military Ethics (ISME), Springfield, Va., 2007. http://www.usafa.edu/isme/ISME07/Cook07.html/. Accessed Sept. 24, 2009.

Cox, Harvey G. "The 'New Breed' in American Churches: Sources of Social Activism in American Religion." *Daedalus* 96 (Winter 1967).

Davidson, Donald. "Christian Ethics And The Military Profession." *Military Chaplains' Review* (Fall 1986).

Davidson, Donald L. "Religious Strategists: The Churches and Nuclear Weapons." *Military Chaplains' Review* (Spring 1984).

Dobosh, William J., Jr. "Coercion in the Ranks: The Establishment Clause Implications of Chaplain-Led Prayers at Mandatory Army Events." *Wisconsin Law Review* 6 (2006).

[General Commission on Chaplains and Armed Forces Personnel]. "Armed Forces Chaplains: *All Civilians?* (A Feasibility Study)." *Chaplain* 29 (Spring Quarter 1972).

Gilbert, Chaplain Bertram. "On Prayers in Jesus' Name." *Military Chaplains' Review* (Fall 1987).

Gilbert, Chaplain Bertram C. "Value Education." *Military Chaplains' Review* (Apr. 1972).

Gilbert, Rev. Dr. Bertram C. "The Scandal and the Glory." *Military Chaplains' Review* (Spring 1984).

Heinl, Robert D., Jr. "The Collapse of the Armed Forces." *Armed Forces Journal* 108 (June 7, 1971).

Hunter, Chaplain Charlotte E., Commander, U.S. Navy, Defense Equal Opportunity Management Institute. "The Ethics of Military Sponsored Prayer." Paper presented at a meeting of the International Society for Military Ethics (ISME), Springfield, Va., 2007. http://www.usafa.edu/isme/ISME07/Hunter07.html/. Accessed Sept. 14, 2009.

Hutcheson, Richard G., Jr. "The Chaplain and the Structures of the Military Society." *Chaplain* 24 (Nov.–Dec. 1967).

———. "The Chaplaincy in the Year 2000." *Military Chaplains' Review* (Summer 1976).

Hutcheson, R. G., Jr. "Should the Military Chaplaincy Be Civilianized?" *Christian Century* 90 (Oct. 31, 1973).

Hyatt, Chaplain (MG) Gerhardt W. "Men of Faith." *Military Chaplains' Review* (Spring 1975).

Hyatt, Gerhardt W. "A Pastoral-Prophetic Ministry to Leaders." *Chaplain* 30 (Fall Quarter 1973).

[Hyatt, Gerhardt W.] "The Special Ministry of the Chief of Chaplains, an Interview with Chaplain (MG) Gerhardt W. Hyatt, US Army Ret." *Military Chaplains' Review* (Summer 1978).

Johnson, James T. "Just War in the Thought of Paul Ramsey." *Journal of Religious Ethics* 19 (Fall 1991).

Johnson, Kermit D. "How Army Chaplains View Themselves, Their Work and Their Organization." *Chaplain* 34 (Second Quarter 1977).

———. "A New Stage: Beyond 'In-House' Ethical Issues." *Military Chaplains' Review* (Spring 1982).

[Johnson, Kermit D.] "Of Peace and Policy." *Sojourners* 12 (Oct. 1983).

Kaplan, Julie B. "Military Mirrors on the Wall: Nonestablishment and the Military Chaplaincy." *Yale Law Journal* 95 (May 1986).

Knight, David E. "Supreme Six." In *Vietnam: The Other Side of Glory,* edited by William R. Kimball. New York: Ballantine Books, 1988.

Kosmin, Barry A., and Ariela Keysar. "American Religious Identification Survey Summary Report March 2009." Hartford, Conn.: Trinity College, 2009. http://www.americanreligionsurvey-aris.org/reports/ARIS_Report_2008.pdf/. Accessed Jan. 5, 2011.

Kosmin, Barry A., Egon Mayer, and Ariela Keysar. "American Identification Survey, 2001." Graduate Center of the City Univ. of New York, 2001. http://www.gc.cuny.edu/faculty/research_briefs/aris.pdf/. Accessed Jan. 5, 2011.

Kuehne, Wayne E. "Faith and the Soldier: Religious Support on the Airland Battlefield." U.S. Army War College Military Studies Program paper. Carlisle Barracks, Pa., 1988.

Lamback, Sam, Dan Garrett, and John Brinsfield. "The Ministry in the Military." *Reflection: Journal of Yale Divinity School* 67 (Jan. 1970).

Leroe, Robert G. "Faith Sharing in the Military." *Military Chaplains' Review* (Fall 1991).

Lewy, Guenter. "The Punishment of War Crimes: Have We Learned the Lessons of Vietnam?" *Parameters* 9 (Dec. 1979).

Loveland, Anne C. "Character Education in the U.S. Army, 1947–1977." *Journal of Military History* 64 (July 2000).

———. "Evangelical Proselytizing at the U.S. Air Force Academy: The Civilian-Military Controversy, 2004–2006." *Journal of Ecumenical Studies* 44 (Winter 2009).

———. "From Morale Builders to Moral Advocates." In *The Sword of the Lord: Military Chaplains from the First to the Twenty-First Century,* edited by Doris L. Bergen. Notre Dame: Univ. of Notre Dame Press, 2004.

Lupu, Ira C., and Robert W. Tuttle. "Instruments of Accommodation: The Military Chaplaincy and the Constitution." *West Virginia Law Review* 110 (Fall 2007).

Malone, Jerry E. "The Chaplain as an Advocate of Religious Freedom." *Military Chaplains' Review* (Fall 1983).

Marty, Martin E. "Foreword." In *Understanding Church Growth and Decline, 1950–1978,* edited by Dean R. Hoge and David Roozen. New York: Pilgrim Press, 1979.

——. "Protestantism Enters Third Phase." *Christian Century* 78 (Jan. 18, 1961).

Massanari, Ronald L. "The Pluralisms of American 'Religious Pluralism.'" *Journal of Church and State* 40 (Summer 1998).

Mathis, Rick D. "Constitutional Guidelines for the Military Chaplain Evangelist and Chaplaincy." *Military Chaplains' Review* (Fall 1991).

Miller, Luther D. "The Chaplains in the Army." *Army and Navy Journal* 85 (Sept. 20, 1947).

——. "Post-War Program." *Army and Navy Journal* 85 (Sept. 20, 1947).

Miller, Luther G. [*sic*] "Moral Effect of Military Service." *Army and Navy Chaplain* 16 (July–Aug. 1945).

National Association of Evangelicals. "The National Association of Evangelicals Statement on Religious Freedom for Soldiers and Military Chaplains, February 7, 2006." http://mca-usa.org/wp-content/uploads/2009/09/NAEStatement.pdf/. Accessed Feb. 21, 2011.

National Conference on Ministry to the Armed Forces. "The Covenant and The Code of Ethics for Chaplains of the Armed Forces." http://www.ncmaf.org/policies/PDFs/CodeofEthics-NCMAF.pdf/. Accessed Mar. 16, 2011.

Resnicoff, LCDR A. E., CHC, USN. "Prayers that Hurt: Public Prayer in Interfaith Settings." *Military Chaplains' Review* (Winter 1987).

Riker-Coleman, Erik Blaine. "Reflection and Reform: Professionalism and Ethics in the U.S. Army Officer Corps, 1968–1974." Chapel Hill, [N.C.], 1997. http://www.unc.edu/~chaos1/reform.pdf/. Accessed Mar. 31, 2012.

Rogers, General Bernard W. "The Challenges of the Chaplaincy." *Military Chaplains' Review* (Fall 1977).

Rosen, Richard D. "*Katcoff v. Marsh* at Twenty-Two: The Military Chaplaincy and the Separation of Church and State." *University of Toledo Law Review* 38 (Summer 2007).

Schweiker, Kenneth J. "Military Chaplains: Federally Funded Fanaticism and the United States Air Force Academy." *Rutgers Journal of Law and Religion* 8 (Fall 2006).

Smylie, James H. "American Religious Bodies, Just War, and Vietnam." *Journal of Church and State* 11 (Autumn 1969).

Spector, Ronald H. "The Vietnam War and the Army's Self-Image." In *The Second Indochina War: Proceedings of a Symposium Held at Airlie, Virginia, 7–9 November 1984,* edited by John Schlight. Washington, D.C.: Center of Military History, United States Army, 1985.

Stricklin, Gil A. "Evangelism in the Ranks." *Military Chaplains' Review* (Fall 1991).

Vickers, Robert. "The Military Chaplaincy: A Study in Role Conflict." *Military Chaplains' Review* (Spring 1986).

Wacker, Grant. "Uneasy in Zion: Evangelicals in Postmodern Society." In *Evangelicalism and Modern America,* edited by George Marsden. Grand Rapids, Mich.: William B. Eerdmans, 1984.

Wildhack, William A., III. "Navy Chaplains at the Crossroads: Navigating the Intersection of Free Speech, Free Exercise, Establishment, and Equal Protection." *Naval Law Review* 51 (2005).

Books

Abercrombie, Clarence L., III. *The Military Chaplain.* Beverly Hills, Calif.: Sage, 1976.

Ackerman, Henry F. *He Was Always There: The U.S. Army Chaplain Ministry in the Vietnam Conflict*. Washington, D.C: Office of the Chief of Chaplains, Department of the Army, 1989.

Ahlstrom, Sydney E. *A Religious History of the American People.* 2 vols. Garden City, N.Y.: Doubleday, 1975.

Appelquist, A. Ray, ed. *Church, State and Chaplaincy: Essays and Statements on the American Chaplaincy System.* N.p.: General Commission on Chaplains and Armed Forces Personnel, 1969.

Autry, Jerry. *Gun-Totin' Chaplain.* San Francisco: Airborne Press, 2006.

Bailey, Beth. *America's Army: Making the All-Volunteer Force.* Cambridge: Belknap Press of Harvard Univ. Press, 2009.

Bowers, Curt. *Forward Edge of the Battle Area: A Chaplain's Story.* Kansas City: Beacon Hill Press of Kansas City, 1987.

Brinsfield, John W., Jr. *Encouraging Faith, Supporting Soldiers: The United States Army Chaplaincy, 1975–1995.* Washington, D.C.: Office of the Chief of Chaplains, Department of the Army, 1997.

Brinsfield, John W., Jr., and Kenneth E. Lawson. *A History of the United States Army Chaplaincy: Taking Spiritual Leadership to the Next Level: The Hicks Years, 2003–2007.* Washington, D.C.: Office of the Chief of Chaplains, Department of the Army, 2007.

Broyles, William, Jr. *Brothers in Arms: A Journey from War to Peace.* New York: Alfred A. Knopf, 1986.

Budd, Richard M. *Serving Two Masters: The Development of American Military Chaplaincy, 1860–1920.* Lincoln: Univ. of Nebraska Press, 2002.

Carroll, Jackson W., Douglas W. Johnson, and Martin E. Marty. *Religion in America: 1950 to the Present.* San Francisco: Harper & Row, 1979.

Cincinnatus [Cecil B. Currey]. *Self-Destruction: The Disintegration and Decay of the United States Army during the Vietnam Era*. New York: W. W. Norton, 1981.

Cox, Harvey G., Jr., ed. *Military Chaplains: From Religious Military to a Military Religion*. New York: American Report Press, 1971.

Davidson, Donald L. *Nuclear Weapons and the American Churches: Ethical Positions on Modern Warfare*. Boulder, Colo.: Westview Press, 1983.

DeBenedetti, Charles, and Charles Chatfield. *An American Ordeal: The Antiwar Movement of the Vietnam Era*. Syracuse, N.Y.: Syracuse Univ. Press, 1990.

Drazin, Israel, and Cecil B. Currey. *For God and Country: The History of a Constitutional Challenge to the Army Chaplaincy*. Hoboken, N.J.: KTAV Publishing House, 1995.

Dulany, Joseph P. *Once a Soldier: A Chaplain's Story*. N.p.: n.p., 1971.

Falabella, J. Robert. *Vietnam Memoirs: A Passage to Sorrow*. New York: Pageant Press International, 1971.

Gallup, George Jr., and Jim Castelli. *The People's Religion: American Faith in the 90's*. New York: Macmillan, 1989.

Gilbert, James. *Redeeming Culture: American Religion in an Age of Science*. Chicago: Univ. of Chicago Press, 1997.

Groh, John E. *Facilitators of the Free Exercise of Religion: Air Force Chaplains, 1981–1990*. Washington, D.C.: Office of the Chief of Chaplains, USAF, 1991.

Gushwa, Robert L. *The Best and Worst of Times: The United States Army Chaplaincy, 1920–1945*. Washington, D.C.: Office of the Chief of Chaplains, Department of the Army, 1977.

Hall, Mitchell K. *Because of Their Faith: CALCAV and Religious Opposition to the Vietnam War*. New York: Columbia Univ. Press, 1990.

Hartle, Anthony E. *Moral Issues in Military Decision Making*. 2d ed., rev.; Lawrence: Univ. Press of Kansas, 2004.

Hauser, William M. *America's Army in Crisis: A Steady Erosion in Civil-Military Relations*. Baltimore: Johns Hopkins Univ. Press, 1973.

Haworth, Larry. *Tales of Thunder Run*. Eugene, Ore.: ACW Press, 2004.

Herberg, Will. *Protestant-Catholic-Jew: An Essay in American Religious Sociology*. 1955. 2d ed., Garden City, N.Y.: Anchor Books, 1960.

Honeywell, Roy J. *Chaplains of the United States Army*. Washington, D.C.: Office of the Chief of Chaplains, Department of the Army, 1956.

Hopkins, Samuel W., Jr. *A Chaplain Remembers Vietnam*. Kansas City, Mo.: Truman Publishing, 2002.

Hunter, James Davison. *Culture Wars: The Struggle to Define America*. N.p.: Basic Books, 1991.

Hutcheson, Richard G., Jr. *The Churches and the Chaplaincy.* Atlanta: John Knox Press, 1975.

———. *Mainline Churches and the Evangelicals: A Challenging Crisis?* Atlanta: John Knox Press, 1981.

Johnson, Haynes, and George C. Wilson. *Army in Anguish.* New York: Pocket Books, 1972.

Johnson, James D. *Combat Chaplain: A Thirty-year Vietnam Battle.* Denton: Univ. of North Texas Press, 2001.

Johnson, Kermit D. *Realism and Hope in a Nuclear Age.* Atlanta: John Knox Press, 1988.

Jorgensen, Daniel B. *The Service of Chaplains to Army Air Units, 1917–1946.* Washington, D.C.: U.S. Government Printing Office, n.d.

Kettler, Earl C. *Chaplain's Letters: Ministry by "Huey" 1964–1965, the Personal Correspondence of an Army Chaplain from Vietnam.* Cincinnati: Cornelius Books, 1994.

Lang, Daniel. *Casualties of War.* New York: McGraw-Hill, 1969.

Loveland, Anne C. *American Evangelicals and the U.S. Military, 1942–1993.* Baton Rouge: Louisiana State Univ. Press, 1996.

Mahedy, William P. *Out of the Night: The Spiritual Journey of Vietnam Vets.* New York: Ballantine Books, 1986.

McCullough, Douglas B., John W. Brinsfield, and Kenneth E. Lawson. *Courageous in Spirit, Compassionate in Service: The Gunhus Years, 1999–2003.* Washington, D.C.: Office of the Chief of Chaplains, Department of the Army, 2003.

Murphy, Andrew R. *Prodigal Nation: Moral Decline and Divine Punishment from New England to 9/11.* New York: Oxford Univ. Press, 2009.

Newby, Claude D. *It Took Heroes: One Chaplain's Story and Tribute to Combat Veterans and Those Who Waited for Them.* 2d ed., rev. Bountiful, Utah: Tribute Enterprises, 2000.

Roof, Wade Clark, and William McKinney. *American Mainline Religion: Its Changing Shape and Future.* New Brunswick, N.J.: Rutgers Univ. Press, 1987.

Scharlemann, Martin H. *Air Force Chaplains, 1961–1970.* Washington, D.C.: U.S. Government Printing Office, 1975.

Settje, David E. *Faith and War: How Christians Debated the Cold and Vietnam Wars.* New York: New York Univ. Press, 2011.

———. *Lutherans and the Longest War: Adrift on a Sea of Doubt About the Cold and Vietnam Wars, 1964–1975.* Lanham, Md.: Lexington Books, 2007.

Silk, Mark. *Spiritual Politics: Religion and America Since World War II.* New York: Simon and Schuster, 1988.

Stover, Earl F. *Up from Handymen: The United States Army Chaplaincy, 1865–1920.* Washington, D.C.: Office of the Chief of Chaplains, Department of the Army, 1977.

Stromberg, Peter L., Malham M. Wakin, and Daniel Callahan. *The Teaching of Ethics in the Military.* Hastings-on-Hudson, N.Y.: Hastings Center, 1982.

U.S. Army War College. *Study on Military Professionalism.* Carlisle Barracks, Pa.: U.S. Army War College, June 30, 1970.

Venzke, Rodger R. *Confidence in Battle, Inspiration in Peace: The United States Army Chaplaincy, 1945–1975.* Washington, D.C.: Office of the Chief of Chaplains, Department of the Army, 1977.

Wuthnow, Robert. *The Restructuring of American Religion: Society and Faith Since World War II.* Princeton, N.J.: Princeton Univ. Press, 1988.

Theses and Dissertations

Beasley, Joseph Hodgin. "Implications of Teaching Ethics: The West Point Experience." Ph.D. diss., Univ. of North Carolina, 1985.

Brinsfield, John Wesley, Jr. "Developing a Ministry of Teaching the History of Ethics and World Religions at the United States Military Academy, West Point, New York." D.Min. thesis, Drew Univ., 1983.

Whitt, Jacqueline Earline. "Conflict and Compromise: American Military Chaplains and the Vietnam War." Ph.D. diss., Univ. of North Carolina at Chapel Hill, 2008.

INDEX

Page numbers in **boldface** refer to illustrations.